Ulrich Meier-Westhues
Karsten Danielmeier
Peter Kruppa
Edward P. Squiller

Polyurethanes

Coatings, Adhesives and Sealants

2nd Revised Edition

VINCENTZ
Providing Professional Information

Cover: Onionastudio, Adobe Stock

Bibliographische Information der Deutschen Bibliothek
Die Deutsche Bibliothek verzeichnet diese Publikation in der Deutschen Nationalbibliographie; detaillierte bibliographische Daten sind im Internet über http://dnb.d-nb.de abrufbar.

Ulrich Meier-Westhues, Karsten Danielmeier, Peter Kruppa, Edward P. Squiller
Polyurethanes: Coatings, Adhesives and Sealants, 2nd Revised Edition
Hanover: Vincentz Network 2019
European Coatings Library
ISBN 3-86630-782-9
ISBN 978-3-86630-782-7

Vincentz Network, Plathnerstr. 4c, 30175 Hanover, Germany
T +49 511 9910-033, F +49 511 9910-029, books@european-coatings.com

Discover further books from European Coatings Library at:
www.european-coatings.com/shop

Layout: Vincentz Network, Hanover, Germany

European Coatings Library

Ulrich Meier-Westhues
Karsten Danielmeier
Peter Kruppa
Edward P. Squiller

Polyurethanes

Coatings, Adhesives and Sealants

2nd Revised Edition

Foreword

A world without polyurethanes is hard to imagine these days. Polyurethane chemistry has become an established technology worldwide in various industrial applications. For innovative coatings, adhesives and sealants, it plays a significant role and has been in many cases the key to new technology implementation. Polyurethanes enable innovation and technological advancement and help to develop products that meet market needs, devising efficient and environmentally friendly manufacturing processes and asserting a position in the global competitive environment.

In 2007 the first edition of the textbook Polyurethanes for Coatings, Adhesives and Sealants was published with the purpose to give a comprehensive overview of the potential offered by polyurethane chemistry. It was a time, when the significance of solvent-free, water-borne or UV-cured systems had steadily increased and reached a considerable importance. While solvent-borne formulations occupied center stage for many years before, polyurethane chemistry suddenly demonstrated that environmental improvements, better quality and economy were not mutually exclusive, but could be synergistic.

The book became an established standard work for professionals and students in technical and commercial areas, who desired the know-how and background of polyurethane related topics for coatings, adhesives and sealants.

Now, 11 years later, we once more deal with this topic and analyze the current role of polyurethane chemistry for coatings, adhesives and sealants, taking the development within the past decade into account. Based on our findings we updated the book, added new technical developments, extended the global view and included increasingly important topics such as sustainability and digitalization. Additionally, we refreshed citations and reduced content in areas deemed to be less relevant as they had been eleven years ago.

Polyurethane chemistry appears to have become even more established in the last decade and has continued to develop. Some of the trends foreseen in the previous edition have come to fruition, e.g., the trend towards water-borne systems. Furthermore, new, even more specialized systems and application technologies have been developed. This widened the applications fields and pushed the performance of the PU technology even further. Examples include the first bio-based PU raw materials which are entering the market, new catalyst systems which have been developed, or the digital printing process as new application technology that makes its way into the market place. Due to the adaptability of polyurethanes, modern systems are able to address the more stringent market, legislation and sustainability demands of today, and it is anticipated that polyurethanes will be able to do this in the future. The enormous diversity and possible combinations of polyurethane raw materials result in an impressive breadth of properties enabling the development of specific, customized solutions. The potential of further development is by no means exhausted.

Although the book has been revised, we took care to maintain the original intent of providing a comprehensive overview that allows newcomers to the industry an understanding of the principles and the potential of the polyurethane chemistry in the applications of coatings, adhesives and sealants, as well as, give the experienced specialist the option to refresh their knowledge and inspire the interested reader to think about how new innovations in polyurethanes can solve the problems of the future. The book opens with an introduction to polyurethane chemistry and technology, followed by a discussion of the many different applications, their current significance and their future prospects. Like the first edition of the book, the second one was also created in a global team approach involving experts in their respective fields for each chapter including experts of the Covestro group of companies, as well as, partners from universities and institutes. We would like to thank all of them for their contribution and patience. The book would not have been possible without their commitment and dedication to the PU technology.
The following authors contributed to the respective chapters in alphabetical order:

- Chapter 1: Karsten Danielmeier, Peter Kruppa, Ulrich Meier-Westhues, Edward P. Squiller
- Chapter 2: Florian Golling, Mareen Sandrock, Diethelm Rappen, Florian Stempfle, Dagmar Ulbrich
- Chapter 3: Evgeny Avtomonov, Karsten Danielmeier, Piet Driest, Christoph Eggert, Veronika Eilermann, Lyubov Gindin, Stephanie Goldfein, Florian Golling, Hans Georg Grablowitz, Dorota Greszta-Franz, Christoph Irle, Hans-Josef Laas, Michael Ludewig, Dieter Mager, Frank Richter, Myron Shaffer
- Chapter 4: Dirk Achten, Piet Driest, Wolfgang Fischer, Charles Gambino, Florian Golling, Michael Hilt, Martin Melchiors, Raul Pires, Torsten Pohl, James W. Rawlins, Eva Tejada Rosales, Christoph Thiebes, Jörg Tillack, Robert Wade, Jan Weikard
- Chapter 5: Thomas Baeker, Beate Baumbach, Kurt Best, Richard Shen, Ellen Chu, William Corso, Theivanayagam Deivaraj, Sebastian Doerr, Ulrich Freudenberg, Florian Golling, Hans Georg Grablowitz, Scott Grace, Tanja Hebestreit, Annette Hüttner, Christoph Irle, Jinqi Li, Markus Mechtel, Thomas Michaelis, Hiroshi Morita, Ahren Olson, Contardo Pafumi, Joe Pierce, Torsten Pohl, Steven Reinstadtler, Thomas Schüttler, Edward P. Squiller, Srba Tasic, Eva Tejada Rosales, Christoph Thiebes, Rainer Trinks, Robert Wade, Jan Weikard, Andreas aus der Wieschen, Todd Williams, Karl H. Wuehrer, Anson Xue, Sherry Yang, Mary Ye, Frank Zhang, Steven Zhu
- Chapter 6: Dirk Achten, Wolfgang Arndt, Beate Baumbach, Jörg Büchner, Jeff Dormish, Winfried Jeske, Anand Khot, Martin Melchiors, Rainer Trinks

- Chapter 7: Jay Johnston, Christoph Thiebes
- Chapter 8: Dirk Achten, Friedrich-Karl Bruder, Sebastian Doerr, Chris Koppenborg, Thomas Rölle, Sophie Viala, Marc-Stephan Weiser
- Chapter 9: Péter Krüger, James A. Thompson-Colón, Thomas P. Fäcke
- Chapter 10: Frithjof Hannig, David Harrison, Linda Liu, Lisa Marie Nespoli, Joachim Petzoldt, Jürgen Schrot
- Chapter 11: Eric Bischof, Péter Krüger, Lydia Simon, Edward P. Squiller, Daniel Steinke
- Chapter 12: Karsten Danielmeier, Peter Kruppa, Ulrich Meier-Westhues, Edward P. Squiller

Most notably we like to thank James W. Rawlins, Wolfgang Fischer and Michael Hilt for their contribution in the respective chapters and Piet Driest, Christoph Eggert, Veronika Eilermann, Lennart Gehrenkemper, Silke Köster, Birgit Schäfer, Tina Stockhausen as well as Jan Sütterlin for their support in managing diverse challenges with regard to the writing of the book.

Leverkusen, Germany, January 2019

Ulrich Meier-Westhues
Karsten Danielmeier
Peter Kruppa
Edward P. Squiller

Contents

BUILDING BRIDGES FROM
INNOVATION TO SUSTAINABILITY.
WHY NOT?
Through curiosity, courage and creativity Covestro has become a world leader in high-tech polymer raw materials for products from many areas of everyday life. Ever since Otto Bayer and his team discovered polyurethane chemistry in 1937, we have been pushing boundaries through groundbreaking innovations to enhance the efficiency, quality, and environmental footprint of our polymers. Nowadays, we work closely with our customers to develop innovative and sustainable products along the entire value chain.
What may we develop for you?
www.covestro.com
covestro

1 Introduction

1.1 Historical aspects

Coating, bonding and sealing are techniques that humankind has used for many centuries. For a long time, only natural resins, oils and fats were used for these purposes.

Shellac, a natural resin secreted by the scaly lac insect, has been used in India as a weather-resistant coating for surfaces. The word *lacquer* in English is derived from the Sanskrit word *laksha*, which means one hundred thousand and describes the unimaginably large number of insects required to produce shellac lacquers. Later it was learned that lacquer resins could also be obtained from other sources, e.g. by boiling down wood oil.

Animal tissue, especially bones and hide, were the basis for glues and adhesives in many applications for a long time. The makers of high-quality glues were called *Kellepsos* in ancient Greece. During the Middle Ages development was largely static. The invention of the printing press by *Johannes Gutenberg* then led to a new and rapidly growing need for adhesives in the emerging bookbinding industry.

The development of synthetic resins began in the early 20th century driven by the oil industry and the emerging downstream industry with their related products. This laid the foundation for the production of coatings, adhesives and sealants of vastly improved quality and in volumes. Polyurethanes were discovered in 1937 when *Heinrich Rinke* produced 1,6-hexamethylene diisocyanate (HDI) and *Otto Bayer* developed the diisocyanate polyaddition process[1–3]. Initial research in this new field of polymer chemistry in the 1940s focused on polyurethane fibers, while the first polyurethane foams were produced a little later.

Figure 1.1: Otto Bayer – Inventor of polyurethane chemistry[4]

Fifty years ago, the first polyurethane coatings were developed. Otto Bayer and his team discovered that the technical properties of alkyd resins could be improved through modification with diisocyanates. However, the real conquest of the coatings sector by polyurethanes only began with the development and industrial use of low-monomer polyisocyanates. The first products were based on toluene diisocyanate. Due to the aromatic nature of the base isocyanate, these tend to yellow on exposure to light and can therefore only be used for interior applications or in primers.

The range of applications was broadened later with the introduction of products based on aliphatic diisocyanates, initially hexamethylene diisocyanate. "Desmodur" N was the first product of this type and was launched in the early 1960s by Bayer AG. Gradually, the two-component (DD) coatings prepared by combining polyisocyanates ("Desmodur") with polyols ("Desmophen") replaced the traditional alkyd coatings, first in the coating of large vehicles. The driving force was the quality of the coatings which, even when dried under mild conditions, matched the performance of coatings which had been baked. This is important when coating large vehicles (airplanes, rail wagons and buses) as their size makes baking impossible.

In the 1970s, it was found that the quality of automotive refinish coatings could be substantially improved with the help of polyurethane chemistry. By adding polyisocyanates based on isophorone diisocyanate (IPDI) to the medium oil-based alkyd resins mainly used at that time, the hardness, overcoatability and gasoline resistance of the resulting coatings could be improved significantly. Today, two-component polyurethane coatings have almost completely replaced alkyd resin chemistry in this segment.

The broad range of applications for polyurethanes in coatings was quickly recognized. Other examples of applications include wood finishing, corrosion protection and construction, as well as textile coating. Another advance has been the development of two-component metering technology. The breakthrough in automotive OEM finishing occurred in the mid-1980s, and since this time, plastic coatings have become a further domain for polyurethanes.

Polyurethane adhesives came onto the market in the 1950s with the development of the hydroxyl polyurethanes and the first trifunctional isocyanate crosslinker, "Desmodur" R. The early 1960s also saw the development of plasticizer-resistant hydroxyl polyurethanes, which laid the foundation for the success of these products in shoe manufacture.

Solvent-free polyurethane reactive adhesives have been used since the 1970s, first in automotive production, and then in the manufacture of laminated films and sandwich elements. They were later joined by reactive sealants. Since the 1990s, polyurethane-based reactive and waterborne adhesives have gained significant market share in automotive, construction, furniture and shoe production.

The process of substituting traditional technologies in coatings, adhesives and sealants with polyurethane is ongoing, and can be observed occurring around the world. Against a background of increasingly demanding quality requirements, ever more stringent environmental legislation, and cost optimization of the end-product manufacturing processes, there has been growth in the use of low-solvent, solvent-free, waterborne and radiation-curing formulations of one- and two-component polyurethane systems[5–9]. Bearing in mind current concomitant developments, polyurethanes will continue to gain further importance. Their spectrum of use will thus expand beyond the established applications into other new areas.

1.2 Definition of scope

This book describes the use of polyurethane raw materials for coatings, adhesives and sealants in selected application areas. Topics covered include applications in the wood and furniture industry, the automotive sector, construction, the broad area of metal coatings including corrosion protection, the shoe industry, and plastic coating and bonding. Also discussed are the manufacture of laminated films and the coating of textiles, leather, glass and paper. In addition to describing the chemical and technical principles involved, the issues of occupational hygiene and sustainability associated with the handling of polyurethane coatings, adhesives and sealants are covered. The broad spectrum of applications is evidence of one key property of polyurethanes: the versatility resulting from their chemistry which is also exploited in segments other than coatings, adhesives and sealants.

For example, polyurethane raw materials are used in the manufacture of foams: rigid foams for insulation (construction industry, refrigerators) or energy-absorbing components in automobile interiors (instrument panels); integral skin foams, e.g. for furniture and medical applications; flexible foams for upholstery, mattresses and packaging materials. Other applications for polyurethanes are found in the manufacture of versatile elastomers for the footwear and electrical industries, thermoplastic urethanes, e.g. for sports and leisure equipment, and polyurethane elastic fibers for stretch fabrics.

This book does not intend to address all these applications. Detailed information can be found in other sources[10–12].

1.3 References

[1] DRP 728 981 (1937) I.G. Farben

[2] O. Bayer, Angew. Chem. 59 (1947) 257

[3] D. Dieterich, Chemie in unserer Zeit, 24 (1990) 135

[4] Archiv Bayer AG, Leverkusen

[5] Brock, Groteklaes, Mischke, European Coatings Handbook, 2nd editon (2010) Vincentz Network ISBN: 978-3-86630849-7

[6] Goldschmidt, Streitberger, BASF Handbook on Basics of Coating Technology, 3rd edition, BASF Coatings AG, Münster, Vincentz Network, Hannover, 2018, ISBN 978-3-86630-336-2

[7] W. Brockmann, P. L. Geiss, J. Klingen, B. Schroeder et al., Adhesive bonding: materials, applications and technology, Wiley-VCH, Weinheim 2009

[8] R. M. Evens, Polyurethane Sealants, CRS Press Inc. Lancaster 1993

[9] K. L. Mittal, A. Pizzi, Handbook of Sealant Technology, CRS Press 2009

[10] G. Avar, U. Meier-Westhues, H. Casselmann, D. Achten in K. Matjaszewski, M. Möller (eds.), Polymer Science: Vol 10, 411–441, Elsevier BV, 2012

[11] H.-W. Engels, H.-G. Pirkl, R. Albers, R. W. Albach, J. Krause, A. Hoffmann, H. Casselmann, J. Dormish, Angew. Chem., 125, 9596–9616, 2013

[12] G. Oertel, Polyurethane Handbook, Hanser Gardner Publications, 1994, ISBN-13: 978-156990-157-1

2 Economic aspects and market analysis

2.1 Introduction and definitions

When looking at the economic importance of coatings, adhesives, and sealants, there is a significant market of formulated products with 64 million tons or 170 billion € globally in 2017 (see Figure 2.1). The formulated product always describes the ready to use material as it will be applied to the end product – being it a wall, car or industrial good. The formulated product contains the so called resin, binder or film forming component and further additives, pigments, fillers and organic solvents or water. This market is split roughly 75:25 for coatings versus adhesives and sealants with 49 versus 15 million tons of material (see Figure 2.1).

With regard to the resin within the formulated product, several market studies and also the authors of this book refer to the resin supply form. This includes not only the mon-

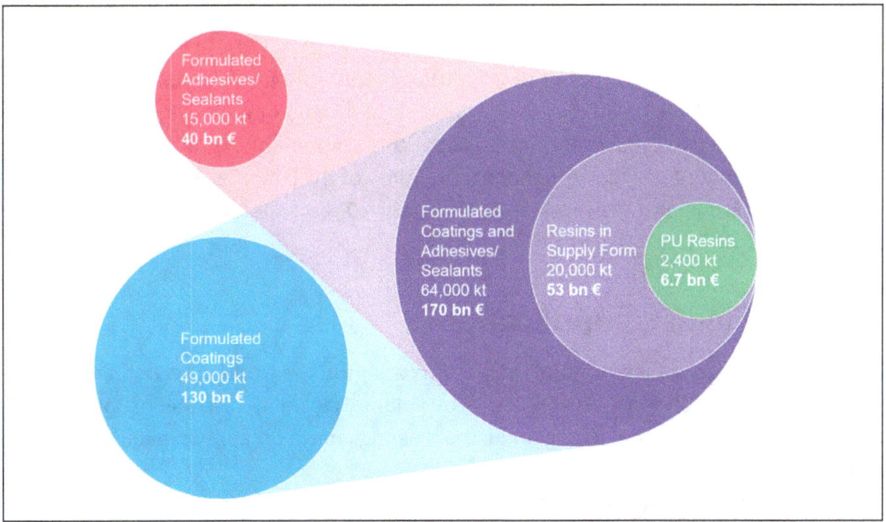

Figure 2.1: Overview world consumption 2017 for coatings, adhesives and sealants[1]

omeric, oligomeric or polymeric resin material itself, but also a certain share of organic solvent or water that is usually included in the commercially available products.

Other market data sources use the term of dry resin to describe only the active resin material without any solvents that will eventually evaporate in the final film forming process during application of the formulated product.

Within this book the term polyurethane (PU) chemistry is used to describe markets, which comprise all resin materials that are crosslinked by urethane or urea groups. In addition, the term also covers all materials that are indirectly based on isocyanates or the reaction products thereof. This means, that the isocyanate is not necessarily reacted in the last crosslinking step, but might be used as a precursor (see Chapter 3). In general, polyurethane resins consist of aliphatic polyisocyanates, aromatic polyisocyanates, prepolymers, polyurethane dispersions (PUDs), polyols and specific diamines.

Based on these definitions, polyurethane based resins have a market size of 2.4 million tons or 6.7 billion €. This represents a 12 % share of the overall resins market in supply form. Therefore, polyurethane chemistry is usually recognized as a specialty market where the final application demands for increased performance, e.g. durability or quality of appearance.[1]

2.2 Coatings

The world consumption of coatings for industrial and architectural applications (i.e. the total production of formulated coating products sold) amounted to 49,000 kilotons in 2017 (see Figure 2.2). This represented global consumption of 16 million tons of coating resins in supply form, corresponding to a value of 44 billion €. Polyurethane resins account for 1.6 million tons representing roughly 10 % of the overall resins market in supply form.[2]

2.2.1 Fields of application

There are two general fields of application for coatings in the world: The architectural applications account for approximately 66 % by volume, including dispersions for architectural coatings, whereas the general industrial coatings make up approximately 34 % (see Figure 2.3).

The architectural market is dominated by interior wall applications, i.e. wall emulsions/dispersions with 51 %. This segment is followed by exterior wall coatings with 24 %. After wood coatings with 15 %, floor coatings (8 %) and roof coatings (2 %) hold the smallest share (see Figure 2.3). Within this market, polyurethane technology only holds a very small share and thus the architectural applications are not further considered within this book with regard to market data.

The industrial market accounts for 34 % of the coating market and can be divided into seven different fields (see Figure 2.4, see page 25) with the largest, named metal, representing roughly 40 % of the market share. This application is followed by protective and marine with 22 %. Industrial wood and automotive account for 14 % and 13 %, respectively. They are followed by agricultural, construction, earthmoving equipment (ACE) and plastics with 4 % each. The smallest segment is transportation with 3 %.[4] Other markets include textile, leather, glass and paper coating. The market size for textile and leather coating sums up to approximately 5,100 kilotons of resins in supply form, of which 50 % is based on polyurethane technology.[5] This market is not reflected in the following graphs but will be discussed in detail in the respective chapter.

The weighting is different if one compares polyurethane coatings to the total formulated industrial coatings market

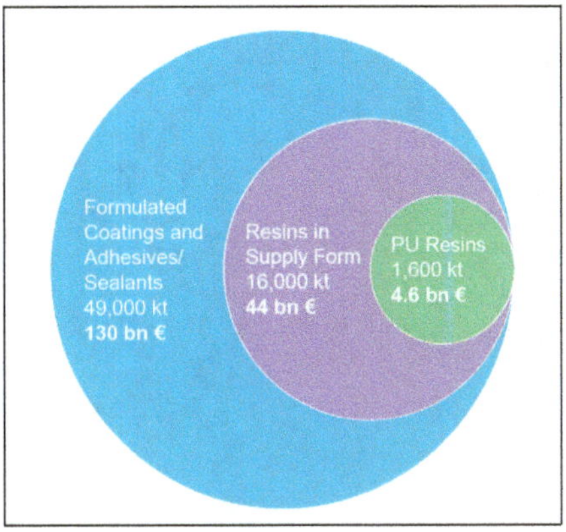

Figure 2.2: Overview world consumption 2017 for industrial and architectural coatings[2]

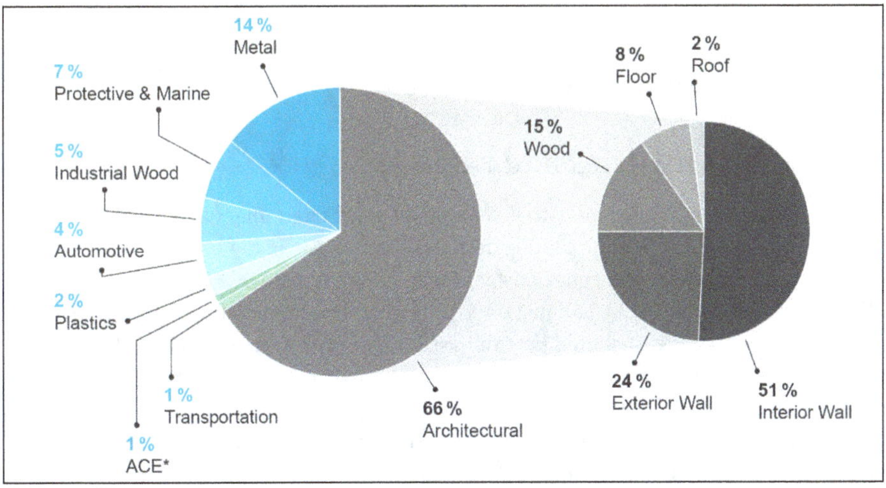

Figure 2.3: Formulated coatings volumes per application 2017 (industrial and architectural): 49,000 kt in total[3]

**ACE: Agricultural, Construction, Earthmoving Equipment*

Table 2.1: Formulated polyurethane coatings volume per application vs. the total formulated industrial coatings market[4]

Application	Total industrial coatings market per application [%]	Polyurethane industrial coatings market per application [%]
Automotive	13	21
Transportation	3	8
Industrial wood	14	24
Protective & marine	22	16
Plastics	4	15
Agricultural, construction, earthmoving equipment	4	6
Metal	40	10
	100	**100**

(i.e. the sum of all coatings types) in terms of share of technologies (see Table 2.1): Industrial wood, automotive, plastics, and ACE have a higher or even significantly higher market share in the polyurethane market compared to the total formulated industrial coatings market. This clearly shows that high-quality polyurethane resins are preferred for applications with more demanding requirements in terms of quality and durability – and these are found accordingly in a higher price segment.

2.2.2 Chemistries and regions

The world consumption for industrial coatings is distributed unequally (see Figure 2.5). More than half of the industrial coating demand is consumed in the Asia-Pacific (APAC) region, while slightly more than one quarter is used in EMEA, namely Europe (21 %) and the Middle East and Africa (4 %), and just under one quarter is going to the Americas (NAFTA and LATAM) with 23 %. It should be emphasized here that APAC's share of global industrial coatings consumption has grown in the last five years (2012 to 2017) to more than 50 % and is expected to continue to rise.

The volume of formulated solvent-borne coating products summed up to 65 % over all technology classes (see Figure 2.6), whereas water-borne coatings, including electrocoats for cathodic electrodeposition (CED), make up 17 % of the formulated coatings market. The remaining share is distributed among powder coatings (14 %), UV coatings (3 %), and others (1 %).

The volume distribution by chemistry type shows that the traditional systems, such as alkyd resins, acrylics, and polyesters, including others, are still of great importance (see Figure 2.7). In 2017, these product groups accounted for just over 51 % of world consumption. Polyurethane coatings have a market share of 20 %.

The main components of polyurethane coating resins are polyisocyanates, polyurethane dispersions, and other polyurethane materials, such as prepolymers and polyols based on polyacrylates, polyesters, polyethers as well as specific diamines. The 1.6 million tons of polyurethane coating resins used worldwide are distributed as shown in Table 2.2 (see page 26):
The key isocyanates used in coatings resins are polyisocyanates, prepolymers and further derivatives. Aliphatic materials are based on hexamethylene diisocyanate (HDI), isophorone

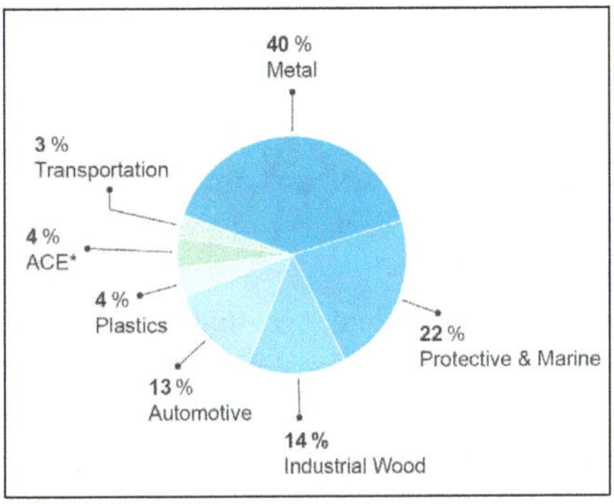

Figure 2.4: Formulated industrial coatings volume per detailed areas of application in 2017, around 17,000 kt[4]

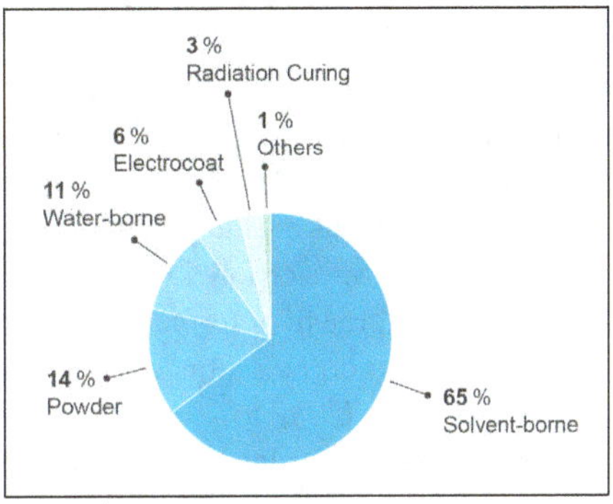

Figure 2.6: Formulated industrial coatings volumes per technology 2017, around 17,000 kt[4]

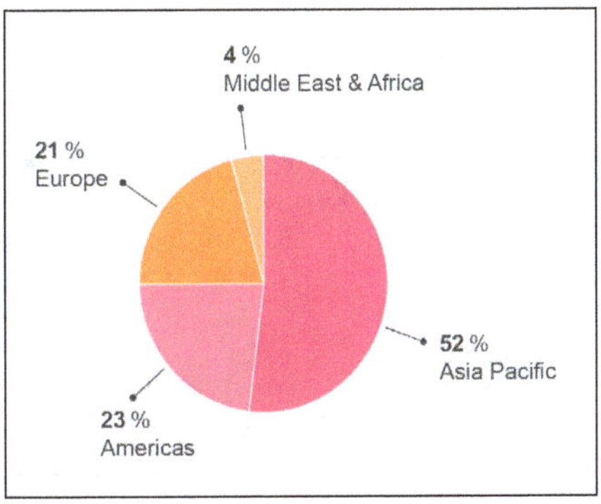

Figure 2.5: Formulated industrial coatings volumes per region 2017, around 17,000 kt[4]

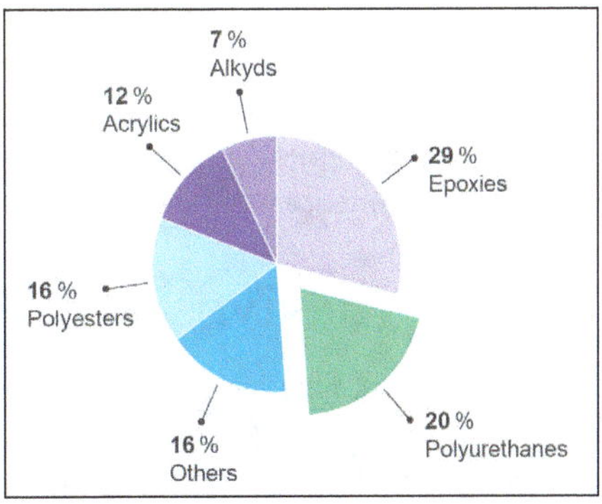

Figure 2.7: Formulated industrial coatings volumes per chemistry 2017, ~ 17,000 kt[6]

Table 2.2: Components of polyurethane coating resins in supply form (volumes in kt)[7]

	Resins in supply form [kt]	**Relative share [%]**
Aliphatic polyisocyanates	210	13
Aromatic polyisocyanates	130	8
Polyurethane dispersions	170	11
Polyols and prepolymers	1,090	68
Polyurethanes total	**1,600**	**100**

diisocyanate (IPDI) or dicyclohexylmethane diisocyanate (H_{12}MDI). Aromatic isocyanates are based on toluene diisocyanate (TDI) or diphenylmethane diisocyanate (MDI).

2.2.3 Market forecasts

For the coming years (2017 to 2021), annual market growth of 3.1 % is forecasted for formulated industrial coatings, which represents a higher growth rate than the 2.1 % observed from 2012 to 2017. It is expected that growth in APAC will be above global average at 4.1 %, whereas the European and American markets are expected to grow at around 1.8 % and 2.1 % respectively.

The market growth of formulated industrial coatings for different coating technologies reveals that water-borne coatings and electrocoats grew above regional market average in the past five years (2012 to 2017). This trend is expected to continue for APAC and Americas. For Europe, electrocoats should continue to grow above average, while water-borne coatings will grow at average market rates.

Polyurethanes are expected to experience an increasing annual growth of approximately 2.7 % in the upcoming years (2017 to 2021) compared to the past (2012 to 2017), which grew at 2.2 %. This is driven by APAC and Middle East and Africa with roughly 4 % annual growth. For polyurethanes, the market growth was driven by different technologies, depending on the region. Most of the growth was accounted for by solvent-borne and powder coatings in APAC, while in Europe water-borne and UV coatings posted stronger growth. In the Americas, UV and solvent-borne coatings grew above average.[4]

2.3 Adhesives and sealants

In 2007, the world adhesives and sealants industry had a volume of approximately 11.8 million tons and was worth some 30 billion €.[8] After a decrease of about 4 % during the financial crisis in 2008/2009 and the subsequent economic recovery in the following years the volume and value growth in this industry sector averaged around 3 %. In 2017, world demand of formulated adhesives and sealants reached approximately 15 million tons, which according to the latest estimates represent a value of approximately 40 billion € (see Figure 2.8).[9] This total volume of formulated adhesives and sealants corresponds to a global consumption of approximately 3.8 million tons of resins in supply form in 2017. The total world market of all polyurethane resins in supply form for adhesives and sealants amounts to approximately 800,000 tons.[10]

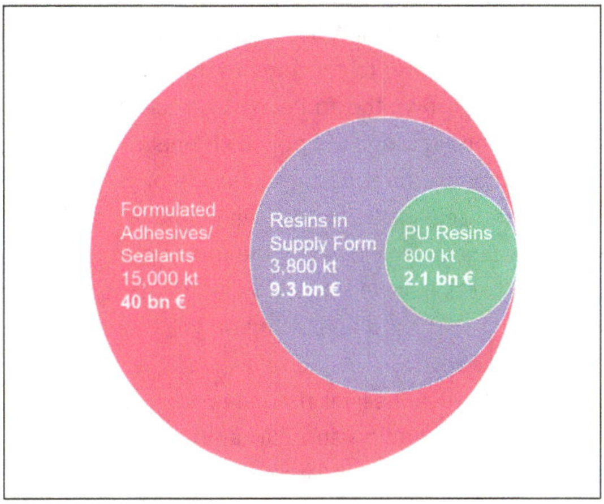

Figure 2.8: World consumption for adhesives and sealants in 2017[11]

2.3.1 Fields of application

The main field of application for adhesives and sealants is the packaging segment (approximately 37 %), followed by the construction industry (24 %), assembly (13 %), woodworking (10 %) and transportation (8 %). Consumer/do-it-yourself (DIY) applications and footwear have a share of 5 % and 3 % respectively (see Figure 2.9).

Considering the polyurethane-based adhesives and sealants, a slightly different picture is observed. Of the total market of approximately 1.4 million tons, around 30 % is attributed to the packaging segment, 19 % to construction, 16 % to transportation and 11 % to footwear. In these market segments, polyurethane-based adhesives and sealants are typically preferred for applications with more demanding requirements in terms of quality and durability.[12]

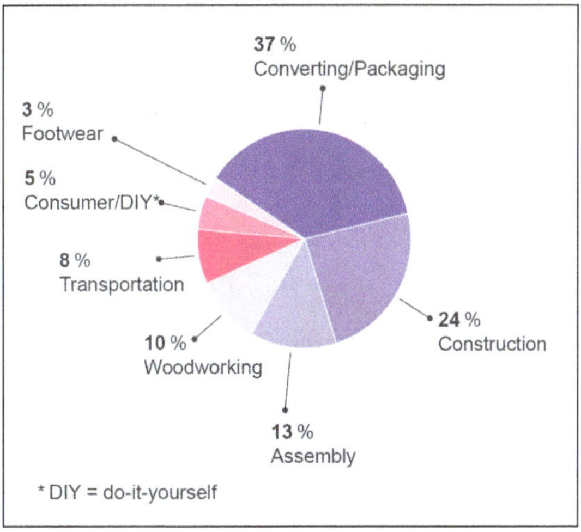

Figure 2.9: Worldwide volume demand of formulated adhesives and sealants by application, around 15,000 kt in total[9]

2.3.2 Chemistries and regions

The regional distribution of adhesives and sealants consumption is fairly balanced between Europe, Middle East and Africa (EMEA), Americas (North and South), and the Asia-Pacific region (APAC), each accounting for approximately one third of the total market volume (see Figure 2.10). APAC in particular has seen above-average growth in recent years, and this is expected to continue in the upcoming years.

Breaking down the adhesives and sealants market into the underlying material classes shows the broad range of applied chemistries. In addition to the polyurethane-based materials, natural products and a multitude of polymer materials can be used to formulate adhesives and sealants systems (see Figure 2.11).

In recent years, however, the more complex application profiles in adhesives and sealants processes have led to a growing demand for high-quality systems. In particular, polyurethane systems are benefiting due to their outstanding property profile compared with other materials.

Overall the adhesives market is dominated by water-borne systems due to large applications in construction and packaging (see Figure 2.12). Solvent-borne adhesives are still traditionally used in some applications with a volume market share of 14 %. New systems of water-borne materials and formulations have been developed to ensure compliance with VOC (volatile organic compounds) workplace guidelines.

The main components of polyurethane resins for adhesives and sealants are isocyanates, polyisocyanates and prepolymers as well as polyols such as polyesters or polyethers. The most important isocyanates, polyisocyanates and prepolymers for adhesives and sealants resins are based on the aromatic diphenylmethane diisocyanate (MDI) and toluene diisocyanate (TDI), as well as on the aliphatic hexamethylene diisocyanate (HDI), isophorone diisocyanate (IPDI), dicyclohexylmethane diisocyanate ($H_{12}MDI$), and the corresponding derivatives of these diisocyanates.

Polyesters may have an aliphatic, aromatic or mixed (araliphatic) structure, vary in chain length, and exhibit amorphous or crystalline properties. Polyethers can be varied to yield the desired property profile through the choice of starter or chain-building monomers.

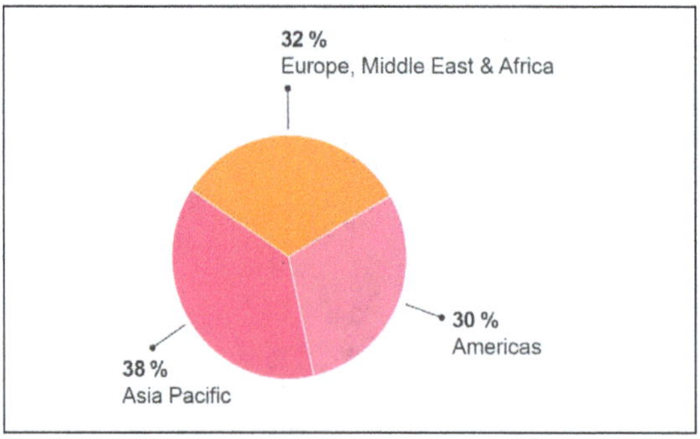

Figure 2.10: Worldwide demand for adhesives and sealants by region (volume of formulated product) 2017, around 15,000 kt in total[9]

2.3.3 Market forecasts

In the upcoming years (2017 to 2021) the global market for adhesives and sealants resins in supply form is expected to grow at least 4.0 % per year compared to a global growth rate of 3.5 % in the past five years (2012 to 2017). In the APAC region growth is expected to be above global average at up to 4.9 %, whereas Europe and American markets are forecasted at 3.7 % and 3.9 %, respectively.[9] Polyurethane materials in general are expected to grow at least with the market. Here, besides the construction area, converting and packaging as well as footwear will be the main growth drivers.[12]

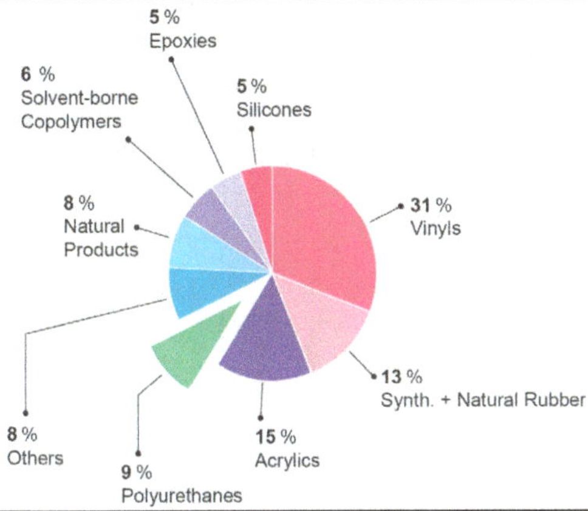

Figure 2.11: Worldwide demand for adhesives and sealants by material type (volume of formulated product) 2017, around 15,000 kt in total[9]

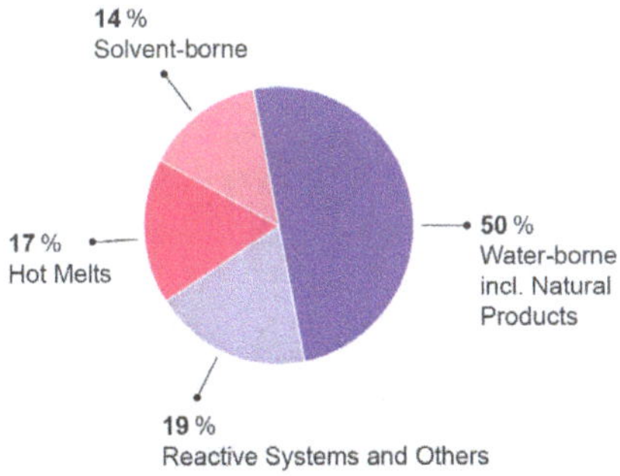

Figure 2.12: Worldwide demand for adhesives and sealants by technology (volume of formulated products) 2017, around 15,000 kt in total[9]

2.4 References

[1] "Irfab" Global Industrial Coatings Markets (GICM) and Global Architectural Coatings Market (GACM) 2015, 2017, 2018 PRA World Ltd, CHEM Research 2018, Orr & Boss 2017, Covestro assumptions
[2] "Irfab" Global Industrial Coatings Markets (GICM) and Global Architectural Coatings Market (GACM) 2015, 2017, 2018 PRA World Ltd, Orr & Boss 2017, and Covestro assumptions
[3] "Irfab" Global Industrial Coatings Markets (GICM) and Global Architectural Coatings Market (GACM) 2015, 2017, 2018 PRA World Ltd
[4] "Irfab" Global Industrial Coatings Markets (GICM) 2017, 2018 PRA World Ltd.
[5] Covestro estimate 2018
[6] "Irfab" Global Industrial Coatings Markets (GICM) 2017, 2018 PRA World Ltd. and Covestro assumptions
[7] Orr & Boss 2017
[8] CHEM Research, Market Review – FEICA Conference Izmir Sept. 13, 2013 and Covestro assumptions
[9] CHEM Research 2018
[10] Orr & Boss 2017. CHEM Research estimates the market to be up to 5.5 million tons of resin in supply form
[11] CHEM Research 2018, Orr & Boss 2017 and Covestro assumptions
[12] Orr & Boss 2017 and Covestro assumptions

3 Chemical principles

3.1 Diisocyanates

Polyurethanes are obtained by the reaction of polyfunctional polyols (in coatings formulations often referred to as component A) and polyfunctional isocyanates (component B). The latter are synthesized by oligomerization of monomeric diisocyanates. The diisocyanates are usually prepared on an industrial scale by liquid or gas phase phosgenation of their corresponding primary amines, and subsequent removal of the excess of monomeric isocyanates (see Figures 3.1 and 3.2).[1, 2]

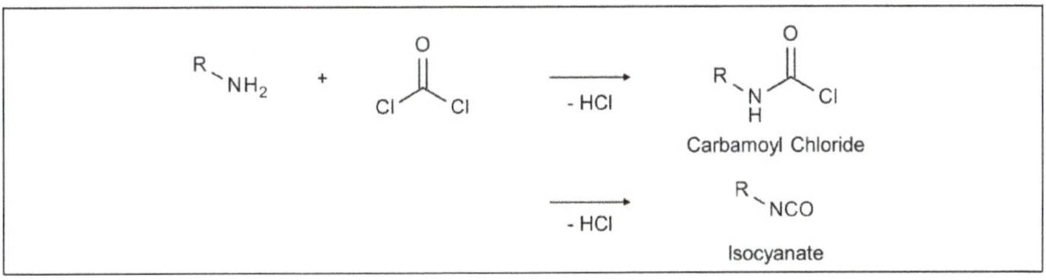

Figure 3.1: Manufacture of isocyanates by phosgenation of primary amines

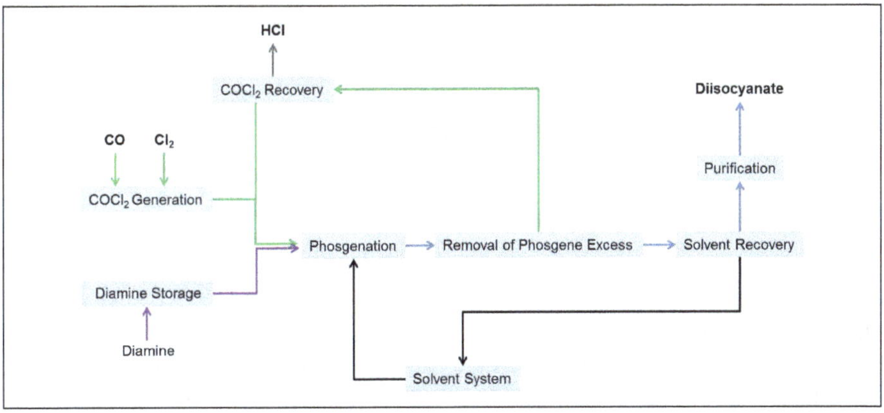

Figure 3.2: Principal technical procedure of manufacture of diisocyanates by phosgenation

Table 3.1: Industrial diisocyanates

	Manufacturers/trade names
Hexamethylene diisocyanate (HDI) 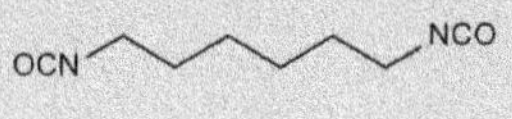	Covestro: "Desmodur" H BASF: "Basonat" H" Vencorex: "Tolonate" Wanhua: "Wannate" HDI Asahi KASEI: "Duranate" 50M-HDI
Isophorone diisocyanate (IPDI) (3-Isocyanatomethyl-3,5,5-trimethylcyclohexyl isocyanate) 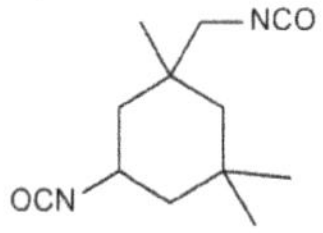	Covestro: "Desmodur" I Evonik: "Vestanat" IPDI Wanhua: "Wannate" IPDI
Bis-(4-isocyanatocyclohexyl)methane ($H_{12}MDI$) 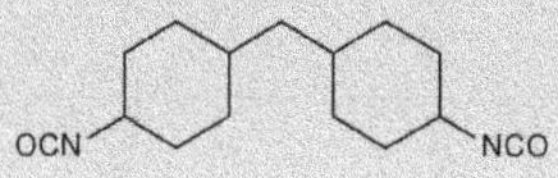	Covestro: "Desmodur" W Evonik: "Vestanat" $H_{12}MDI$ Wanhua: "Wannate" HMDI
2,4- and 2,6-Toluene diisocyanate (TDI) 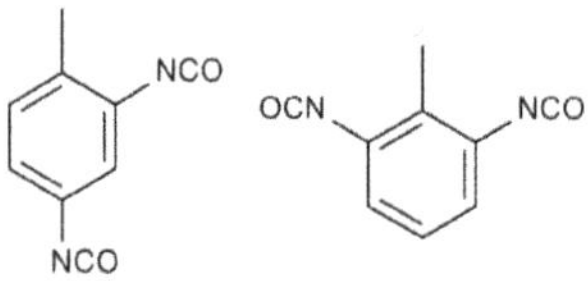	Covestro: "Desmodur" T BASF: "Lupranat" T Dow Chemical: "Voranate" T Vencorex: "Scuranate" T
Diphenylmethane 4,4'- and/or 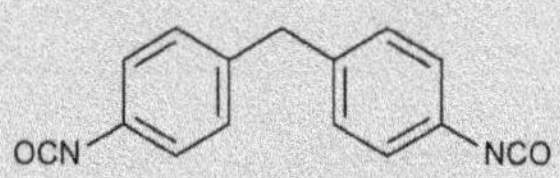 -2,4'-diisocyanate (MDI) 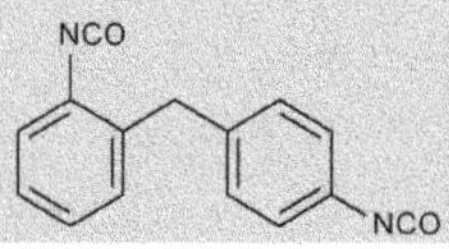	Covestro: "Desmodur" 44 M BASF: "Lupranat" M Dow Chemical: "Isonate" Huntsman: "Suprasec" Wanhua: "Wannate" MDI and/or Covestro: "Desmodur" LS 2424

Alternatively, for selected aliphatic diisocyanates, phosgene-free manufacturing processes have been developed. One such process involves the reaction of an amine and an alcohol with a urea to give a urethane that is then split at elevated temperatures to yield an isocyanate. This process is also used in the industrial production of some diisocyanates, such as bis-(4-isocyanatocyclohexyl) methane (see Figure 3.3).[3]

The standard commercial polyisocyanates used in coatings and adhesives are all derived from just a few diisocyanates with aliphatically, cycloaliphatically or aromatically bound isocyanate groups.[4] The most important diisocyanates that are available on an industrial scale are summarized in Table 3.1.

In addition to these, a number of other monomeric diisocyanates for the manufacture of specialty coating and adhesive raw materials have been described. However, these have yet to achieve widespread industrial significance. Pentamethylene diisocyanate (PDI) is the most recent example.[4] This product, in contrast to petrochemical based diisocyanates, is produced from a renewable feedstock with improved carbon footprint versus hexamethylene diisocyanate (HDI). Table 3.2 shows some examples of diisocyanate specialities.

With the exception of MDI which has accorded special status on account of its low vapor pressure, other monomeric diisocyanates have a significant volatility. For occupational health reasons, monomeric diisocyanates are generally not used as coating and adhesives raw materials. To overcome the hazardous potential related to diisocyanate monomers and to achieve low volatility, they must first be converted into higher molecular weight polyisocyanates, using suitable modification reactions like the formation of water-borne polyurethane dispersions, UV curable resins, prepolymers, and polyisocyanate crosslinkers. When necessary, the removal of the monomeric diisocyanates is done as part of the production process of these modified products. These components and different classes of polyurethane resins will be covered in subsequent chapters.

3.2 Isocyanate reactions

The most important type of reactions involving isocyanates is the addition of compounds containing active hydrogen atoms, especially polyols, polyamines, and to some much lesser extent thiols and carboxylic acids. Table 3.3 provides an overview of basic reactions of the isocyanate group.

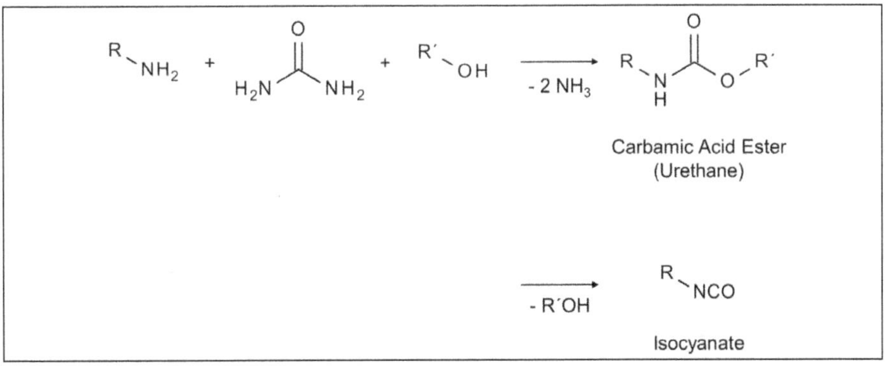

Figure 3.3: Manufacture of organic isocyanates via the urea route

Table 3.2: Diisocyanate specialties (selection)

	Manufacturer/trade name
Xylylene diisocyanate (XDI) 1,3-bis(isocyanatomethyl)benzene 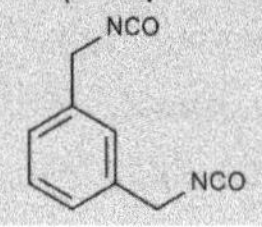	Mitsui Chemicals: "Takenate" 500
Hydrogenated xylylene diisocyanate (H_6-XDI) 1,3-bis(isocyanatomethyl)cyclohexane 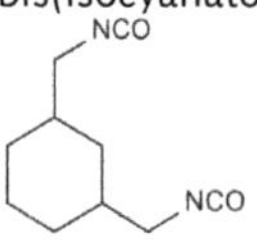	Mitsui Chemicals: "Takenate" 600
2,2,4- and 2,4,4-Trimethyl-1,6-diisocyanatohexane 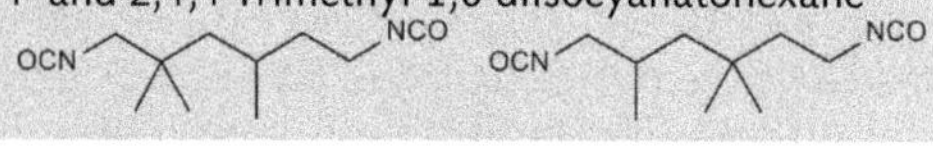	Evonik: "Vestanat" TMDI
Tetramethylxylylene diisocyanate (TMXDI)[5] 1,3-bis(2-isocyanatoprop-2-yl)benzene 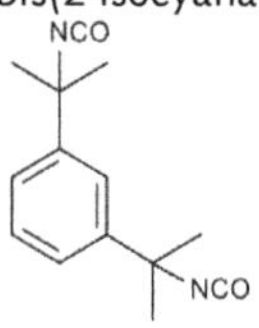	Allnex: TMXDI
1,5-Pentamethylene diisocyanate 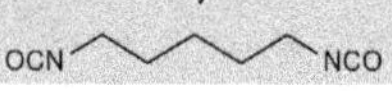	Covestro: "Desmodur" P Mitsui Chemicals: "STABiO"
Triphenylmethane-4,4',4''-triisocyanate 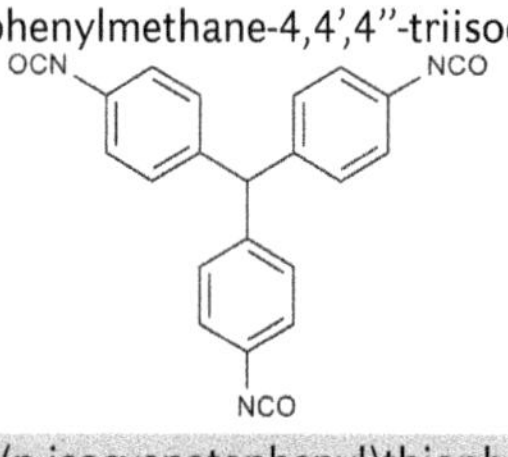	Covestro: "Desmodur" RE
Tris(p-isocyanatophenyl)thiophosphate 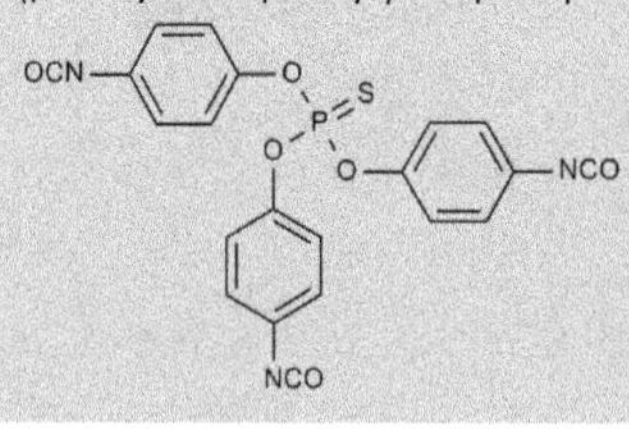	Covestro: "Desmodur" RFE

Table 3.3: Schematic reaction principles of isocyanates

Reaction with	Reaction equation
Alcohol to urethane	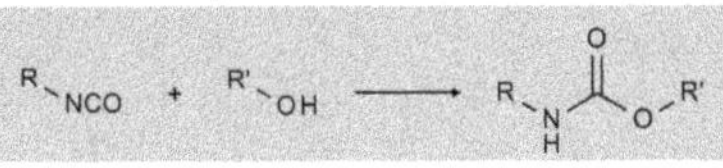
Urethane to allophanate	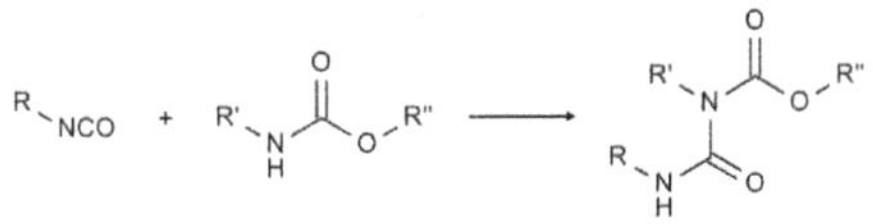
Amine to urea	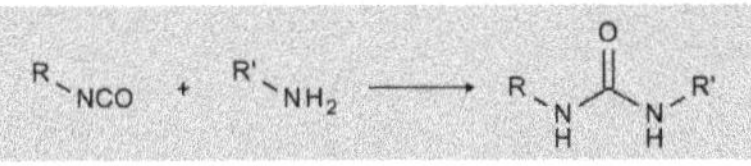
Water to urea	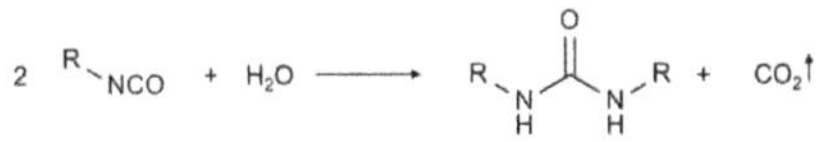
Urea to biuret[7]	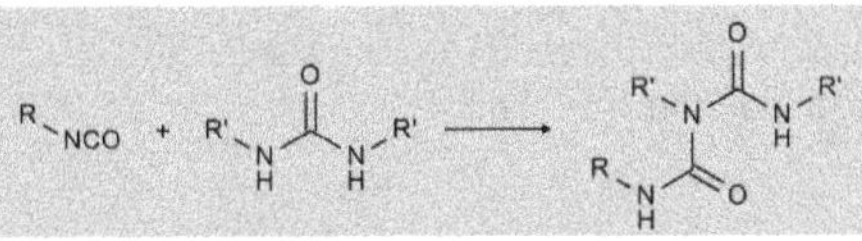
Carboxylic acid to amide[8a]	
Amide to acyl urea	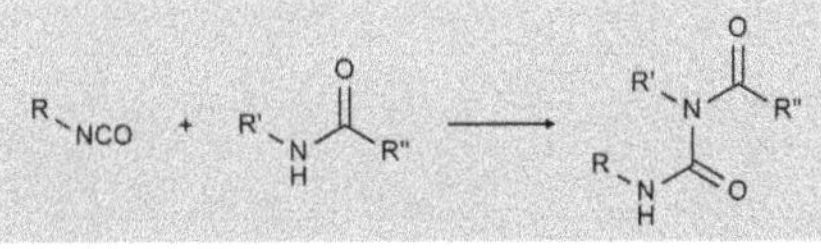
Anhydride to imide	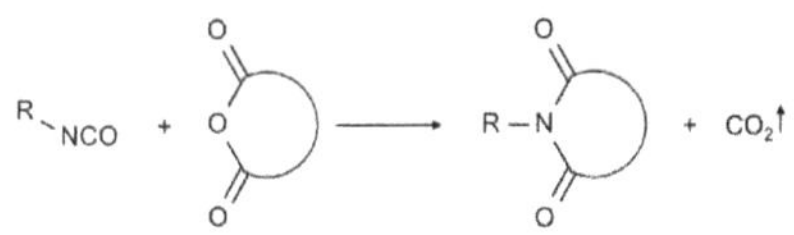
Epoxide to oxazolidone	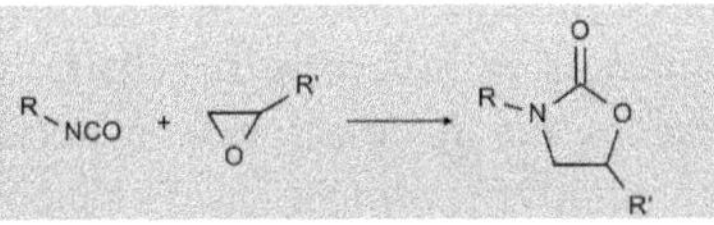
Oxime to oxime carbamate	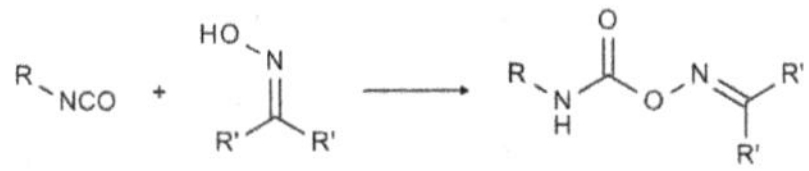
Carbon dioxide to oxadiazinetrione[8b]	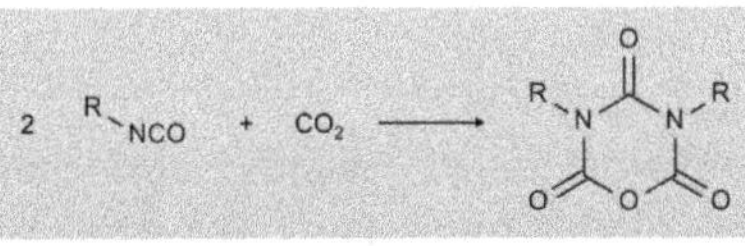

Table 3.4: Schematic (cyclo)polymerization of isocyanates

Formation of	**Reaction equation**
Uretdione[11] (dimer)	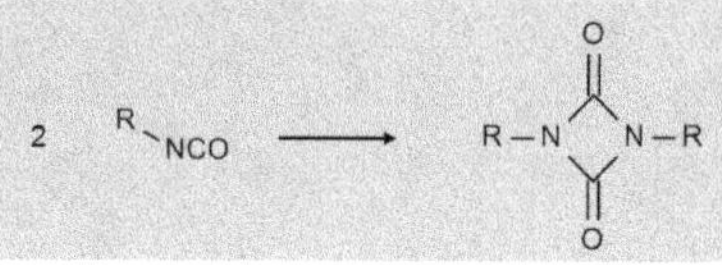
Isocyanurate[12] (trimer)	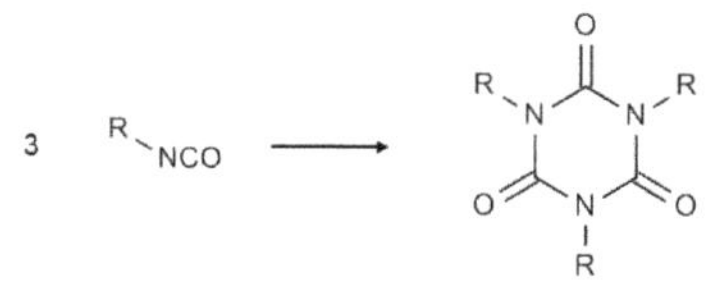
Iminooxadiazinedione[13] (asym. trimer)	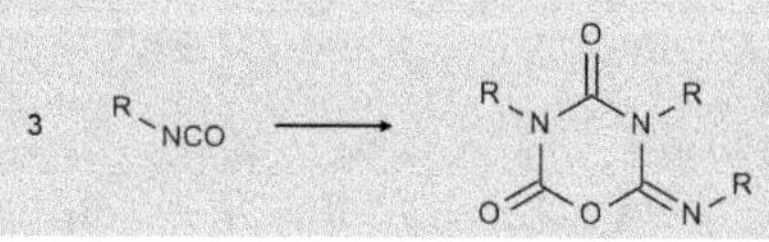
Linear polymer (α-"Nylon")	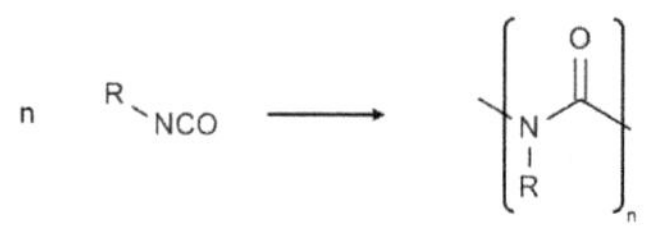
Carbodiimide[14]	$2\ R{-}NCO \longrightarrow R{-}N{=}C{=}N{-}R + CO_2$
Uretonimine[14]	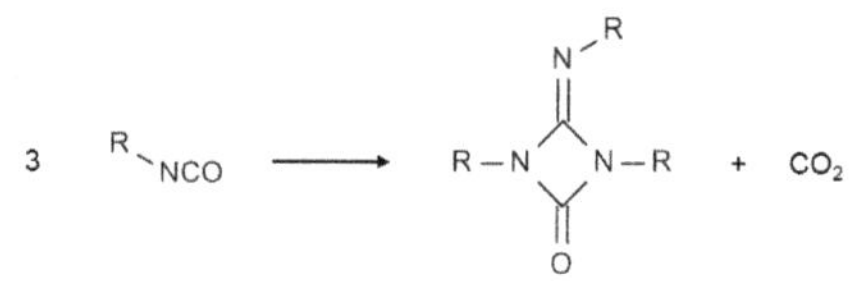

Isocyanate groups may also react with each other. The reaction products shown in Table 3.4 are formed with varying selectivity. The reactions shown here by way of example are used in the manufacture of oligomeric polyisocyanates for coating and (less pronounced) adhesive raw materials. The reactions that form isocyanate functional prepolymers, biurets and cyclooligomers such as trimers and dimers (see Table 3.4), are of particular significance. Isocyanate-functional prepolymers are generally produced via urethanization from diisocyanates with hydroxy-functional co-reactants. The properties of the prepolymers can vary widely as a function of the type of co-reactant, molecular weight, and functionality.

The aforementioned reactions also allow the crosslinking of oligomeric isocyanates with suitable co-reactants. For example, polyisocyanate and polyol react to form polyurethane, or polyisocyanate and polyamine or water react to form polyurea, with the formation of a linear or branched connection.

The kinetic rates of these reactions vary widely and depend on the type of isocyanate used and, on the structure, and type of the co-reactant. Aromatic isocyanates, for example, are far more reactive with alcohols than aliphatic isocyanates. Among the latter, a decline in reactivity from primary through secondary to tertiary NCO groups can also be observed[9] although steric or catalytic influences can alter the typical reactivity profile. Suitable catalysts vary from application to application and must be selected specifically for each use.[10]

The reaction of isocyanates with primary or secondary amines is very fast and urea forms spontaneously even at room temperature. At higher temperature, optionally in the presence of suitable catalysts, urea and urethanes react with excess isocyanate to form biurets or allophanates respectively. Special catalysts are required for the oligomerization of aromatic and aliphatic isocyanates, and a number of tailor-made NCO-functional oligomers are available for industrial use.

3.3 Polyisocyanates

A further argument in favor of converting diisocyanates to polyisocyanates – apart from the occupational health aspect already mentioned – is the increase in isocyanate functionality per molecule that can be achieved. The reactions described above can provide polyisocyanates with average functionality in the commercial mixture greater than two which provides a three-dimensional crosslinked network. This is desirable for a high crosslink density in the cured film, which helps ensure high durability in the final product. Likewise, the characteristic properties of the base isocyanate are retained in the polyisocyanate derivative, e.g. the lightfastness of aliphatics compared with the tendency to yellow of aromatics. Polyisocyanates based on cycloaliphatic diisocyanates yield hard and sometimes brittle coatings, which can be flexibilized if necessary by proper choice of the polyol. In contrast, polyisocyanates based on linear diisocyanates such as HDI or PDI result in flexible films, which can be modified with a wide range of polyols to meet various requirements in terms of flexibility and hardness. As far as the reactivity of the polyisocyanates is concerned, the same principles that applied to diisocyanates also apply to polyisocyanates. Reactivity is influenced by steric and electronic effects and by the nature of the carbon atom the NCO group is attached to (primary, secondary or tertiary), and can be fine-tuned with catalysts.

3.3.1 Derivatization

The reaction principles shown in Figures 3.4 and 3.5 have gained industrial significance in the derivatization of diisocyanates.

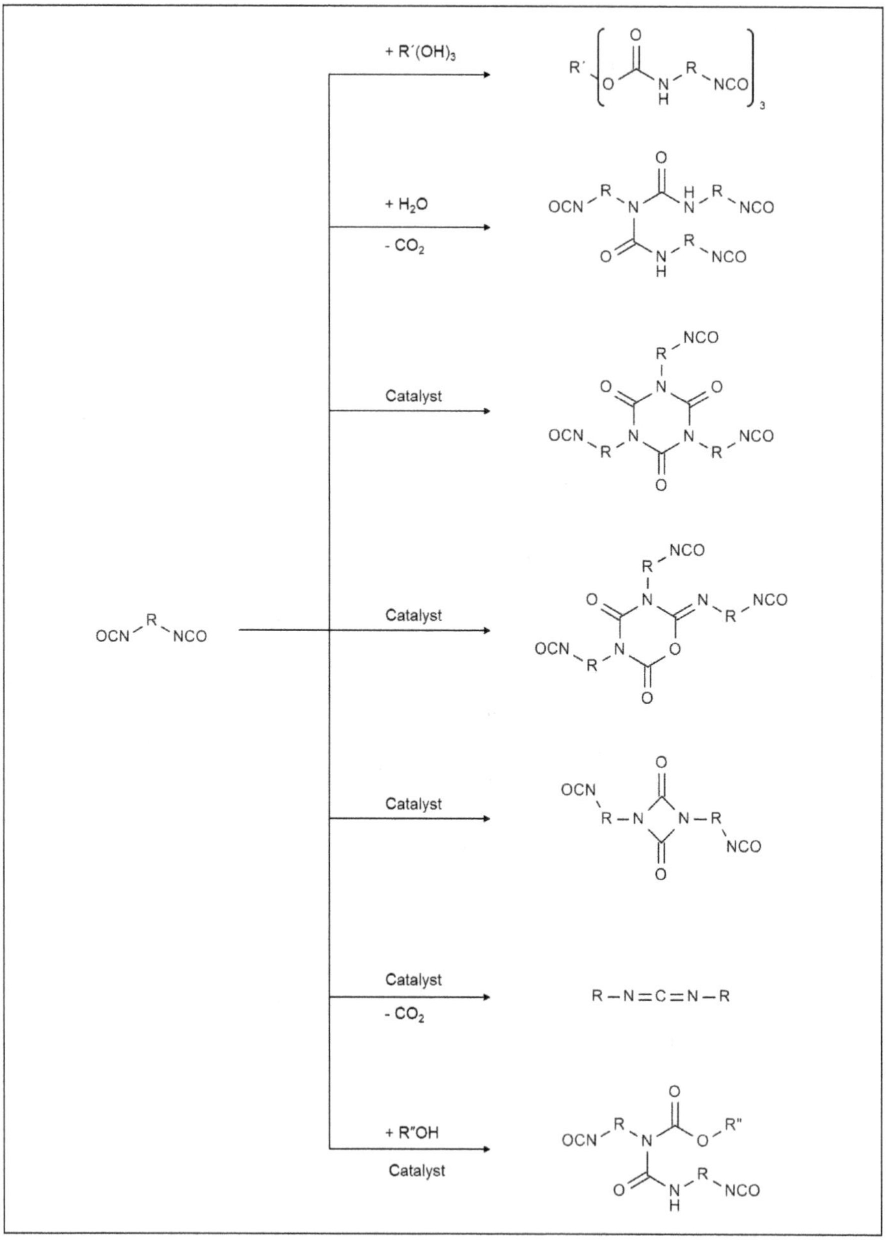

Figure 3.4: Schematic manufacture of polyisocyanates from monomeric diisocyanates

The following product classes are of special interest for coating and to some degree for adhesive and sealant applications:

- Isocyanate functional urethanes and prepolymers obtained by reaction with polyols (e.g. "Desmodur" L, "Polurene" AD, "Desmodur" E, "Ucopol" M 33)
- Biurets obtained, e.g. by reaction of three diisocyanate molecules with one water molecule[7, 11] (e.g. "Desmodur" N 100, "Basonat" HB 100, "Tolonate" HDB, "Wannate" HB 100)
- Isocyanurates based on HDI (e.g. "Desmodur" N 3300, "Basonat" HI 100, "Wannate" HT) and most recently based on PDI (e.g. "Desmodur" eco N 7300, "STABiO" D-370N) as well as iminooxadiazindiones (e.g. "Desmodur" N 3900) obtained by catalytic trimerization of diisocyanates[11, 15]
- Uretdiones (e.g. "Desmodur" N 3400) obtained by catalytic dimerization of diisocyanates[11, 15]
- Allophanates or combinations of isocyanurates and allophanate structures obtained by reaction with alcohols in the presence of suitable catalysts or at very high temperature (e.g. "Desmodur" 3580, Desmodur" XP 2580, "Bayhydur" 305, "Tolonate" X-Flo 100)[16]
- Mixtures of different hardeners including hardeners containing hybrids of aliphatic and aromatic diisocyanates (e.g. "Desmodur" HL) and combinations of different crosslinking technologies are frequently used in dedicated applications explaining the variability of the polyurethane

The manufacturing process does not yield these polyisocyanates as uniform compounds – such as pure isocyanurate triisocyanates – but as oligomer mixtures exhibiting a molecular weight distribution. In the case of isocyanurates, the smallest oligomer in the mixture

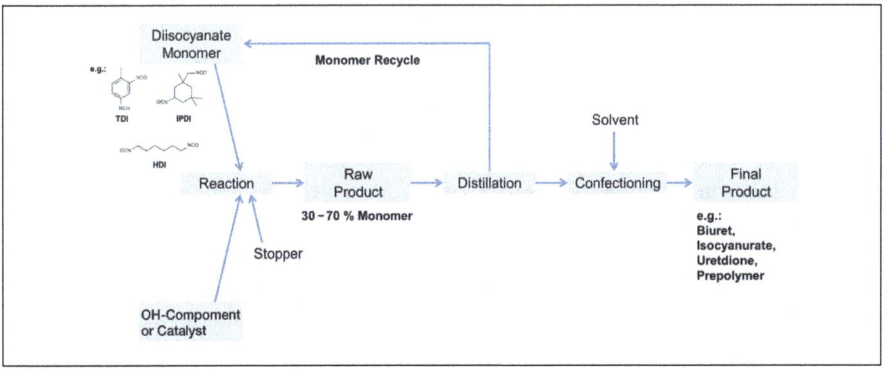

Figure 3.5: Principle technical procedure of manufacture of polyisocyanates from monomeric diisocyanates

Table 3.5: Influence of conversion rate to properties of trimers (exemplified)

HDI trimer (examplified)	Solid content [%]	Viscosity at 23 °C [MPa]	NCO functionality	Drying time	Oligomer distribution [%]		
					n = 3	n = 5	n = 7
High functionality trimer	100	15,000	> 4.5	++	~ 30	< 20	~ 50
Standard trimer	100	3,000	> 3.7	+	~ 50	< 20	~ 30
Low viscosity trimer	100	1,200	> 3.2	0	~ 70	~ 15	< 15

Table 3.6: Aliphatic coating polyisocyanates (selection)

Type	Manufacturer	Trade name
HDI biuret	Covestro	"Desmodur" N 100
	Vencorex	"Tolonate" HDB
	BASF	"Basonat" HB 100
	Wanhua	"Wannate" HB-100
HDI trimer	Covestro	"Desmodur" N 3300
	Covestro	"Desmodur" N 3600
	Vencorex	"Tolonate" HDT
	BASF	"Basonat" HI 100
	Asahi Kasei	"Duranate" TKA-100
	Wanhua	"Wannate" HT-100
HDI dimer	Covestro	"Desmodur" N 3400
IPDI trimer	Covestro	"Desmodur" Z 4470
	Evonik	"Vestanat" T 189
	Vencorex	"Tolonate" IDT 70 B
	BASF	"Basonat" IT 170 B

consists of three monomer units (n=3) of the applied diisocyanate. The next higher oligomer is formed by adding two further monomer units resulting in the pentamer (n=5), followed by the heptamer (n=7) by adding another two monomer units.

This is explained by the fact that not only the monomeric base diisocyanates, but also the polyisocyanates which are formed continue to react, yielding higher molecular weight structures. In order to prevent the formation of highly branched, highly polymeric products, only some of the isocyanate groups are reacted and the unreacted excess of diisocyanate is subsequently removed from the reaction mixture, typically by distillation.

The average molecular weight and the molecular weight distribution of the oligomeric polyisocyanate mixtures, obtained in this way, are directly related to the degree of conversion of the base monomers. This allows for the customization of product properties, such as equivalent weight, average isocyanate functionality and viscosity within certain ranges.

A higher degree of conversion leads to a higher average molecular weight, which increases the product viscosity and the isocyanate functionality. A lower degree of conversion leads to a higher amount of lower molecular weight oligomers, and thus yields of particularly low viscosity, somewhat reduced in functionality but high isocyanate content. Table 3.5 gives an exemplified example. The reactions used in the manufacture of polyisocyanates can be customized to produce specific coating properties. For example, using a high degree of conversion, which corresponds to having higher molecular weight oligomers, the speed of physical drying (initial drying) of the coating film can be increased. By contrast, using a lower degree of conversion to produce lower molecular weight polyisocyanates provides a route to low volatile organic compounds (VOC) and solvent-free coatings. This is getting more important as there is more pressure globally on reducing VOCs.

Table 3.7: Aromatic coating polyisocyanates (selection)

Type	Manufacturer	Trade name
TDI adduct	Covestro Sapici Benasedo Synthesia	"Desmodur" L "Polurene" AD "Hartben" 75 "Synthane" P 80
TDI trimer	Covestro Sapici Benasedo Synthesia	"Desmodur" IL "Polurene" IR "Hartben" SV 100 "Synthane" R 50
TDI/HDI trimer	Covestro Sapici Benasedo	"Desmodur" HL "Polurene" OK.D "Hartben" AM 29
TDI prepolymer	Covestro Sapici Benasedo	"Desmodur" E 1361 "Ucopol" M "Hartben" MC 35
MDI prepolymer	Covestro Baxenden Dow Chemical	"Desmodur" E 21 "Trixene" SC "Isonate" 181

3.3.2 Monomer separation

Strict color requirements are set for polyisocyanate crosslinkers used for coatings. To remove the excess diisocyanate, which has not reacted, only extremely mild processes (vacuum) can be used, which have no detrimental effect on the color of the product. This is usually done via different short-path distillation processes that provide for a short residence time of the product at the elevated temperature necessary for efficient monomer removal.

To ensure that handling of the respective polyisocyanates is not hazardous, the content of volatile monomeric diisocyanates in the polyisocyanates used in coating and adhesive applications must be reduced to a residual content of less than 0.1 to 0.5 % by weight depending on the product. Even after prolonged storage, this threshold may not be exceeded. More modern product lines such as "Desmodur" ultra and "Bayhydur" ultra lines or "Polurgreen" line reduce the residual monomer content to less than 0.1 % by weight.[11] The polyisocyanates and any side products formed during their manufacture must therefore be stable to splitting back into the free diisocyanates.

Tables 3.6 and 3.7 list some of the standard commercial polyisocyanates available for the formulation of coatings.

3.4 Prepolymers

Prepolymers based on aliphatic or aromatic diisocyanates play an important role in polyurethane chemistry. They are regarded as intermediates of the isocyanate polyaddition. Depending on the approach, NCO- or OH-functional prepolymers can be synthesized, since the reactivity of the diisocyanate and the alcohol is (almost) independent of their molar ratio. This enables to generate targeted average molecular weights with the desired end groups and a statistical distribution of components in the prepolymer. The described synthetic route renders the preparation advantageous to other polymerization procedures, e.g. of olefins..[1]

NCO-terminated prepolymers are manufactured by reacting excess diisocyanates with long-chain polyols (see Figure 3.6), especially polyether polyols, but also polyester and polycarbonate diols, with the excess monomeric diisocyanate removed if necessary. These products have found widespread applications for polyurethanes since the isocyanate groups shows a high reactivity towards various nucleophiles (e.g. OH, NH, SH). This approach can be used for the synthesis of prepolymers with hard or soft segments. Slower reacting polyols, having secondary hydroxyl groups, can also be utilized using this method. Common applications for NCO-terminated prepolymers would be for one-component moisture-curing coatings and adhesives as well as two-component PU systems.

Hydroxy-terminated prepolymers are synthesized from NCO-functional prepolymers by addition of a molar excess of diol or by chain extension with amino alcohols. The OH-functional prepolymer products are often used for the manufacture of polyurethane dispersions (PUDs). Additionally, prepolymers can be used in thermally activated PU systems, as well as silane-terminated prepolymers (STPs).

3.4.1 Water content of the raw materials

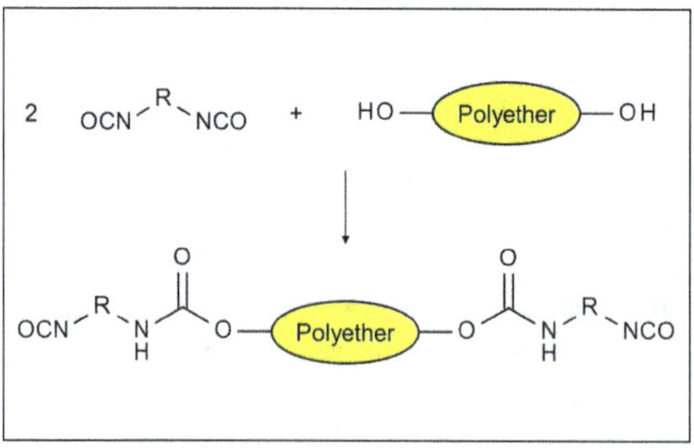

Figure 3.6: Manufacture of polyether based prepolymers

The molecular weight of water is very low: one mole (18 g) reacts with two moles NCO groups. Therefore, the water content of all raw materials used for manufacturing NCO prepolymers must be very low (< 0.05 %) since a small amount of water consumes a large amount of NCO groups. In the case of filled NCO prepolymers the moisture content of the fillers should therefore also be kept at minimum. Urea groups are formed from the reaction of NCO groups with water, and the overall stability of the prepolymer is often negatively affected by these urea groups. During storage, particularly under warm conditions, urea groups react with free NCO groups to produce higher functionality biurets, which results in an increase in prepolymer viscosity in addition of the increase by the ureas itself. Consequently, the polyols and fillers should be dried before use. Unmodified polyisocyanates are also effected by the presence of moisture, since it results in an increased content of dissolved carbon dioxide. Dissolved gases are easily removed by applying a vacuum in the initial phase of prepolymer manufacture. If this is not done, foaming may occur during subsequent application of the prepolymer.

3.4.2 Stability of NCO prepolymers

Stable prepolymers can be stored at room temperature for an extended period of time and can also be handled for short periods above the recommended storage temperature, without significant change in NCO content and viscosity. In the case of unstable prepolymers, a considerable decrease in the NCO value and a marked increase in viscosity are observed. Aromatic polyisocyanate-based prepolymers should not be subjected to temperatures above 90 °C during production, because higher temperatures trigger side reactions (formation of allophanates and biurets, trimerization reactions), which can negatively affect the storage stability of the prepolymers, the viscosity and the molecular structure of the systems.

3.5 Thermally activated polyurethane crosslinkers

Storage-stable, long shelf-life formulations, which have high-end polyurethane performance properties as a result of chemical crosslinking would be an ideal situation. Considering all the conditions of polyurethane crosslinking described previously, this appears to be an unsolvable issue.

However, thermally activated polyurethane crosslinkers open a door to solve that paradox. Chemically blocking the reactive isocyanate group in a given polyisocyanate results in a crosslinker that can be formulated into a stable system with standard co-reactants and additives. Chemical crosslinking is initiated after application by curing at elevated temperature. Chemically speaking, the blocking agent (Bl) can be cleaved at elevated cure temperature, allowing the NCO-reactive co-reactant to replace the blocking agent resulting in a cured polyurethane system (see Figure 3.7). As the blocking agent is in most cases released special care needs to be taken with appropriate industrial hygiene measures and careful studies of the respective material safety data sheet. Figure 3.8 displays a list of typical blocking agents for isocyanate groups.

Thermally activated polyurethane crosslinkers are being used as the hardeners in one-component polyurethane baking coatings. A selection of standard commercial products is shown in Table 3.8.

The advantages of thermally activated polyurethane crosslinkers for end users are their ease of handling, since no mixing equipment or expertise is involved to get the correct ratio of ingredients. One-component polyurethane coatings with adequate storage stability can be formulated without difficulty. However, the advantage of easy handling is balanced with the limitation that the systems require elevated cure temperatures to reach full properties. The type of blocking agent used is of significance for a number of reasons. Reactivity, thermal yellowing and other specific coating properties, including cost, are significantly affected by the choice of blocking agent. Most commonly used are butanone oxime, dimethylpyrazole

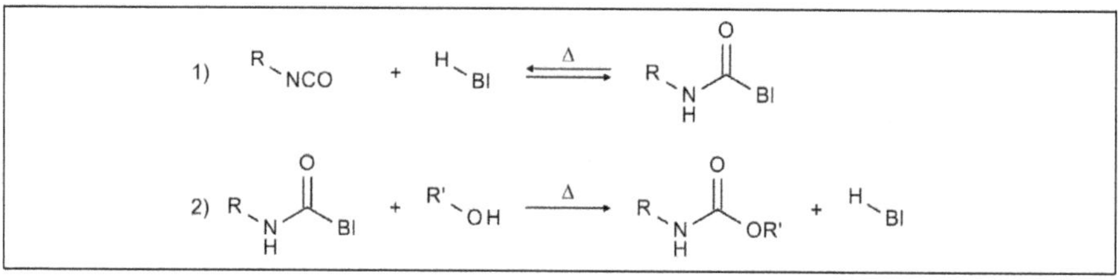

Figure 3.7: Principle of one-component thermally activated polyurethanes, (Bl = Blocking agent)

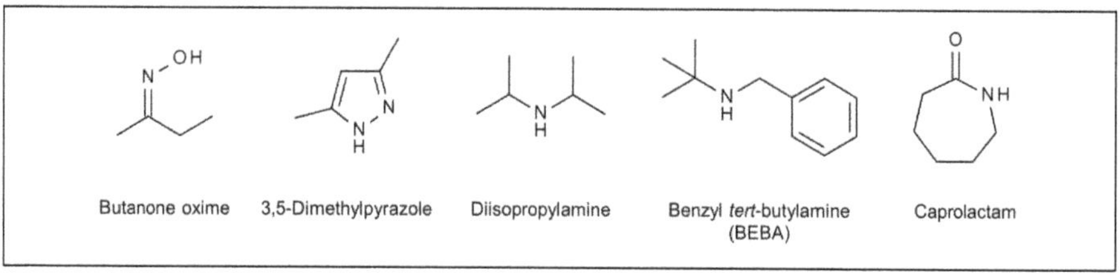

Figure 3.8: Exemplary selection of blocking agents for isocyanate groups

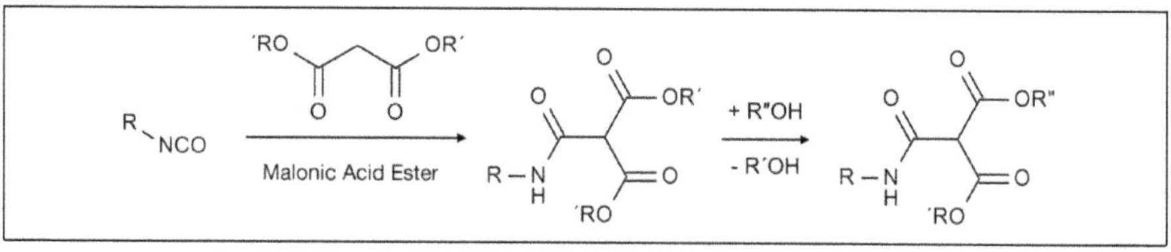

Figure 3.9: Blocking and crosslinking reaction with malonic acid esters

Table 3.8: Selection of thermally activated polyurethane crosslinkers

Blocking agent	**Diisocyanate-based**	**Manufacturer**	**Trade name**
Butanone oxime	HDI	Covestro	"Desmodur" BL 3175
		Covestro	"Desmodur" VP LS 2257
		Vencorex	"Tolonate" D2
		Mitsui Chemicals	"Takenate" 160 N
	IPDI	Covestro	"Desmodur" BL 4265
		Evonik	"Vestanat" B 1370
		Mitsui Chemicals	"Takenate" 140 N
	"Desmodur" W	Covestro	"Desmodur" VP LS 2117
Dimethylpyrazole (DMP)	HDI	Covestro	"Desmodur" VP LS 2253
		Covestro	"Desmodur" VP LS 2376
		Covestro	"Desmodur" PL 350
		Baxenden	"Trixene" BI 7982
	IPDI	Covestro	"Desmodur" PL 340
		Baxenden	"Trixene" BI 7951
Malonic ester	HDI/IPDI	Covestro	"Desmodur" BL 3475
	MDI	Covestro	"Desmotherm" 2170
Diisopropylamine	HDI/IPDI	Covestro	"Desmodur" VP LS 2352
Diisopropylamine/ malonic ester	HDI	Covestro	"Desmodur" BL 3370
Dimethylpyrazole/ malonic ester	HDI	Baxenden	"Trixene" BI 7992
	IPDI	Baxenden	"Trixene" BI 7990
Diisopropylamine/ triazole	IPDI	Covestro	"Desmodur" VP LS 2114
ε-Caprolactam	HDI	Covestro	"Desmodur" BL 3272
		Baxenden	"Trixene" BI 7981
	TDI	Covestro	"Desmodur" BL 1100
	IPDI	Covestro	"Desmodur" VP LS 2078
Isononylphenol	TDI	Covestro	"Desmocap" 11
			"Desmocap" 12

and ε-caprolactam (see Figure 3.8). The corresponding thermally activated polyurethane crosslinkers are obtained by chemical reaction of the blocking agent with a given polyisocyanate. The blocking reaction is done at elevated temperatures; in most cases it runs smoothly without catalyst addition. Curing of the coating includes liberation of the blocking agent and chemical reaction with the co-reactant to form a urethane bond. However, blocking of polyisocyanates with malonic acid ester represents a special case. The use of crosslinkers blocked with malonic acid ester results in transesterification (see Figure 3.9).

Cleavage reactions

One-component polyurethane coatings are formulated to react at various temperatures. The minimum temperature required for a fast crosslinking reaction strongly depends on the blocking agent, as well as the polyisocyanate backbone. The reactivity of thermally activated polyurethane crosslinkers with polyols has been studied by dynamic mechanical analysis (DMA) on model systems formulated with various blocking agents that are readily available on an industrial scale[18] (see Table 3.8). The typical curing temperatures for selected blocking agents and the influence of a catalyst are shown in Table 3.9.

The results obtained confirm the experience that thermally activated PU crosslinkers blocked with ε-caprolactam require high curing temperatures, while in contrast, those blocked with malonic ester can be crosslinked at significantly lower temperatures. While it is most common to use such one-component PU systems for metal coatings, malonic ester based crosslinkers can be used for coating thermally sensitive substrates, like plastics. What is interesting with respect to the blocking agent is the catalytic effect of DBTL (dibutyl tin dilaurate), which substantially reduces the initial crosslinking temperature. DBTL has been shown to be a less suitable catalyst only in the case of malonic ester, which has been attributed to the different crosslinking mechanism (see Figure 3.9). In this special case, the use of a titanium compound has proven beneficial.[19]

Using thermogravimetric analysis (TGA) on model systems, the effect of the polyol on the deblocking temperature has been found to be quite significant (see Table 3.10). Whereas DMA measures the curing (crosslinking) of the one-component polyurethane system, TGA records the release of the blocking agent by weight loss. The deblocking process can be tracked and the deblocking temperature determined. The underlying chemical procedure is identical in both cases, but different parameters are measured.

Comparison of Tables 3.9 and 3.10 shows that the crosslinking temperatures measured by DMA for the non-catalyzed system is of the same magnitude as the deblocking temperature measured by TGA (see Table 3.10) for the analogous one-component polyurethane system. Of particular interest in these systems, the deblocking temperatures of the pure blocked polyisocyanates (i.e. in the absence of polyol co-reactants) are unexpectedly high. They are in the range of the cleavage temperature of aliphatic urethanes, which are approximately 240 °C.

The reactivity sequence for the blocking agents revealed by DMA and TGA measurements has been confirmed in systems other than the model system used. The only difference is smaller fluctuations in the crosslinking temperatures when using other aliphatic polyisocyanates or polyols. Corresponding data for blocked aromatic polyisocyanates are not widely available. However, in the light of past experience, it is expected that their initial

Table 3.9: Typical curing temperature and influence of a catalyst (in this case dibutyl tin dilaurate (DBTL)) for various blocking agents (DMA method)

Blocking agent	**Initial crosslinking temperatures [°C]**	
	with 1 % DBTL, based on the solid resin	**without DBTL catalysis**
ε-Caprolactam	163	175
Diethylamine	144	173
Butanone oxime	137	156
3,5-Dimethylpyrazole	112	158
Diisopropylamine	115	136
BEBA	100	-
Imidazole	98	123
Malonic acid diethyl ester	-	104

Table 3.10: "Polyol effect" on the deblocking initiation temperature for various blocking agents (TGA method)

	Initial deblocking temperature [°C]	
Blocking agent	**with polyol**	**without polyol**
ε-Caprolactam	177	234
Butanone oxime	157	220
3,5-Dimethylpyrazole	167	220
Diisopropylamine	140	220

Table 3.11: Curing states of one-component polyurethane coatings as a function of the baking time, characterized by T_g of the coating film and residual blocking agent in the coating film

	Stoving temperature 140 °C			
	Stoving time 30 min		**Stoving time 60 min**	
Blocking agent	**T_g [°C]**	**Residual amount of blocking agent [%]**	**T_g [°C]**	**Residual amount of blocking agent [%]**
3,5-Dimethylpyrazole	73	14	80	5
Butanone oxime	68	27	77	12
For comparison: corresponding 2-component polyurethane coating	80	-	80	-

crosslinking temperatures will be around 10 °C lower. Model studies in the literature confirm our internal work. Table 3.11 shows the analysis of three one-component clear coat systems in which different blocking agents were used and a two-component polyurethane reference system.

In general, thermally activated polyurethane crosslinkers per se are relatively heat-stable compounds suitable for one-component coating. Only in the presence of a co-reactant are feasible crosslinking temperatures achieved, which can be lowered further by catalysis. Effects on the coatings performance such as yellowing, effects of blocking agent which remains in the film are described in the literature.[20–22] An extensive overview about thermally activated polyurethane crosslinkers was published from *Wicks* and *Wicks*.[23]

3.6 Hydrophilically-modified polyisocyanates

Aqueous two-component polyurethane systems represent a low-VOC (volatile organic compounds) alternative to conventional solvent-borne coatings in a large number of application areas. In these two-phase systems, the use of a suitable aqueous polyol dispersion and, in particular, the choice of polyisocyanate crosslinker is critically important to the quality of the resulting coating.

Basically, there are two options for crosslinking aqueous polymer dispersions [24]:

- The use of hydrophobic, preferably low-viscosity polyisocyanates, such as those also used in conventional two-component polyurethane coatings.
- The use of hydrophilically modified, self-emulsifying polyisocyanates specially developed for this purpose.

Whereas low-viscosity hydrophobic polyisocyanates can only be incorporated into aqueous dispersions with the concomitant use of suitable organic co-solvents and high shear (e.g. using dispersion apparatus), the use of hydrophilically-modified polyisocyanates enables the production of a homogeneous and co-solvent-free mixture of binder and crosslinker by simple manual stirring.

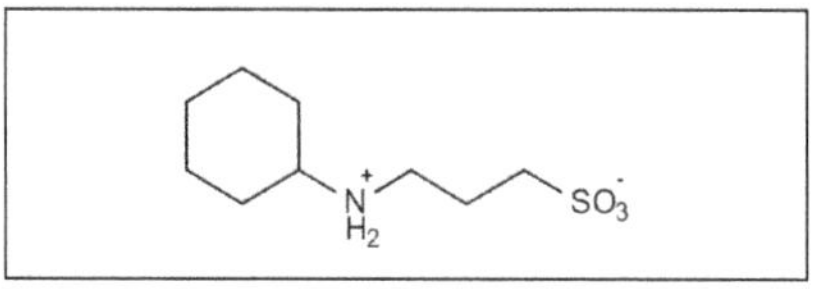

Figure 3.10: 3-(Cyclohexylamino)-1-propane sulfonic acid

Various approaches have been used in the manufacture of hydrophilic polyisocyanates. In addition to the use of external emulsifiers[25, 26], which sometimes exhibit surface defects associated with their migration through the coating film, the partial reaction of hydrophobic polyisocyanates with hydrophilic co-reactants has proven especially effective.

Highly active emulsifier molecules are incorporated into the polyisocyanate crosslinker. In this way, coatings can be produced that meet the highest quality requirements.

For example, the simple reaction of aliphatic or cycloaliphatic polyisocyanates, such as HDI or IPDI trimers, with a small amount of a monofunctional polyethylene oxide polyether alcohol[27, 28] leads to polyisocyanate mixtures containing non-ionic emulsifiers of the polyether urethane type (see Figure 3.11, structure [a]). Polyisocyanates hydrophilically-modified in this way can be easily emulsified in water using a manually driven stirrer, without

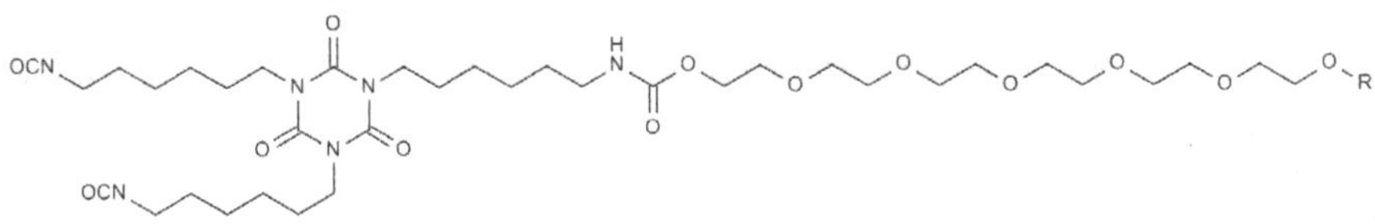

[a] Nonionic Hydrophilic Modification: Emulsifier Molecule of the Polyether Urethane Type (e.g., "Bayhydur" 3100, "Basonat" HW 2100, "Desmodur" DA-L or "Desmodur" DN)

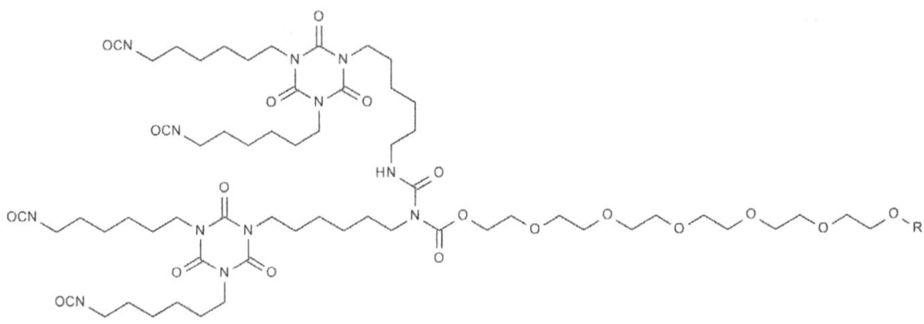

[b] Nonionic Hydrophilic Modification: Emulsifier Molecule of the Polyether AllophanateType (e.g., "Bayhydur" 304, or "Bayhydur" 305)

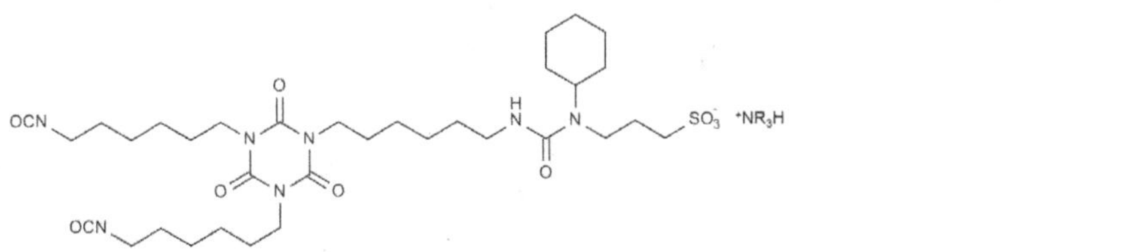

[c] Ionic Hydrophilic Modification: Emulsifier Molecule of the SulfonateType (e.g., "Bayhydur" XP 2487/1, or "Bayhydur XP" 2547 or "Bayhydur XP" 2655)

Figure 3.11: Hydrophilically-modified HDI polyisocyanurates – emulsifier structures (idealized)

the application of high shear forces. Water-emulsifiable polyisocyanates of this type (e.g. "Bayhydur" 3100, "Bayhydur" 401-70, "Basonat" HW 2100) are today established as the standard NCO crosslinkers for most aqueous coating and adhesive applications.

However, the chemical hydrophilization of polyisocyanates via urethanization with polyether alcohols always leads to a reduction in the average isocyanate functionality and thus yields coating films with lower crosslinking density, which usually becomes apparent through lower chemical resistance.[24]

This effect spurred the development of improved polyether-modified polyisocyanates in which the hydrophilic polyether chains are each linked with two polyisocyanate molecules (see Figure 3.11, emulsifier structure [b]) by means of allophanatization.[29] In the case of these second-generation non-ionic hydrophilically-modified polyisocyanates, the polyether modification increases rather than reduces the isocyanate functionality. As a smaller amount of polyether alcohol also suffices to achieve the desired dispersibility of these products, hydrophilic polyisocyanates containing a polyether allophanate emulsifier (e.g. "Bayhydur" 304, "Bayhydur" 305) yield coating films with much better water and chemical resistance than those produced with the standard crosslinkers described above.

Despite their broad market acceptance for many applications, all polyether-modified polyisocyanates have one fundamental disadvantage. When used as the crosslinker in aqueous two-component polyurethane coatings, the polyether content required to ensure satisfactory dispersibility results in a relatively longer drying time and gives the coatings permanent hydrophilicity. The latter is the reason why polyether-modified polyisocyanates have not been accepted for applications requiring the highest levels of resistance, e.g. in automotive base coats or for anti-graffiti coatings. These disadvantages were overcome with the development of special, ionically modified polyisocyanates.

Figure 3.12: Example of a polyisocyanate production plant[115]

Aliphatic polyisocyanates react with 3-(cyclohexylamino)-1-propanesulfonic acid (CAPS) (see Figure 3.10), a zwitterionic amino sulfonic acid, in the presence of tertiary neutralizing amines under mild conditions.[30] The sulfonate urea derivatives that are produced are excellent emulsifiers (see Figure 3.11, structure [c]).

Irrespective of the salt groups, CAPS-modified polyisocyanates have good storage stability, are free from turbidity, and yield highly dispersed emulsions in water, even with a relatively low sulfonate group content. A range of ionically modified polyisocyanates is now available and can be used in the formulation of environmentally friendly, aqueous two-component polyurethane coatings of high quality for a broad spectrum of applications. Following the trend to reduce the use of fossil resources, an anionically hydrophilized polyisocyanate for two-component water-borne systems has been developed based on a trimer of pentamethylen diisocyanate (PDI), the diisocyanate produced from renewable feedstock. This polyisocyanate ("Bayhydur" eco 7190), which has a biomass content of 66 %, gives similar performance to the established fossil-based HDI polyisocyanate hardeners.[31]

Coatings formulated with sulfonate modified polyisocyanates are often as good as conventional, solvent-borne systems in terms of drying, curing and chemical resistance. [32–33] Due to new legislation mandating further reductions in VOCs, and the very high quality of the resulting coatings, these crosslinkers will see increased usage in the future.

3.7 Co-reactants for polyisocyanates

The properties of polyurethane coatings and adhesives are primarily determined by the choice of which polyisocyanate and co-reactant to use. Solid content, surface wetting, adhesion to the substrate, drying speed, gloss, elasticity/hardness, resistance to chemicals and hydrolysis, and cost-effectiveness are some of the important parameters. Therefore, in past years, substantial research has been focused on the development of new and improved resin systems as co-reactants for polyisocyanates.

The most important class of co-reactants consists of hydroxyl group containing polymers, such as polyacrylate, polyester, and polyether polyols. More specialized building blocks for polyurethane coatings and adhesives are polycarbonate-, polycaprolactone- and polyurethane-based polyols. Co-reactants containing unmodified amino groups show much higher reactivity towards isocyanates, and thus are used only in applications that require this unique property.

In addition to solvent-borne co-reactants for low and high-solid polyurethane formulations, water based polyurethane dispersions have experienced significant growth in the past decades. More and more suppliers are starting up around the globe, particularly in the Asian market. Polyurethane dispersions can be used as one-component systems, without any additional crosslinker, but also as components for two-component water-borne (WB) systems that require some amount of a crosslinker, such as a polyisocyanate. Also, hydroxyl group containing polyacrylate and polyester dispersions as co-reactants for two-component WB PU systems have gained greater importance over past years.

The following contributions describe the principal routes for the synthesis of key co-reactants for polyisocyanates. Their special significance in individual applications is addressed in later chapters.

3.7.1 Solvent-borne polyacrylate polyols

The term polyacrylate polyols covers co-polymers of acrylic and/or methacrylic acid esters – typically ethyl acrylate, butyl acrylate and methyl methacrylate in combination with hydroxyl group bearing monomers such as 2-hydroxyethyl methacrylate, 2-hydroxyethyl acrylate, hydroxypropyl methacrylate, or 4-hydroxybutyl acrylate. Other co-monomers such as styrene, vinyl esters, or maleates, are also frequently used (see Figure 3.13). Another way of generating hydroxyl groups is by appropriate chemical modification of polyacrylates initially containing no hydroxyl groups.[34]

Manufacturing process

Industrial preparation of polyacrylate polyols for solvent-based two-component PU technology is carried out by a free radical polymerization of the aforementioned monomers. This polymerization takes place in either organic solvents or directly in the monomers themselves, which is referred to as bulk- or mass-radical polymerization.[35–37]

Initiators used in the preparation of polyacrylate polyols are azo-compounds or peroxides, e.g. azo-bis-isobutyro nitrile or t-butyl peroxy-2-ethyl hexanoate.

Application of polyacrylic polyols in two-component SB PU formulations is a mature technology and a substantial part of polyol manufacturing takes place for captive use, which allows coating companies to control costs. Aqueous polyacrylic polyols, which are used in two-component WB PU, are discussed in Chapter 3.8.2.

Properties

The properties of polyacrylate polyols, and of the polyurethane coatings based on them, are largely governed by the choice of co-monomers, hydroxyl content, molecular weight and glass transition temperature (T_g).[38] The molecular weight of a polyacrylate can be influenced significantly by the polymerization temperature, the type and amount of initiator used, and the addition of chain transfer agents, such as thiols.[39, 40] T_g can be influenced by the appropriate selection of hardening and softening monomers and their dosage during polymerization, whereas the hydroxyl content is directly related to the amount of hydroxy monomer used.

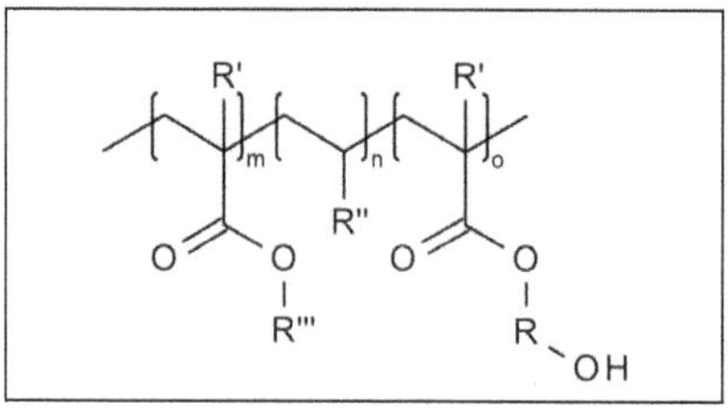

Figure 3.13: Basic structure of a typical polyacrylate polyol

(R = Alkylene ; R' = H or Me; R'' = Ph; R''' = H or Alkyl)

Applications

Polyacrylate polyols occupy a significant position in polyurethane coatings and adhesives technology. As a result of the ongoing development of these products, the market now offers a broad range of resins for use in a wide variety of applications, as shown by the following examples:

- Wood/furniture coating with a need for fast drying properties.
- Automotive refinish, with its special requirements in terms of quality and cost-effectiveness.
- Automotive OEM, with its requirement for long-term resistance properties such as gloss retention.
- Plastic coatings that require high flexibility and good chemical resistance.
- Corrosion protection that requires good long-term resistance properties in topcoats.

3.7.2 Polyester polyols

Polyester polyols are produced via a polycondensation reaction comprised of di- and polycarboxylic acids along with a molar excess of polyfunctional alcohols (polyols) (see Figure 3.14). A wide range of primary polycarboxylic acids and their anhydrides are available on an industrial scale for the manufacture of polyester polyols. These compounds include: aromatic acids, e.g. phthalic and isophthalic acid; aliphatic acids, e.g. adipic and maleic acids; as well as cycloaliphatic acids, e.g. tetrahydrophthalic acid and hexahydrophthalic acid.

Monomeric polyfunctional alcohols used as raw materials are: aliphatic diols, e.g. ethane diol, 1,2-propanediol, 1,6-hexanediol, neopentyl glycol; trifunctional polyols, e.g. glycerol and trimethylolpropane; and cycloaliphatic alcohols, e.g. 1,4-cyclohexanedimethanol. In addition, monofunctional alcohols or mono-carboxylic acids, e.g. 2-ethyl hexanol or 2-ethyl hexanoic acid can be used as chain terminators.

As in the case of polyacrylic polyols, many polyesters are produced by coating and adhesive manufacturers for captive use.

Over the past decade, there have been considerable efforts to develop polyesters based on renewable raw materials, with the first thousand-ton scale plants for the production of those raw materials (mainly di-acids and diols) built only in the last few years. Some of these monomers are bio-succinic acid, bio-butane diol, and bio-1,3-propane diol. [41-43]

Reaction process

As previously mentioned, polyester polyols are generally produced by a polycondensation reaction between polycarboxylic acids and polyols. This reaction generates water, which

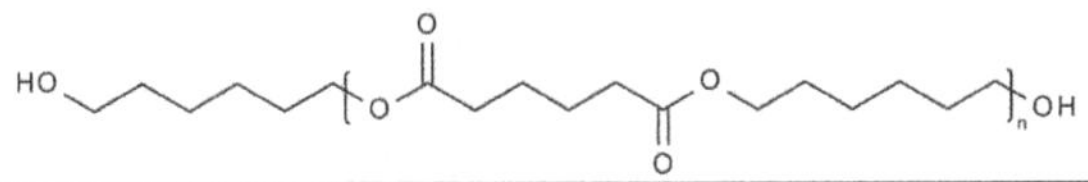

Figure 3.14: Example of a structure of a linear polyester polyol based on 1,6-hexane diol/adipic acid

is removed during the reaction. There are two primary, commercially-used processes to produce polyesters. The first is the melt condensation process, in which the components are reacted in a melt at approximately 160 to 260 °C. The reaction is performed either in a vacuum or in a stream of inert gas to remove water. The second process is an azeotropic esterification that is performed in the presence of an organic solvent which forms an azeotrope with water and acts as a carrier to remove it from the reaction.

The polycondensation reaction can be accelerated by increasing the reaction temperature, increasing the flow rate of the inert gas used to strip the generated water, reducing the pressure, and/or the use of catalysts, which is particularly advantageous towards the end of the esterification process. Suitable catalysts include metal derivatives, such as tin compounds, bismuth carboxylates, titanium alkoxy derivatives, or lithium hydroxide, to name a few.[44]

Polyester polyols can be used as building blocks to produce polyurethane-based binders; for instance, both hydroxyl-functional polyurethanes for two-component WB systems and high molecular polyurethanes are used in polyurethane dispersions. When used directly as a binder for coatings and adhesives, the polyester polyol is used mainly as a solvent-free viscous resin, but it can also be dissolved in a solvent or used as a dispersion in water.[45]

Properties and applications

The properties of polyester polyols are determined by the choice of raw materials, molecular weight and functionality.[46] Generally speaking, saturated polyester polyols are easily pigmented, and yield coatings with good weather stability and high gloss. This makes them suitable for a broad range of applications, from vehicle finishing (including aircraft) to general industrial coatings (machinery).[47]

Polyester polyols are also broadly used in the adhesive applications.

One of the disadvantages of polyesters as a co-reactant in polyurethane coatings and adhesives is their limited hydrolytic stability.

Unsaturated polyesters that are often used in putties are not in the focus of this overview.

3.7.3 Polyether polyols

Polyether polyols are typically used as co-reactants in specific applications and as building blocks for a wide range of products including adhesives, spray elastomers, and polyurethane dispersions, to name a few. Polyether polyols are formed by the addition of alkylene oxides, e.g. ethylene oxide or propylene oxide, to polyfunctional starter molecules (see Figure 3.15). Usually, polyvalent alcohols or amines are used as starter molecules. The addition of alkene oxide is usually performed in an alkaline medium with potassium hydroxide as the base or by using a metal-complex catalyst such as double-metal cyanide catalyst based on zinc hexa(cyano)cobaltate.[48, 49]

The latter method is characterized by less side reactions, e.g. water elimination, and can be used for the synthesis of polyols of high molecular weight, while keeping the desired hydroxyl group functionality ("Acclaim" polyol technology). In contrast, the level of allyl group formation instead of the desired hydroxy group increases significantly using KOH-catalyzed technology. As a result, this process is limited to the production of lower molecular weight polyols (molecular weight approximately 2000 g/mol).

Molecular weights of polyether diols are ranging from 154 to 18,000 g/mol. Trifunctional polyols span from short-chain ethers with molecular weights from 300 to 725 g/mol to long-chain polyether triols with molecular weights from 3,000 to 6,000 g/mol. High molecular weight polyols lead to polyurethane polymers with exceptional good mechanical and dynamic properties, which are important for sealant applications.

Key properties of polyethers, such as T_g, viscosity, hydrophilicity, and compatibility with polyisocyanates, can be controlled by varying the ratio of ethylene oxide to propylene oxide as well as the amount of starter molecules.

Due to their low viscosity, polyether polyols are also used in solvent-free coating systems. However, their use is restricted to applications with limited UV exposure due to their poor

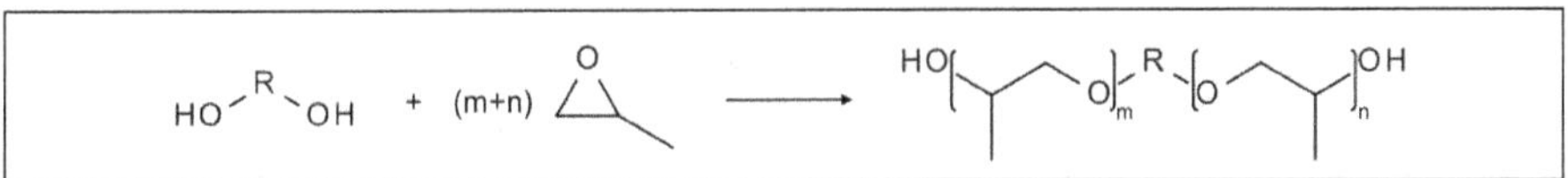

Figure 3.15: Manufacture of a polyether polyol

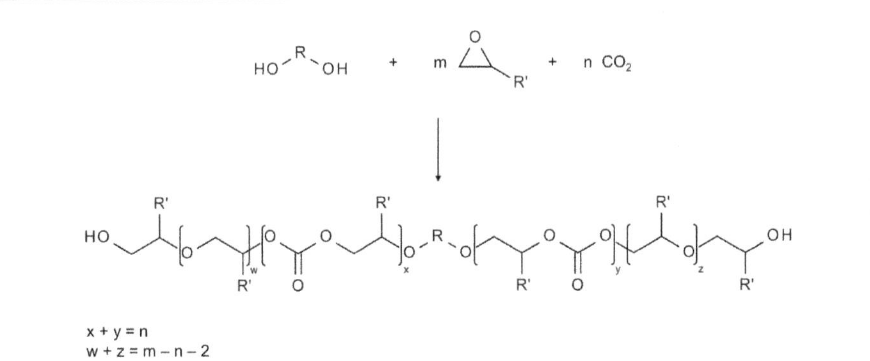

Figure 3.16: Co-polymerization of alkylene oxides with CO_2

R = backbone of the starter molecule; R' = H, CH_3 or C_2H_5, etc.)

weather stability, a consequence of oxidative polyether chain degradation. On the other hand, these systems are characterized by unique hydrolytic resistance and mechanical stability. For these reasons, they are often used in the construction sector for coatings on mineral substrates, such as concrete.

In recent years, co-polymers of alkylene oxides that incorporate CO_2 have been developed, with the amount of incorporated CO_2 varying between 20 to 40 %.[50] The incorporation of CO_2 not only provides increased mechanical properties and reduces sensitivity towards oxidation, but also opens up the possibility of using a green-house gas as a raw material for polyurethanes (see Figure 3.16). Compared to polyalkylene oxide-based polyols, CO_2-containing polyethers show an increased viscosity.

3.7.4 Polycarbonate polyols

Polycarbonate polyols are esterification products formally derived from the reaction of carbonic acid with polyols and thus represent a special subgroup of the polyester polyol family. In practice, the carbonate structure is introduced by trans-esterification using dialkyl carbonates as carbonyl donors.[51] Linear aliphatic polycarbonates[52] are used both as binders in high quality polyurethane coatings and in the production of polyurethane binders, especially polyurethane dispersions.[53] The most commonly used hexane diol-based polycarbonate diols (see Figure 3.17) offer a much improved hydrolytic resistance compared to other polyester polyols and very good weather stability and toughness compared to polyether polyols, thus combining the best properties from both polyol classes. However, due to their complex production process they come at a comparatively high cost.
1,6-Hexane diol-based polycarbonate diols are solids at room temperature and generally give solid or high viscosity products after conversion with isocyanates. Liquid polycarbonate polyols based on branched diols or mixtures of different diols are available, but they

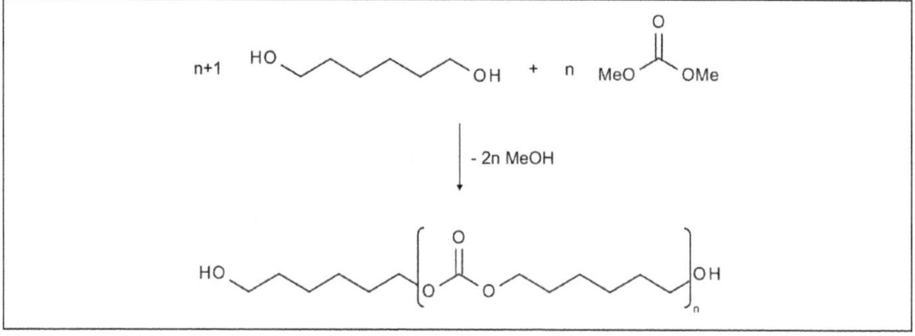

Figure 3.17: Synthesis of a polycarbonate diol based on 1,6-hexane diol

Table 3.12: Major providers of polycarbonate polyols

Producer	Trade names	Product examples (M_n in [g/mol])
Asahi KASEI	"Duranol"	T 6002 (Mn 2000) G 4672 (Mn 2000) T 5652 (Mn 2000)
Caffarao	"Ravecarb"	106 (Mn 2000) 292 (Mn 2000) 107 (Mn 1850)
Covestro	"Desmophen"	C 1100 (Mn 1000) C 1200 (Mn 2000) C 2100 (Mn 1000) C 2202 (Mn 2000) XP 2613 (Mn 2000) XP 2716 (Mn 650)
Cromogenia	"Ropol"	PC 12 (Mn 2000) PC 22 (Mn 2000)
Daicel	"Placcel"	CD 220 (Mn 2000) CD 220PL (Mn 2000)
Kuraray	"Polyol"	C-2050 (Mn 2000)
Sumitomo Chemical	"Durez"-ter	S 2000-55 (Mn 2000) S 2001-120 (Mn 1000)
UBE	"Eternacoll"	UH-200 (Mn 2000) BH-200 (Mn 2000) PH-200 (Mn 2000)

provide a slightly lower toughness in the final application. Another liquid product group is comprised of mixed ε-caprolactone-hexane diol carbonate polyesters which sacrifices some of the excellent hydrolytic resistance in the final product. Table 3.12 summarizes major providers of polycarbonate polyols.[54]

3.7.5 Polycaprolactone polyols

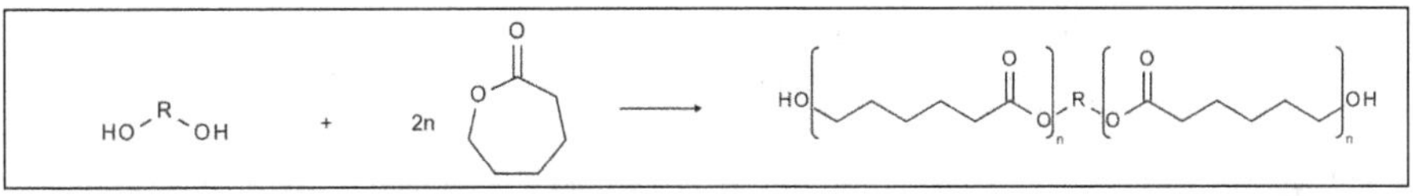

Figure 3.18: Synthesis of ε-caprolactone-based polyester polyols

Polycaprolactone polyols are produced by the ring-opening polymerization of ε-caprolactone and thus are another subgroup of the polyester polyol family (see Figure 3.18). Suitable starter molecules are polyfunctional alcohols e.g. ethylene glycol, 1,2-propanediol, glycerol and trimethylolpropane.[55–57] The process yields low-viscosity products with defined functionality.

These polyesters are used either as sole binders or as reactive thinners in solvent-free or high-solids, two-components coatings.

Polycaprolactone polyols are also used as polyol building blocks in manufacturing high molecular weight polyurethanes. An additional use of the ring-opening polymerization of ε-caprolactone is for the modification of higher molecular weight polyols, e.g. polyacrylate polyols.[58] The technical advantages of polycaprolactone polyols are their flexibility and weather stability, as well as a very low viscosity compared to other polyester polyols. The main applications for these materials are in two-component polyurethane coatings for plastics substrates within the automotive sector and in solvent-free coatings for construction applications. Disadvantages of this class of co-reactants are a slightly reduced hydrolysis resistance and quite high cost.

3.7.6 Solvent-based polyurethane polyols

Polyurethane polyols are produced using the diisocyanate polyaddition process in which diisocyanates are reacted with an excess of diols and/or polyols.[59–61]

Due to their high viscosity, these products are currently of little importance in the field of solvent-borne coating systems but have found a significant use in adhesive applications (see Chapter 6.4).

3.7.7 Polyamines

Primary and secondary amino groups usually react rapidly with isocyanate groups to form polyureas. Therefore, polyurea coatings can only be applied using two-component application equipment that is specially developed for this purpose. Typically, polyurea coatings use polyether amines in combination with aromatic isocyanate prepolymers. Aliphatic prepolymers have only played a subordinate role up to now. The main applications of polyurea coatings are coatings for concrete systems or systems requiring heavy-duty corrosion protection.[62, 63]

The reactivity of the amino group can be significantly reduced by "blocking". The most important blocking reactions for amines are those with aldehydes to form aldimines and with ketones to form ketimines (see Figure 3.19).[64–66]

In contact with water, atmospheric humidity for example, the blocking reaction is reversed to produce the starting amines, which are then available to react with the isocyanate groups.[67, 68]

In the absence of water crosslinking can also be achieved, especially under the influence of heat. Direct reaction of ketimines and aldimines with isocyanates results in a variety of products, depending on the structure of the reactants and the reaction conditions.[64, 65]

Reacting amino alcohols with aldehydes or ketones (see Figure 3.20), along with the removal of water, yields 1,3-oxazolanes (oxazolidines), another class of blocked amines.[69, 70]

These compounds can then be reacted with diisocyanates via available hydroxyl groups to form difunctional "blocked" amines (bisoxazolidines). As in the case of aldimines and ketimines, upon exposure to atmospheric moisture the oxazolidine group will hydrolyze to the original amino alcohol which then reacts with the respective electrophilic hardener.

Low molecular weight blocked diamines have been particularly important as reactive thinners for high-solid polyurethane coatings, e.g. for automotive refinish. However, in recent years, their importance has diminished due to the release of aldehydes and/or ketones to the environment as a result of the aforementioned hydrolysis reaction.

Polyaspartic acid esters

Polyaspartic acid esters are a special class of polyamines having hindered secondary amino groups. They can be produced either by the addition of primary (cyclo)aliphatic diamines to maleic or fumaric acid di(cyclo)alkyl esters[71–73] (see Figure 3.21), or by the addition of primary (cyclo)aliphatic amines to unsaturated oligoesters or polyesters.[74]

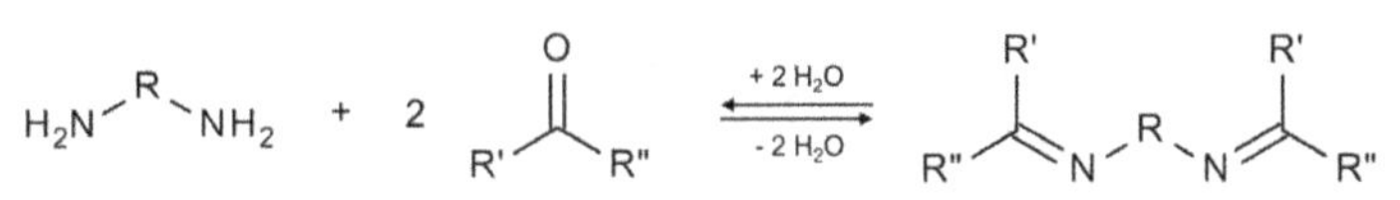

Figure 3.19: Aldimine and ketimine formation as blocking reactions for amines

(R = (cyclo)alkylene; R' = (cyclo)alkyl; R" = (cyclo)alkyl, H)

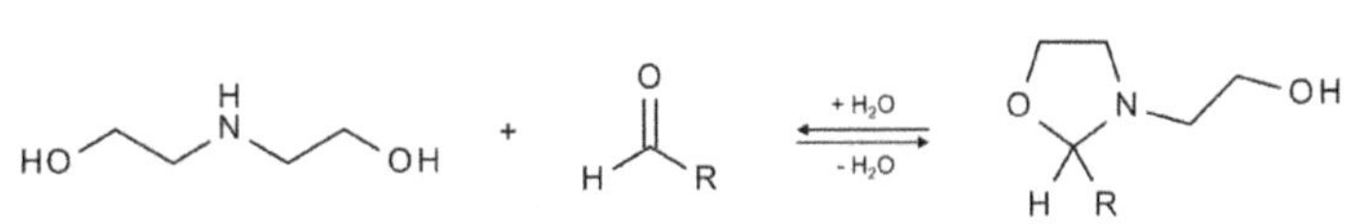

Figure 3.20: Formation of 1,3-oxazolanes

R = (cyclo)alkyl

Table 3.13: Largest manufacturers of polyaspartic acid esters

Company	**Trade name**	**Product examples**
Covestro	"Desmophen"	NH 1420 (standard product) NH 1220 (higher reactivity) NH 1520 (lower reactivity) NH 2850 XP (flexibility, medium reactivity)
Feiyang	"Feispartic"	F 420 (standard product) F 220 (higher reactivity) F 520 (lower reactivity) F 2850 (medium reactivity) + further grades available
Wanhua	"Wanelast"	984 (standard product) 982 (higher reactivity) 985 (lower reactivity) + further grades available
Evonik/Air Products	"Amicure"	IC 20 (standard product) IC 322 (satin finish) IC 40 (lower reactivity) IC 166 (high reactivity, direct to metal coatings IC 133 (lower reactivity, direct to metal coatings) + further grades available

Polyaspartic acids ester-based on diethyl maleate and cycloaliphatic diamines are solvent-free, low-viscous substances with secondary amino groups, whose reactivity is reduced by electronic and steric effects to such an extent that they can be used to formulate highly reactive, two-component polyurea coatings with an adequate pot life.

This adjustable reactivity enables the formulation of fast-drying coatings for highly productive, modern coating concepts. These systems are of particular interest for use in automotive refinish, large vehicle and industrial coatings, protective coatings, gelcoats in

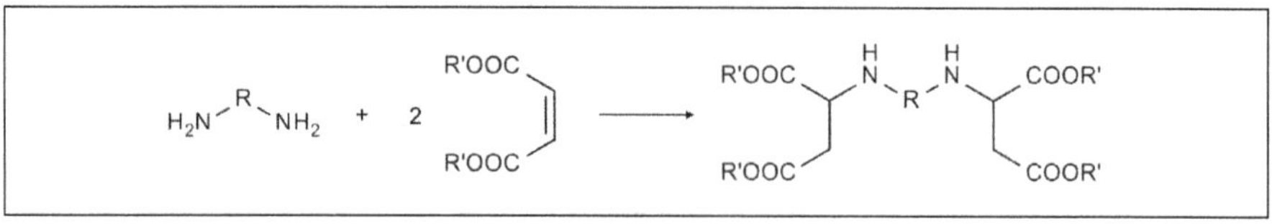

Figure 3.21: Manufacture of diaspartic acid esters

(R = (cyclo)alkylene; R' = (cyclo)alkyl)

wind-mill blades, and construction. The polyaspartic acid esters can be used either as the main binder or as a reactive thinner.[75–77]

The most appreciated characteristics of polyaspartic acid ester-based coatings (in the following polyaspartic coatings) include:

- quality (excellent appearance, long lasting performance),
- efficiency (rapid curing, fewer coats needed, ease of application),
- environmental benefits (100 % or high-solids systems, low energy consumption).

Due to the properties mentioned above, polyaspartic coatings have enjoyed an increased interest by the coatings market, and as a result several manufacturers of polyaspartic acid esters have emerged. Some of them are listed in Table 3.13.[54]

3.8 Aqueous dispersions

Against the backdrop of ever-increasing and more stringent environmental and occupational health regulations, aqueous dispersions offer a wide range of properties for coatings and adhesives, while maintaining low to zero VOC (volatile organic compounds) level. For this reason, aqueous dispersions have gained significant importance for use in two-component water-borne as well as one-component water-borne polyurethane technology. Another important reason to use aqueous dispersions of polymers is the potential to tune molecular weight, and therefore, performance of the polymers across a wide range of properties without losing process efficiency. Unlike solvent-borne systems, even at a very high molecular weight of the polymer itself the viscosity of the dispersed system is governed by the colloidal interactions of the polymer particles in water, and the viscosity stays much lower than for solutions of high molecular weight polymers in organic solvents.

Substantial technical development and significant investments into manufacturing assets are taking place in this area. In particular, increased activity in Asia has led to the appearance of several players in the area of WB PU technology.

Table 3.14 summarizes the largest producers of polyurethane dispersions and water-based polyols, which are the most important binders for one-component and two-component WB PU systems.[54]

3.8.1 Polyurethane dispersions

An aqueous polyurethane dispersion (PUD) consists of discrete polyurethane or more commonly, polyurethane/polyurea polymer particles, suspended in a continuous water or water/organic solvent media. Typically, the polymer is characterized as high molecular weight and linear. However, some branching can be introduced during synthesis as well. The polymer can be crosslinked via post-addition of a wide variety of crosslinking agents.

Table 3.14: Major producers of binders for coatings and adhesives in one- and two-component water-borne polyurethane technology

Producer	**Typical trade names**	**Technology/Application**
Alberdingk Boley	"Alberdingk": PU dispersions U-grades, PUR MATT-grades, AC-grades	PU dispersions PAC polyol dispersions for 2K WB PUR
Allnex	"Daotan", "Ebecryl", "Macrynal"	PU dispersions, UV-cure PU dispersions UV-cure urethane acrylates PAC polyol dispersions for 2K WB PUR
BASF	"Luhydran", "Luphen", "Basonat"	PAC polyol dispersions for 2K WB PUR PU dispersions for adhesive applications Water-miscible polyisocyanate cross-linkers
Bond Polymers	"Bondthane"	PU dispersions for leather, textile, adhesive, inks, and industrial applications
Covestro	"Bayhydrol", "Bayhytherm", "Baybond", "Impranil", "Dispercoll" U, "Baycusan", "Baymedix", "Bayhydur"	1K and 2K WB PU dispersions, PAC polyol dispersions for 2K WB PUR Water-based thermo-activated cross-linker dispersions Water-miscible polyisocyanate cross-linkers Products for coatings, adhesives, medical, cosmetics and textiles
DIC	"Hydran", "Vondic" , "Burnock"	PU dispersions Water-miscible cross-linkers
DSM	"NeoRez", "NeoPac", "NeoCryl"	PAC polyols, PU dispersions PAC-PU hybrid dispersions for coatings
Hauthaway	"Hauthane"	PU dispersions for various applications
Huafon	"JF"	PU dispersions for synthetic leather applications
Ketien/ SCISKY	"KT"	PU dispersions predominantly for artificial leather applications
Lamberti	"Esacote", "Rolflex"	Non-ionic, anionic, and cationic PU dispersions, water miscible polyisocyanates
Lubrizol	"Sancure", "Turboset", "Permax"	1K and 2K WB PU technology PU dispersions
Stahl	"Permutex", "Permuthane"	PU dispersions for textile and leather applications Polycarbodiimide cross-linker dispersions
Wanhua	"Adwel", "Antkote", "Archsol", "Crysol", "Lacper", "Leasys", "Tekspro", "Wantipro", "Aquolin"	1K and 2K WB polyurethane dispersions Hybrid polyurethane-polyacrylate dispersions Polyacrylic polyol dispersions Water-miscible polyisocyanate cross-linkers

Composition

A broad range of properties of PUDs is achieved through the careful selection of building blocks while designing the polymer. The common building block families are hydroxy-functional compounds, isocyanates, and amines. Hydroxy functional compounds consist of polyols, e.g. polyesters, polyethers, or polycarbonates, and diols, e.g. 1,6-hexane diol, neopentyl glycol, butane diol and ethylene glycol. In the market today, PUDs based on aliphatic isocyanates, e.g. hexamethylene diisocyanate (HDI), isophorone diisocyanate (IPDI), dicyclohexylmethane diisocyanate (H_{12}-MDI), and in some specialties tetramethyl-xylylene diisocyanate (TMXDI)[78] are predominantly used due for their light stability and lower reactivity, which is beneficial during the production process. However, aromatic diisocyanates are used in some special cases, e.g. in adhesive applications and as tie coats in textile coatings. Amine functional compounds are used in PUDs as chain extenders, in order to build high molecular weight and hard block content. In contrast, mono-functional amines or monools can be introduced in order to control molecular weight through chain termination. Tri-functional amines and triols can also be incorporated to introduce branching and thus increasing physical-mechanical properties of the coating or adhesive layer. The building blocks of polyurethane dispersions can also be selected in such a way that the films obtained are biodegradable, e.g. under composting conditions.[79]

Hydroxy-functional PUDs can be synthesized by altering the ratio of NCO/OH-containing reaction components or by introducing chain-terminating, hydroxyl-functional amines. These hydroxyl-functional PUDs are used in two-component, water-borne systems.

PUDs are widely used in one-component and two-component WB PU systems for wood and plastic coatings. Other areas where they also have been used successfully are general industrial coatings, glass coatings, or when soft-feel effects are desired. Semi-crystalline PUDs are very important for the use in thermo-activated adhesives for automotive, furniture, shoe, and other applications.

A polyurethane polymer as a resin component is typically inherently hydrophobic. In order to make it dispersible into water, solubilizing (also referred to as hydrophilizing) agents must be introduced. These solubilizing agents are mainly incorporated into the backbone to ensure proper dispersibility and stability of the resulting dispersion. However, external emulsifiers can also be used. These solubilizing agents can be further characterized as anionic, cationic, or nonionic emulsifiers, depending on their chemical functionality. In PUDs with anionic modification, the polymer can be hydrophilically-modified by incorporating carboxylate or sulfonate groups. These functional groups are introduced into the polymer backbone via building blocks with OH or NH functionality and contain the necessary acid groups. The polymer with incorporated carboxylic groups is then neutralized with bases such as trialkyl amines or alkali hydroxides. PUDs with internal, cationic hydrophilic modification are created by the incorporation of tertiary amino groups followed by neutralization with acid or quaternization with an alkylating agent.[80, 81]

Stabilization

As discussed above, PUDs are colloidal systems and can be stabilized by ionic and/or non-ionic mechanisms. The ionic dispersion is stabilized by the formation of a charged double layer on the surface of each particle, which results in ionic repulsion between particles. This phenomenon is demonstrated in Figure 3.22, see below.

Ionic dispersions are sensitive towards electrolytes and freezing. However, they are stable at high temperatures (> 70 °C), have a sufficiently long shelf-life, and are stable under moderate shear forces.

The stabilizing mechanism of non-ionic dispersions can be explained in terms of entropic or steric repulsion. When the particles approach one another, the degree-of-freedom of the hydrophilic chains in the continuous phase becomes restricted, leading to an unfavorable reduction in entropy. Non-ionic dispersions demonstrate a greater stability with regard to pH changes, lower temperatures, and strong shear, but are sensitive to coagulation at high temperatures. Polyethers with high ethylene oxide content are used as non-ionic, hydrophilic building blocks which are incorporated into the polymer backbone.[82, 83] Both stabilization types can be combined to achieve desirable, synergistic effects.

Greater hydrophilic modification of the polymer yields a smaller particle size of the dispersion. However, in this case, the viscosity of the resulting dispersion tends to increase, due the rapidly increasing number of interactions between the colloidal particles, thus leading to either dispersion of high viscosity or lower solid content. Ultimately, the amount of internal emulsifier in the polymer backbone must be balanced to obtain a stable dispersion with the highest possible solid content.[83, 84]

Additionally, solvent-free PUDs, as any other dispersions, need to be preserved in order to avoid damage caused by bacteria, fungi and algae as those organisms easily grow in water-based systems. Therefore, biocides are often used in order to protect PUDs before they can be applied in coatings and adhesives formulations.[85]

Manufacturing process

Polyurethane dispersions can be manufactured by various manufacturing methods (see Figure 3.23). The most widely-used processes for manufacturing PUDs refer to the acetone and prepolymer mixing processes, which typically require the use of an organic solvent. As acetone is nowadays commonly used in both processes, a clear differentiation between them becomes increasingly difficult.

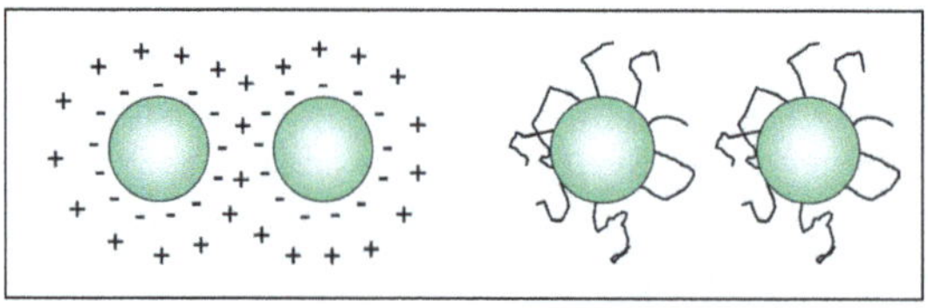

Figure 3.22: Stabilization of colloidal particles in a polyurethane dispersion

In both process modifications an isocyanate/(NCO)

group terminated polyurethane prepolymer with a molecular weight between 5,000 to 15,000 g/mol is synthesized in the first step, which is then chain extended with a polyamine. As a result, final polymer in the polyurethane dispersion also contains urea groups and the chain can be built to a high molecular weight. The most important difference between both processes is the way in which the chain extension step takes place. This step can either be performed in the homogeneous phase or in the heterogeneous phase directly after dispersing the hydrophilized prepolymer into water. Despite the free NCO groups, the polyurethane prepolymer can be dispersed in water since the reaction rate of an aliphatic NCO group with water is rather low. This method is not applicable for aromatic PUDs as aromatic NCO groups are too reactive towards water.

In the universal and rather popular acetone process, a hydrophobic, NCO-terminated prepolymer is first synthesized without any solvent, while maintaining a processable viscosity. Once the desired final NCO content is reached, the prepolymer is dissolved in acetone or another suitable, low boiling solvent, such as, e.g. methyl ethyl ketone. A hydrophilic polyamine like 2-[(2-aminoethyl)amino]ethane sulfonic acid sodium salt, so called AAS-salt (see Figure 3.23), is added with other polyamines to the reaction in the

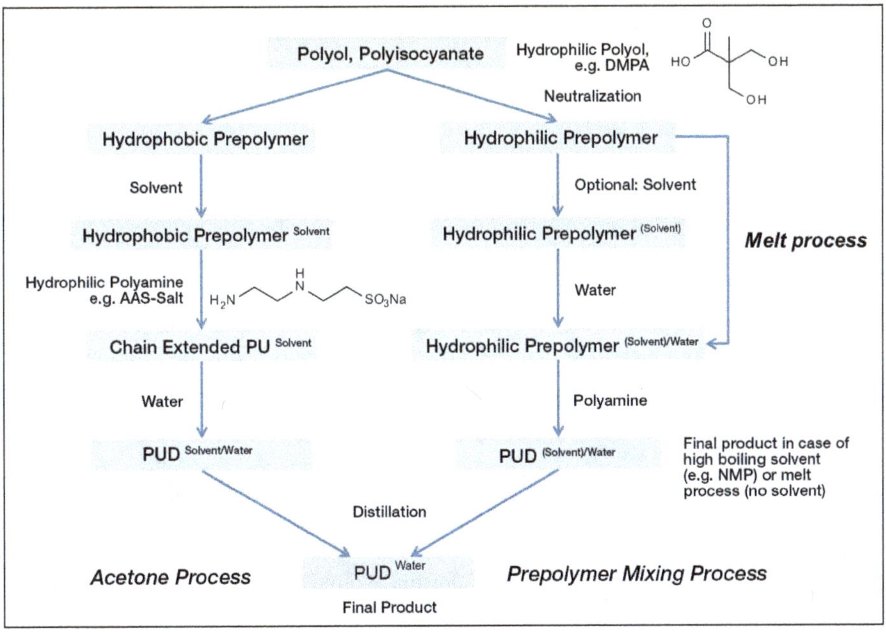

Figure 3.23: Manufacturing of polyurethane dispersions by the acetone or the prepolymer mixing process

subsequent chain extension step. Because the reaction mixture is still a homogeneous solution this step results in a strong viscosity increase. Water is then added to disperse the chain extended and hydrophilized polyurethane polymer, during which a phase inversion occurs. This converts the product from a "water-in-oil" (w/o) emulsion to the final "oil-in-water" (o/w) emulsion. Lastly, the acetone is distilled off and recycled to provide a substantially solvent-free, low-VOC PUD.

In the prepolymer mixing process a hydrophobic, NCO-terminated prepolymer is first modified using a polyhydroxy acid, e.g. dimethylol propionic acid (DMPA), which is then neutralized. The hydrophilic prepolymer can then be directly dispersed in water. In the case of low viscosity, hydrophilic NCO prepolymers having sufficient dispersability, no additional solvent is required to disperse them into water. This special case is referred to as a melt process and it results in a solvent-free, low-VOC PUD (see Figure 3.23). However, typically the prepolymer mixing process requires an additional solvent, such as N-methyl-2-pyrrolidone (NMP) to reduce the viscosity of the prepolymer. Solvents such as NMP are much less used in production due to increasing environmental and health regulations. The dispersed prepolymer is then chain extended with polyamines. Since this chain extension step takes place in the dispersed phase, the viscosity does not increase significantly. When using a low boiling solvent, like acetone, it is removed in the final step by distillation. In both processes, the low boiling solvents can be recycled, which makes these processes more sustainable.

With all the manufacturing processes that are available, a broad range of properties can be achieved. However, each process has its unique advantages and disadvantages. The prepolymer mixing process excludes the use of aromatic isocyanates due to a higher reaction rate of urea formation after dispersing the prepolymer and before the chain extension step. On the other hand, it typically requires less solvent and the polyurethane prepolymer can be more easily branched during the chain extension. The acetone process is mostly utilized to handle high viscosities. The complete polymerization is performed in a homogeneous mixture, which makes this process generally more reliable. Current developments make possible the production of high-solid dispersions, wherein the solid content can be increased even up to 60 % while retaining comparable properties.[86]

Properties and structure

PUDs offer a unique balance of flexibility, hardness, mechanical strength and durability combined with excellent tactile attributes and appearance. In order to design a polyurethane dispersion that will deliver performance specific to a targeted end use, the complex polyurethane structure/property relationship, and morphology, must be understood.

Crystallinity, T_g, degree of crosslinking, and degree of branching are just a few important characteristics when analyzing a polymer's property profile. The varying crystallinity of the building blocks determines the thermal behavior of the final product. A unique property of the polyurethane dispersion polymer is its ability to build up a network via hydrogen bonding between the urea and urethane groups. Once the PU film is stretched, these bonds temporarily break. However, they return once there is no more strain on the film. This phenomenon is referred to as a "self-healing" affect.[87] The PU polymer in a PUD is typically high molecular weight and linear. However, in order to increase the amount of crosslinking via hydrogen bonding or improve hardness and chemical/abrasion resistance, branching can be introduced.

The polyurethane polymer of the PUD is composed of a combination of soft segments and hard segments, which form soft and hard segment domains during film formation. The morphological balance between these soft and hard segments is predominantly responsible for the physical-mechanical properties. This offers unique characteristics to the final coating or adhesive layer, such as high film hardness without brittleness or high flexibility with increased toughness.

The combination of PUDs with other types of dispersions provides further customization of the property profile.[88–91] Anionic dispersions are mainly used due to the similarity of their pH ranges, from 6 to 9, when compared to other commonly used dispersions. The typical mean particle size of an aqueous polyurethane dispersion is between 10 to 500 nm. Compared with other dispersions, the particle size distribution is relatively broad. With increasing hydrophilicity, the particle size decreases, the viscosity increases, and the dispersion becomes more stable. The solid content is likewise affected by the particle size. The average solid content of PUDs is from 30 to 50 %.

The final application of a PUD dictates the molecular weight and other properties of the polyurethane polymer, and the flexibility of the production process allows these targets to be achieved. Molecular weight is adjusted by the inherent molecular weight of the building blocks, the level of chain extension, and the level of chain termination.

There are a multitude of modifications that can be made to the PUD polymer backbone in order to take advantage of their benefits and customize their property profile.[88–91] For instance, when combined with other dispersion polymer technologies, hybrid polyurethane dispersions can be obtained, such as the examples listed below:

- Polyacrylate-polyurethane dispersions,
- Alkyd containing, oxidatively-drying, self-crosslinking dispersions,
- Acrylate containing, radiation-curable dispersions,
- Hydroxy-functional dispersions,
- Thermally activatable, crosslinking polyurethane dispersions , and
- Combinations of hydroxy- and thermally activated isocyanate group containing polyurethane dispersions.

Due to this compatibility, the property profile of the final product can be further tailored to meet the requirements of a wide array of applications.

3.8.2 Polyacrylate dispersions

Hydroxy functional polyacrylate dispersions represent an important class of compounds in the formulation of aqueous two-component WB polyurethane coatings. They experience an increasing interest in the market, especially in Asia. Polyacrylate dispersions can be divided into primary and secondary dispersions. This definition stems from the two ways of manufacturing polyol dispersions. Primary dispersions are produced via emulsion polymerization of acrylate and other vinyl monomers directly in water in the presence of radi-

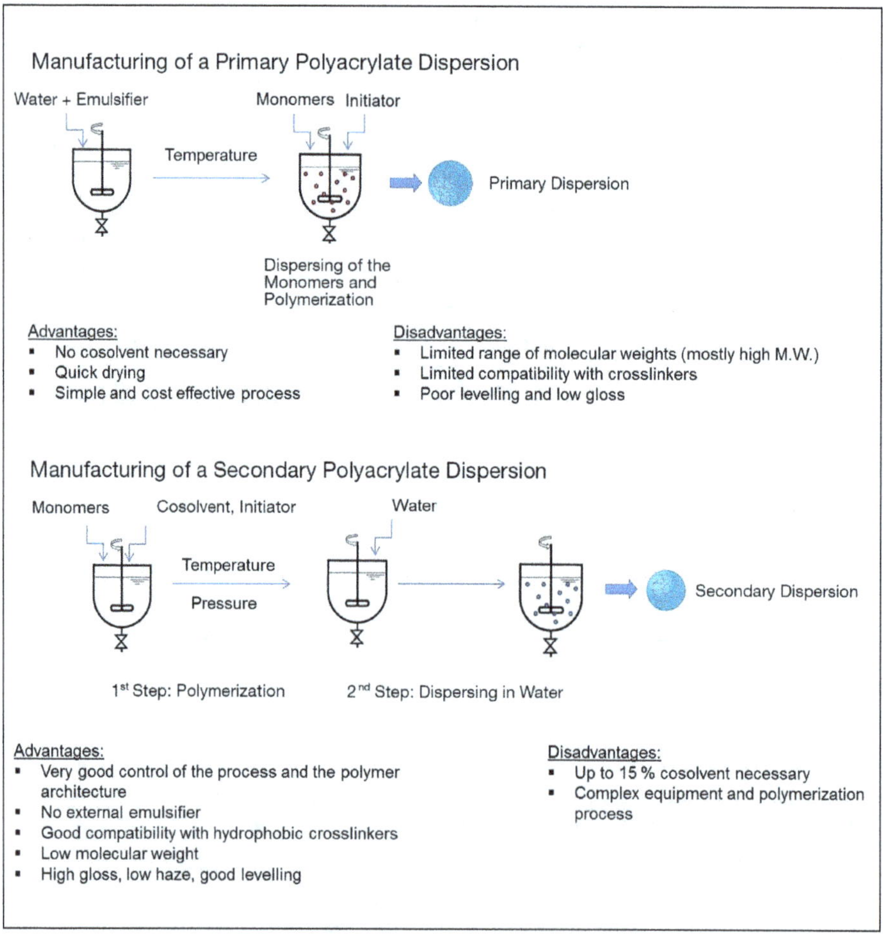

Figure 3.24: Synthetic routes for polyacrylate dispersions

cal initiators and emulsifiers (see Figure 3.24).[92–95] They are characterized by high molecular weights and are usually solvent-free.[96, 97]

On the other hand, secondary dispersions[38, 98–103] are prepared via free radical polymerization of the corresponding monomers, which include carboxy-functional monomers in a suitable solvent or in bulk. Following this step, a neutralizing amine and water are added to the organic solution, whereby a phase inversion takes place with the formation of a dispersion (see Figure 3.24).

Secondary dispersions are typically characterized as having a polymer of moderate molecular weight, which is similar to their solvent-borne analogues, and sometimes contain co-solvents.[101–103]

Secondary polyacrylate polyol dispersions can be used to formulate aqueous two-component polyurethane coating systems that yield high quality coating films. In particular, properties of these films, such as gloss, chemical resistance, and mechanical strength are comparable to the films formed by two-component, solvent-borne polyurethane coatings.[104]

Although secondary polyacrylate dispersions contain only little amount of co-solvent, recent developments have been directed towards a complete elimination of the use of volatile organic solvents. An example of such development is the "NOVOS" technology.[105] Another important trend in the field of polyacrylate dispersions is the use of bio-based monomers, such as, for example, "Visiomer" Terra monomers from Evonik[106] and the Discovery technology from DSM[107], to mention a few.

3.8.3 Hybrid polyurethane/polyacrylate dispersions

Hybrid polyurethane-polyacrylate dispersions (PU-PAC) represent a special type of binders, which consist of polyacrylic and polyurethane chains grafted onto each other. Such hybrid technology is prepared by swelling a polyurethane dispersion with acrylic monomers and a subsequent radical emulsion polymerization. In contrast to physical blends of polyurethane and polyacrylate dispersions, polymers in which are often not compatible with each other, the PU-PAC hybrids allow a coating or adhesive formulator to leverage synergies of both technologies. They can be used in a one-component, physically-drying coating or a crosslinking mechanism can be incorporated to further optimize the hybrid properties.[108]

3.8.4 Polyester polyol dispersions

Using a polyester polyol in its aqueous-supplied form requires the availability of a sufficient amount of hydrophilic groups in the polymer. These hydrophilic groups are generally neutralized acid groups, which can be introduced after the polycondensation reaction by a reaction of the polyester polyol with dicarboxylic acid anhydrides and opening the anhydride ring. However, this often results in polyesters that are susceptible to rapid hydrolysis and can be handled only during a very limited period of time.[109] For this reason, polyester polyol dispersions are often produced by coating companies for captive use.

To produce aqueous polyester resins that are more resistant to saponification, the hydrophilic carboxyl group is introduced during polyurethane dispersion synthesis, which was described in Chapter 3.8.1.[110]

3.9 Urethane acrylates

Urethane acrylates are used in radiation-curing coatings for industrial applications. Compared to other radiation curing raw materials, they provide increased toughness, chemical resistance, weather stability, and chemical diversity, and thus combine the high performance of polyurethane coating systems with the cure rate and efficiency of photopolymerization in many possible applications. Urethane acrylates are available as solutions in reactive thinners (low-viscosity (meth)acrylate esters), as low-viscosity oligomers, as solids for powder coating technologies, or as urethane acrylate dispersions. Isocyanate-functional urethane acrylates are a special group used in dual-cure technology.[111]

3.9.1 Chemistry of urethane acrylates

Urethane acrylates are manufactured by the reaction of hydroxy alkyl acrylates, diisocyanates and polyols, or by the direct addition of hydroxyl alkyl acrylates to polyisocyanates.[111] An alternative that yields especially low-viscosity products consists of the

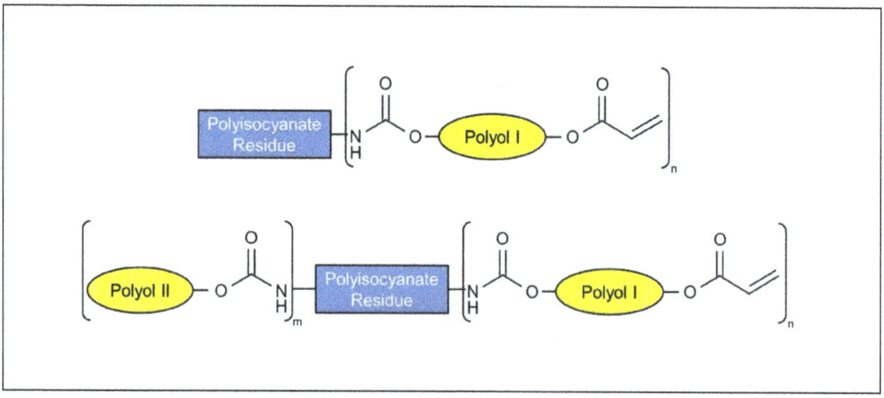

Figure 3.25: Basic structures of urethane acrylates

polyaddition of low molecular weight hydroxyl group containing polyester acrylates with diisocyanates. The idealized basic structures are shown in Figure 3.25.
The following factors can be varied to adjust the coating raw material properties, the UV reactivity of the resulting coating, and the properties of the cured coating film:
- Chemical nature of the di- and polyisocyanates
- Chemical nature of the residual polyol
- Double bond content
- Double bond density, and
- Molecular weight.

The basic structure-property relationships are summarized in Figure 3.26.

While these parameters cannot be varied entirely independent of each other, the structure of the resin can be optimized over a broad range to yield the desired end properties of the coatings. However, the molecular weight will be limited by the corresponding increase of viscosity. This can be counterbalanced by dilution of the final product either with solvents or with reactive thinner. While the first option is environmentally unwanted the second option will dilute many of the favorable properties of urethane acrylates.

Urethane acrylate dispersions in water are largely independent of viscosity effects caused by high molecular weight or hydrogen bond forming urethane groups. Manufactured as shown in Figure 3.23 they can be designed to show new properties like high elasticity, soft touch, multiple curing mechanisms.[112, 114] This variety of possible variations makes urethane acrylates one of the most versatile and useful binder classes in the area of radiation curing.

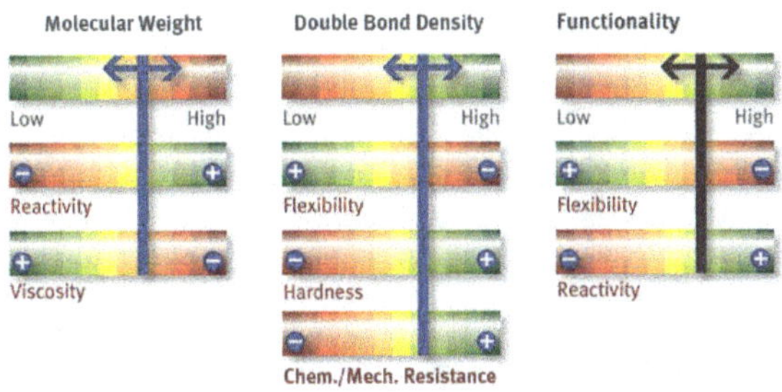

Figure 3.26: Structure-property relationships in urethane acrylates and resulting films [112, 113]

3.10 References

[1] C. Six, F. Richter, Ullmann's Encyclopedia of Industrial Chemistry, Chapter 4, DOI: 10.1002/14356007.a14_611 by Wiley-VCH, Organic Isocyanates, 2002; H.-W. Engels, H.G. Pirkl, R. Albers, R.W. Albach, J. Krause, A. Hoffmann, H. Casselmann, J. Dormish, Angewandte Chemie International Edition, 52, 2013, 9422–9441; G. Örtel, Kunststoffhandbuch 7 (3rd Edition), 1997, p 26

[2] EP 289 840 (1987) Bayer AG

[3] DE 4 231 417 (1992) Hüls AG; DE 4 413 580 (1994) Bayer AG; WO 97/17 323 (1995) BASF AG; H. Blattmann, M. Fleischer, M. Bähr, R. Mühlhaupt, Macromol. Rapid Commun. 35 (2014) 1238–1254

[4] W. Wieczorrek in: Lackharze, D. Stoye, W. Freitag (Hrsg.). Carl Hanser Verlag, München, Wien 1996, p. 183ff.; F. E. Golling, R. Pires, A. Hecking, J. Weikard, F. Richter, K. Danielmeier, D. Dijkstra, Polym. Int. (2018), https://doi.org/10.1002/pi.5665. G. Behnken, A. Hecking, B. Vega-Sanchez, Farbe und Lack, 08 (2015) p. 42–47; G. Behnken, A. Hecking. B. Vega-Sanchez, European Coatings Journal 01 (2016) 46-50; Ikeda et al., AMB Express (2013) 3:67; f) Li et al., Biotechnology and Bioprocess Engineering 19: p. 965–972 (2014); WO16042125 (2016) Covestro Deutschland AG

[5] R. Saxon, R. W. Dexter, G. C. Hewitt, Cellular Polymers 4 (1985) p. 117

[6] EP 749 958 (1995) Bayer AG

[7] DE 1 101 394 (1958) Bayer AG

[8] C. Guertler, K. Danielmeier, Tetrahedron Letters 45 (2004) 2515; EP 379 914 (1989)

[9] H. J. Laas, R. Halpaap, J. Pedain, Farbe und Lack, 100 (1994) p. 330

[10] R. Lomölder, F. Plogmann, P. Speier, Journal of Coatings Technology, 69 (1997) 51–57 Polyurethanes: chemistry and Technology, Part I. (High Polymers, Vol. XVI). J. H. Saunders and K. C. Frisch, Interscience, New York, 1962, p. 129 ff; Ana L. Silva & João C. Bordado (2004) Catalysis Reviews, 46:1, 31–51; L. Thiele, Acta Polymerica 30, 323 (1979); e) EP 249 3948

[11] H. J. Laas, R. Halpaap, J. Pedain, J. prakt. Chem. 336 (1994) 185 Covestro Data Sheet: Order No.: COV00084028; Edition 2018. Sapici: http://www.sapici.it/sapici-launches-us-low-free-monomer-line-wood-coating/

[12] EP 10589 (1978) Bayer AG

[13] DE 196 11 849 (1996) Covestro Deutschland AG (form. Bayer MaterialScience AG)

[14] K. Wagner, K. Findeisen, W. Schäfer, W. Dietrich, Angew. Chem. 93 (1981) p. 855

[15] H. Mundstock, R. Pires, F. Richter, J. Schmitz, Macromol. Symp., 187 (2002) p. 281–292

[16] EP 666 276 (1994) Bayer AG; EP 682 012 (1994) Bayer AG; C. Irle, J. Weikard, High functional, low viscosity hardener, European Coatings Journal 11, 2015, p. 36

[17] Ch. Gürtler, T. Stingl, Farbe und Lack 111 (11/2005), p. 22

[18] Th. Engbert, E. König, E. Jürgens, Einsatz einkomponentiger Einbrennurethanbeschichtungen, Farbe und Lack 102 (1996) p. 51–58

[19] DE 10 109 228 (2002) Bayer AG

[20] EP 615 992 (1993) Miles Inc.; US 5 216 078 (1992) BASF; EP 654 490 (1993) Bayer AG

[21] U. Röckrath et al., Investigations of crosslinking mechanism of etch resistant clearcoats, 22. Int. Conference in Org. Coatings Science and Technology, Athen, June 1996

[22] V. Mirgel, K. Nachtkamp, Farbe und Lack 89 (1983) 928–934; G. A. Sal'nikova, Yu. A. Shmidt, Yu. I. Voladorskaya, Dokl. Akad. Nauk SSSR, 181 (1968) 669;

P. L. Jansse, J. Oil Colour Chem. Assoc., (1989), p. 478
[23] Douglas A. Wicks, Zeno W. Wicks Jr., Progress in Organic Coatings Volume 36, Issue 3, July 1999, Pages 148–172; Douglas A. Wicks, Zeno W. Wicks Jr., Progress in Organic Coatings, Volume 41, Issues 1–3, March 2001, Pages 1–83
[24] R. Pires, H. J. Laas, European Coatings Journal 11 (2001) p. 16–22
[25] WO 97/31960 (1996) Rhodia Chimie
[26] EP 486 881 (1990) BASF AG
[27] EP 206 059 (1985) Bayer AG
[28] EP 540 985 (1991) Bayer AG
[29] EP 959 087 (1998) Covestro Deutschland AG (form. Bayer MaterialScience AG)
[30] WO 01/88006 (2000) Covestro Deutschland AG (form. Bayer MaterialScience AG)
[31] E. Tejada, M. Almató, A. Hecking, H. J. Laas, The first bio-based polyisocyanate for 2K waterborne systems. High performance enabled by nature; PRA's 10th International Woodcoatings Congress, October 25–26, 2016, Amsterdam/The Netherlands; WO16146579 (2016) Covestro Deutschland AG
[32] M. Melchiors, H. J. Laas, M. Mechtel, Sulfonathydrophilierte Polyisocyanate: die nächste Härtergeneration für wässrige Zweikomponenten-Polyurethanlacke; 69. Vortragstagung der GDCh-Fachgruppe „Anstrichstoffe und Pigmente", 22.–24. September 2004, Ulm/Neu-Ulm; M. Mechtel, Ch. Irle, H. J. Laas, M. Melchiors, New Hydrophilic Polyisocyanates for Waterborne 2K-PU-Coatings; The European Coatings Conference: Polyurethanes for High Performance Coatings III, March 18–19, 2004, Berlin/Germany
[33] M. Melchiors, H. J. Laas, T. Münzmay, R. Gertzmann, Graffiti resistant waterborne 2K Polyurethane Coatings, XXVIIIth FATIPEC Congress, June 12–14, 2006, Budapest, Hungary
[34] Technology Backgrounder: "Cardura" E10P Glycidyl Ester; An Easy Process for High Solids Acrylic Resins, literature code HXN-526, Hexion (2017)
[35] D. H. Klein, J. Paint Technol. 42 (1970) 335
[36] W. H. Brown, T. J. Miranda, J. Paint Technol. 36 (2) (1964) p. 92
[37] G. Markert in: Houben-Weyl, Methods of Organic Chemistry, 4th ed., Vol. E20/2, H. Bartl, J. Falbe (Hrsg.). Georg Thieme Verlag, Stuttgart 1987, p. 1156
[38] P. Denkinger in: Lackharze, D. Stoye, W. Freitag (Hrsg.). Carl Hanser Verlag, München, Wien 1996, p. 317
[39] G. Henrici-Olive, S. Olive, Adv. Pol. Sci. 2 (1961) p. 496–577
[40] J. D. Ferry, C. G. Overberger, G. V. Schulz, A. J. Staverman, and H. A. Stuart (editors). Fortschr. Hochpolymer.-Forsch. Springer-Verlag, Berlin 2 (1961) p. 496–577
[41] EP 1882 712 (2006) Mitsubishi Chemical
[42] V. R. Remya, Deepak Patil, V. K. Abitha, Ajay V. Rane and Raghvendra K. Mishra. 2016. Biobased materials for polyurethane dispersions. Chemistry International 2(3) p. 158–167
[43] Evonik Resources Efficiency GmbH, Dynacol Terra EP 480.01, Technical Product Sheet, 2015
[44] A. Steinbüchel (editor) in: Biopolymers Vol. 3, Polyesters II. Properties and Chemical Synthesis. Edited by Y. Doi and A. Steinbüchel. Wiley-VCH, Weinheim 2002, pp. 329–362
[45] A. Sickert, D. Stoye in: Lackharze, D. Stoye, W. Freitag (Hrsg.). Carl Hanser Verlag, München, Wien 1996, p. 51–54
[46] A. Sickert, D. Stoye in: Lackharze, D. Stoye, W. Freitag (Hrsg.). Carl Hanser Verlag, München, Wien 1996, S. 49–51
[47] EP 318 800 (1987) Bayer AG
[48] G. Wegener in: Applied Catalysis A: General 221 (2001) p. 303–335
[49] Mihail Ionescu in: Chemistry and Technology of Polyols for Polyurethanes, 2nd Edition, V. 1, Smithers Rapra, Shawbury, UK, 2016
[50] J. Langanke, A. Wolf, J. Hofmann, K. Böhm, M. A. Subhani, T. E. Müller, W. Leitner, C. Gürtler, Green Chem. 2014, 16, p. 1865–1870

[51] W. Freitag in: Lackharze, D. Stoye, W. Freitag (Hrsg.). Carl Hanser Verlag, München, Wien 1996, p. 103–104
[52] US 2 999 844 (1956) Bayer AG
[53] DE 1 770 245 (1968) Bayer AG
[54] Respective internet homepages of the companies listed in the Tables 3.11, 3.12 and 3.13. Status: 2018
[55] US 2 933 477 (1956) Union Carbide Corp
[56] EP 117 538 (1983) Union Carbide Corp
[57] A. Balas, G. Palka, J. Foks, H. Janik, J. Appl. Polym. Sci. 29 (1984) 2261
[58] EP 279 311 (1987) Bayer AG
[59] DRP 728 981 (1937) I.G. Farben
[60] O. Bayer, Angew. Chem. 59 (1947) 257
[61] G. Oertel in: Kunststoff-Handbuch, G. W. Becker, D. Braun (Hrsg.), 3rd Edition, Vol. 7: Polyurethane. Carl Hanser Verlag, München, Wien 1993
[62] K. Buecking, European Coatings Journal 10 (1998) p. 730–731
[63] A. Perez, J. Reddinger, J.A. Johnston, Modern Paint and Coatings 90 (11) (2000) p. 26–29
[64] D. A. Wicks, P. E. Yeske, Prog. Org. Coat., 30, (1997), p. 265
[65] M. Bock, R. Halpaap, J. Coat. Technol. 59 (755), (1987), p. 313
[66] DE 2 356 213 (1973) Schering AG
[67] DE 1 694 237 (1968) Bayer AG
[68] DE 1 719 121 (1967) Bayer AG
[69] DE 2 446 438 (1974) Bayer AG
[70] US 4 471 102 (1983) Textron
[71] C. Zwiener, L. Schmalstieg, M. Sonntag, K. Nachtkamp, J. Pedain, Farbe und Lack 97 (1991) p. 1052
[72] EP 403 921 (1989) Bayer AG
[73] EP 470 461 (1990) Bayer AG
[74] DE 1 157 773 (1963) Bayer AG
[75] V. Mirgel, M. Sonntag, European Coatings Journal 10/1994, p. 690
[76] C. A. Angeloff, E. P. Squiller, Journal of Protective Coatings and Linings 20 (11) (2003) p. 35–41
[77] G.W. Gardner, Journal of Protective Coatings and Linings 23 (10) (2006) p. 38–47
[78] Brochure by Allnex S.A. LRA1004-EN-WW-0314 (2014)
[79] DE 19 517 185 (1995) Bayer AG
[80] DE 1 495 745 (1963) Bayer AG
[81] DE 2 019 324 (1970) Bayer AG
[82] D. Dieterich in: Houben-Weyl, Methoden der Organischen Chemie, Bd. E20, H. Bartl, J. Falbe (Hrsg.). Georg Thieme Verlag, Stuttgart 1987, p. 1659
[83] J. W. Rosthauser, K. Nachtkamp, Waterborne Polyurethanes, Adv. Ureth. Sie. Tech. 10 (1987) p. 121
[84] R. Satguru, J. McMahon, J. C. Padget, R.G. Coogan, Aqueous Polyurethanes, Journal of Coatings Technology 66 (830) (1994) 47
[85] J. Gillat in: Directory of Microbicides for the Protection of Materials, W. Paulus (Ed.). Springer, Dordrecht 2005, pp. 219–248
[86] C.-F. Chai, Y.-F. Ma, G.-P. Li, Z. Ge, S.-Y. Ma, Y.-J. Luo, Progress in Organic Coatings, 115 (2018) p. 70–85
[87] K. Zhu, Q. Song, H. Chen, P. Hu, Journal of Applied Polymer Science, 135 (2018) APP. 45929
[88] US 4 636 546 (1984) Rohm and Haas Co
[89] DE 3 718 520 (1987) Bayer AG
[90] DE 4 122 265 (1991) Hoechst AG
[91] JP 04 081 405 (1990) Nippon Shokubai KK
[92] S.C. Thickett, R.G. Gilbert, Polymer 48 (2007) p. 6965–6991
[93] H. Fikentscher, H. Gerrens, H. Schuller, Angew. Chem. 72 (1960) p. 856
[94] H. Rauch-Puntigam, Th. Völker, Acryl- und Methacrylverbindungen. Springer Verlag, Berlin 1967
[95] G. Markert in: Houben-Weyl, Methods of Organic Chemistry, 4th ed., Vol. E20/2, H. Bartl, J. Falbe (Hrsg.). Georg Thieme Verlag, Stuttgart 1987, p. 1150
[96] EP 0 739 961 (1996) Rhodia Chimie
[97] EP 0 973 817 (1998) Zeneca Specialities
[98] E. Knappe in: Glasurit-Handbuch: Lacke und Farben, BASF Farben und Fasern AG (Edit.). Vincentz Verlag, Hannover 1984, p. 66–69
[99] EP 358 979 (1988) Bayer AG
[100] DE 4 322 242 (1993) BASF AG
[101] WO 0 039 181 (2000) Akzo Nobel

[102] EP 1 024 184 (2002) Covestro Deutschland AG (form. Bayer MaterialScience AG)
[103] B. Schlarb, M. Gyopar Rau, S. Haremza, Prog, Org. Coat. 26 (1995) p. 207–215
[104] M. Melchiors, M. Sonntag, C. Kobusch, E. Jürgens, Prog. Org. Coat. 40 (2000) p. 99–109
[105] T. Nabuurs, W. J. Soer, W.van Bavel, Eur. Coat. J. 10 (2009) p. 28-31
[106] Press release from Evonik, Essen, 8. April 2014 (www.corporate.evonik.de)
[107] T. Nabuurs, Eur. Coat. J. 4 (2018) p. 20–24
[108] D. Mestach, European Coatings Journal (5), (2007), p. 156–161
[109] E. T. Turpin, J. Paint Technol. 47 (602) (1975) p. 40–46
[110] H. Blum, P. Höhlein, J. Meixner, Farbe und Lack 94 (5) (1988) p. 342–344
[111] For example: PKT Oldring (Ed.) Chemistry Technology of UV-, EB Formulations for Coatings, Inks & Paints, Vol. 2, SITA Technology, London, (1991), pp 73–123
[112] W. Fischer, H. Kuczewski, D. Rappen, J. Weikard, e-5(2005) Conference Proceeding
[113] W. Fischer, Pitture e Vernici-European Coatings 6 (2000) pp. 16–22
[114] E. Tejada Rosales, B. Vega Sanchez, M. Wintermantel, European Coating Journal 11 (2017) pp. 22–28
[115] Photography by Covestro

4 Coating technology principles

4.1 Formulation basics of polyurethane coatings

This chapter will focus on the formulation basics of coatings. Details to formulation of adhesives and sealants will be described in Chapters 6 and 7.

4.1.1 Basics

Typical paint formulations are multicomponent mixtures which can have up to 20 different components. If one keeps in mind that most of the individual components are not pure substances, but polymers which have a molecular size distribution and therefore main and side products, it becomes clear that good paint formulations are rather complex systems which strongly rely on the experience of formulators (increasingly supported by computational power) for their development and manufacturing.
Polyurethane coatings are today marketed in various forms in view of:

- their formulation as one- and two-component coatings,
- curing conditions such as baking (stoving), radiation-curing, room temperature-curing, moisture-curing, and physically drying coatings.

Two-component polyurethane paints typically consist of polyols and polyisocyanate crosslinkers, which are reactive to each other. These systems have a pot life, also called working life or gel time, which is the period for which the two reactive chemicals remain applicable when mixed. Typical two-component polyurethane coatings double their viscosity after being mixed within about 4 to 6 hours and become useless after that.

Such a polyurethane coating formulation contains resins, crosslinkers, pigments, fillers, catalysts additives and thinners. 100 % solids systems with no thinners are also in the market, e.g. in construction and pipe coatings as well as in powder coatings.

The solid part of this type of ready for use paint is typically made of so-called resins or binders, also called "component A", e.g. polyols like polyesters, polyacrylics,

Table 4.1: Example of a typical solvent-borne 2K-PU primer formulation

Component A	**Parts by weight**
Polyacrylate 1	26.00
Polyacrylate 2	7.80
Flow modifier additive	1.31
Defoamer	0.26
Catalyst (DBTL)	0.78
Rheology additive	5.22
Silica filler	0.26
Solvent mixture	8.00
Titanium dioxide pigment	10.10
Iron oxide pigment	0.29
Anticorrosion pigment	5.79
Talc	3.15
Barium sulfate	22.26
Component B	
Polyisocyanate	8.78
	100.00

polyethers, or polycarbonates, and of crosslinkers or hardeners, also called "component B", e.g. polyisocyanates.

In the case of polyurethane primers, base coats, or topcoats, pigments – which can be either organic or inorganic – or anticorrosion pigments will be added. Fillers, e.g. inorganic carbonates or sulfates, give volume to dried paints and may help to dilute cost. Catalysts and additives, such as defoamers, rheology and sagging control agents, levelling agents, antioxidants, UV-absorbents, etc. further, fine tune the formulation.

Thinners are often added to adjust viscosity to a point where the formulated system can be handled accordingly to the application method (e.g. spraying, brushing or rolling). They can be solvents (typically mixtures of aromatic solvents, such as "Solvesso" 100 or 150, aliphatic esters such as methoxypropyl acetate or butyl acetate) or water.

Another group of thinners, also known as reactive diluents, have reactive groups that are chemically incorporated into the paint film upon crosslinking. This approach is rather popular for certain polyurethane systems such as UV coatings, where unsaturated, low viscous oligomers are used to dilute the coating system to spray viscosity and upon UV curing, these oligomers are chemically incorporated into the film.

An example of a typical solvent-borne two-component-polyurethane primer formulation is shown in Table 4.1.

Nowadays, polyurethane coatings are formulated to a large extent with a low solvent content (high-solids), entirely free of solvents, as powder coatings or as water-borne systems, so that they have no difficulty complying with increasing VOC regulations.

The reduced use of solvents in the formulation and application of coatings also results in a significant safety improvement in terms of exposure to solvents at the workplace and also with regard to the fire risk. Another key driver for the increasing popularity of low-solvent coatings is their economic benefits. Especially in times of rising crude oil costs, the future aim will inevitably be to minimize the use of solvents in surface coatings and, if possible, to avoid them completely. This will also make it possible to dispense with cost-intensive, high-energy solvent recycling or post-incineration processes.

When choosing the appropriate thinner for polyisocyanates attention needs to be paid to the fact that it should not contain any NCO-reactive groups (nucleophiles), particularly hydroxyl groups or water. The residual water content of suitable solvents (so called PU grades) for polyisocyanates should not exceed 0.05 %.

One-component solvent-borne coatings based on isocyanate-terminated prepolymers are called moisture cure coatings. They cure after application by means of the reaction of isocyanate groups of the prepolymer with ambient humidity. The humidity will react with isocyanates temporarily forming amines, which in turn, will react with excess isocyanate groups present in the formulation to form urea networks. Since the amount of air humidity is strongly depending on local weather conditions, these systems may vary in their crosslinking speed depending on where they are applied. Special attention has to be paid when formulating moisture cure coatings that only raw materials with no moisture or those with a low water content are used, since the binder reacts with water. In order to keep formulated moisture cure paints stable over time, water-scavengers such as monofunctional isocyanates such as p-toluenesulfonyl isocyanate or triethylorthoformiate are typically added.

Polyurethane coatings are noted for their efficiency in application. Properties such as low-temperature curing, short drying times and early achievement of final properties allow almost immediate practical use and/or further processing of the coated items. There is no doubt that the overall performance of polyurethane coatings is vastly superior to that of traditional technologies. Linked to this is the fact that the much longer maintenance cycles for polyurethane coatings contribute to the sustainable conservation of raw material and energy resources, which, of course, also brings cost benefits for the consumer and adds on to sustainability initiatives along the value chain. From the consumer protection point of view, cured polyurethane coatings are safe as also described in detail in Chapter 10.

The wide range of economical, environmentally friendly polyurethane coating technologies on the market enables users to select products tailored to their individual needs.

4.1.2 Selection of polyurethane raw materials

With respect to the adequate selection of polyurethane raw materials for the formulation of coatings there are a few general concepts that are worth noting.

Only hardeners with very low vapor pressure are in use. In addition to MDI with inherently low vapor pressure, higher molecular weight polyisocyanates such as biurets, trimers, allophanates or prepolymers of the industrially available diisocyanates MDI, TDI, HDI, IPDI, H_{12}MDI and recently PDI are in use. The general concept is described in detail in Chapter 3.

In general coatings based on aromatic polyisocyanates are limited – because of their relatively poor lightfastness – to indoor applications, mainly in the field of wood finishes and primers. Prominent examples for such hardeners are TDI-TMP adducts, which have excellent mechanical and chemical resistances (see Figure 4.1). In construction applications such as floor coatings, isomer mixtures and prepolymers of 2,4´- and 4,4´-MDI are commonly used.

Coatings with aliphatic polyisocyanates in combination with suitable polyols, on the other hand, have excellent weather and light stability and can be used universally. These systems are less reactive than their corresponding aromatic countertypes allowing for their use in manual applications where a short pot life in the range of a few minutes is not acceptable. The most important polyisocyanates for weather-resistant polyurethane coatings today are the HDI biuret and the HDI isocyanurate (see Figure 4.2). Polyisocyanates based on IPDI are also used as crosslinkers but to a much lesser extent. A look at glass transition temperatures (T_g) of HDI biurets, HDI and IPDI isocyanurates serves as a hint to understand the different polymer properties of these hardeners (see Table 4.2).

In addition to the differences in T_g shown in Table 4.2, the different reactivity of the NCO groups in HDI and IPDI also must be considered. HDI has two equal primary NCO groups with similar reactivity, while IPDI has a primary and a secondary NCO group with different reactivity that can be addressed selectively by the choice of the adequate catalyst.[65, 66]

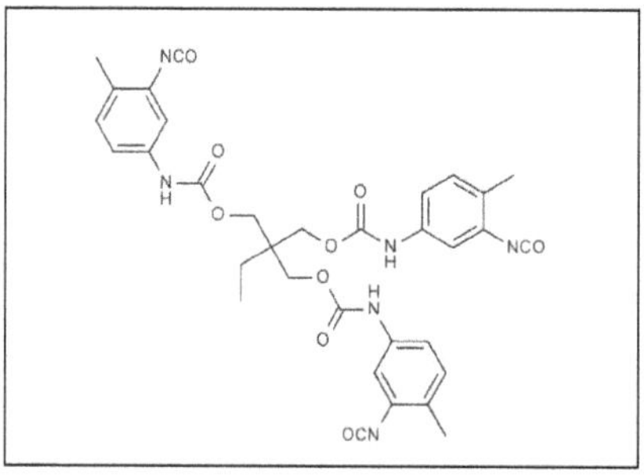

Figure 4.1: Exemplified TMP-adduct of TDI used for the formulation of two-component polyurethane wood finishes and primers

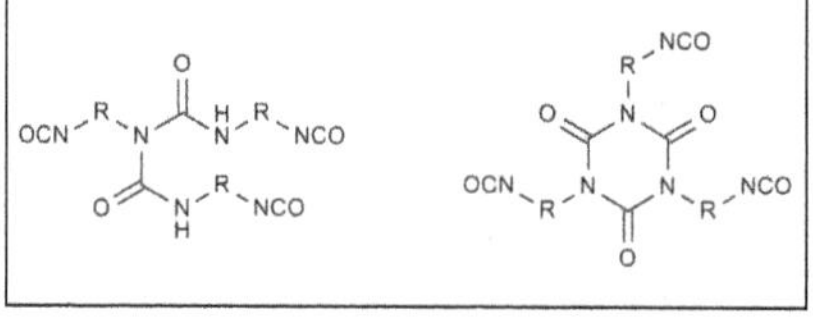

Figure 4.2: HDI biuret and trimer; R = $(CH_2)_6$

HDI-based hardeners are used to crosslink all aliphatic coatings systems and contribute to the well-known outstanding properties of polyurethane coatings. IPDI-based hardeners are used – typically as 10 to 30 % by volume admixture to HDI-polyisocyanates – in order to improve the physical drying properties and to increase the hardness of coatings (higher T_g-value). When adding IPDI hardeners to formulations attention needs to be paid to the elastic properties of the coating which will be compromised as more IPDI hardener is used in a formulation.

HDI-biurets and trimers are both well-established hardeners for aliphatic polyurethane coatings. Biurets show a slightly broader compatibility with hydrophobic, highly hydroxy-functional polyols and are

Table 4.2: Glass transition temperature (T_g) and typical properties and applications of common aliphatic and aromatic polyisocyanates

Hardener type	Glass transition temperature T_g [°C]	Properties and applications
HDI biuret	approx. -60	Standard grade for primers, primer surfacers and topcoats with universal compatibility to polyols
HDI isocyanurate	approx. -60	Standard grade for primers, primer surfacers and topcoats
IPDI isocyanurate	approx. +60	Combination hardener for blends with HDI types
TDI-TMP adduct	approx. +67	Standard grade for primers, and wood topcoats that do not require high levels of lightfastness

thus preferred when such polyols are used. Trimers exhibit lower viscosity and are typically used in modern high-solid and water-borne formulations.

Typical polyol classes used in coatings are polyacrylates, polyesters, polyethers and polycarbonates. Polyacrylates provide paints with fast drying and good hardness, together with outstanding weathering properties (if styrene is not used as a co-monomer). Solvent-borne and water-borne polyacrylates are widely established as the sole resin in one-component paints or as a reaction partner, e.g. to polyisocyanates, in two-component polyurethane industrial coatings applications.

Polyesters show very good wetting properties for pigments and fillers and thus are used to formulate solid color topcoats, primers and base coats. Additionally, polyesters provide very high gloss to paint formulations, but show weaknesses in weathering. Typically, polyesters and polyacrylates are mixed in paint and adhesives formulations in order to provide a good balance of drying and hardness, weathering, gloss and pigment wetting.

Polyethers show very good material properties in applications that do not require top levels of UV resistance (i.e. foams, primers, adhesives or coatings which are not exposed to sunlight). Due to their resistance to saponification they are used for example in combination with aromatic isocyanates to formulate two-component polyurethane floorings and pipe coatings.

More costly polycarbonate diols are used whenever high elasticity, weather stability, and hydrolysis resistance are all required simultaneously.

A typical coating is formulated by first mixing the polyols together with pigments and additives to form the mill base (component A). Because of the high reactivity of the isocyanate groups toward the polyols, the polyisocyanate hardener is not added until shortly before application, which means that the system is applied as a two-component system. Depending on which polyol and hardener are used, pot life, gel time, and drying times can be precisely adjusted, as can various coating properties such as gloss, levelling, scratch resistance, chemical resistance, lightfastness, and weather stability.

When combining polyisocyantes with polyols a chemical reaction between the corresponding isocyanate and the hydroxyl groups will occur. The mixing ratio of polyisocyanate to polyol is therefore a result of the functional groups of both reactants. An easy way of calculating the equivalent amounts of polyisocyanates and polyols in a polyurethane system is by considering the equivalent weights of both components based on their respective supply forms. Equivalent weights of the materials in question are typically delivered by the manufacturers of the raw materials and can be found in technical data sheets or can easily be calculated as follows:

$$EW_{PI} = \frac{42}{\%\, NCO} \cdot 100$$ Equation 4.1

where EW_{PI} = equivalent weight of the polyisocyanate

$$EW_{PO} = \frac{17}{\%\, OH} \cdot 100$$ Equation 4.2

where EW_{PO}: equivalent weight of the polyol

The stoichiometric relation of NCO to OH that equals 1:1 is achieved when one equivalent of polyol is combined with one equivalent of polyisocyanate.

The best mixing ratio between polyisocyanates and polyols that delivers the desired coatings properties should be determined empirically. In the polyurethane paint industry often the term "index" or "indexing" is used. Index is used as a dimensionless value (not a percent), and it is defined as the ratio of the equivalent amount of isocyanate used relative to the theoretical equivalent amount times 100. Above 100 means an excess of isocyanate; less than 100 means a shortage of isocyanate, or an excess of polyol. In practice the 100 factor is not used; more commonly it is written or said as 1.00 rather than 100. In commercial systems usually an over- or an under-indexing is deliberately used. A stoichiometric excess or deficiency of polyisocyanates to polyols is typically used to impact the adhesion properties to substrates or the other coatings layers (intercoat adhesion). An NCO over-indexing will also lead to harder and more resistant films, since the excess NCO will gradually react with air humidity leading to hard polyurea segments.

In general, it can be said that if one assumes a stoichiometric crosslinking (factor 1.00) between a given polyisocyanate and different polyols a higher hydroxyl content of a polyol will lead to harder films with a high resistance to chemicals, whereas polyols with low hydroxyl-content will lead to softer films with a low resistance to chemicals.

Another key factor affecting the overall properties of the coating system is the polyurethane network that is formed. The urethane bonds are resistant to hydrolysis and saponification. Of particular importance is the formation of hydrogen bridge bonds between the urethane groups as displayed in Figure 4.3.

These bonds contribute significantly to the mechanical resistance properties and enable the so-called reflow properties. In practice, this can mean that over time or under the effect of heat, scratched polyurethane coatings are largely able to heal themselves. This principle is used, for example, with automotive clear coats to optimize scratch resistance (see Chapter 5.3).

Low-solvent paints with a content of volatile organic components (VOC) of maximum 420 g/l are called high-solid paints, which have a solid content of at least 55 % by volume. Two-component high-solid polyurethane coatings comply with the current legal regulations on solvent emissions. If the coating system is carefully formulated, the solid content can be increased even further to create completely solvent-free polyurethane coating systems.

High-solid systems need crosslinkers and polyols with low viscosity. Examples of low viscosity polyisocyanate hardeners are the low-viscosity HDI isocyanurates and HDI iminooxadiazinediones – also known as an asymmetric trimer (see Chapter 3). These hardeners are established in high-solid and solvent-free PU coatings. Within the HDI polyisocyanate product range, the asymmetric trimer nowadays represents the optimal combination between low viscosity and high functionality.

Low viscous allophanates and uretdiones are frequently added to polyurethane formulations as reactive thinners to further lower the viscosity and increase the solid content. They have a very low viscosity and a functionality of around 2.5.

Depending on the particular requirements, polyesters, polyacrylates, polyethers, and increasingly also polycarbonate diols are used as high-solid polyols. Through their low molecular weights and narrow molecular-weight distribution, they have relatively low solution viscosity, which allows in turn for a higher solid content at a given application viscosity.

For some years now, amino-functional oligomers have been used with increasing success as reaction partners for polyisocyanates in order to yield very fast drying low VOC coatings systems. Although the reaction produces urea groups rather than urethane groups, such systems are also described in the broader sense as polyurethane coatings.

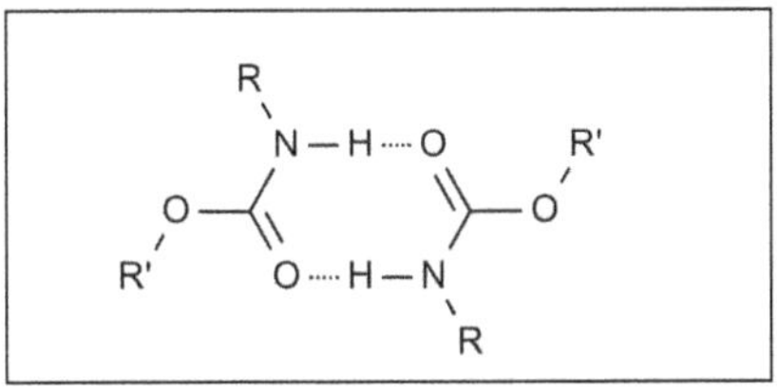

Figure 4.3: Formation of hydrogen-bonds in polyurethane systems

To allow aliphatic polyamines to be used in combination with polyisocyanates, the reactivity of the amino groups must be restricted. Low molecular weight, sterically-hindered diamines have attained particular importance here, particularly polyaspartic acid esters. These are reaction products from maleic acid diethylester with (cyclo)-aliphatic diamines. These solvent-free low-viscosity substances have secondary amino groups with a reactivity weakened by electronic and steric effects. Thus, with aliphatic polyisocyanates, fast drying two-component polyurethane coatings with adequate pot life can be formulated. The reactivity of the polyaspartic acid esters can be adjusted by varying in the molecular structure between the two amine groups in the amine building blocks used to manufacture the polyaspartic acid ester products (see Figure 4.4 and Chapter 3.7).

Because of their good color stability and their high-quality performance, such systems are of particular interest for general metal and protective coatings (Chapter 5.2), automotive refinishing and transportation coatings (Chapter 5.4), plastic coatings (Chapter 5.5) and construction applications (Chapter 5.9). Polyaspartic acid esters enable the formulation of extremely productive coating systems.

Another elegant way of reducing the VOC emission of paints is the use of water-borne coatings. Typical modern systems contain a residual solvent content of 5 to 10 % and a polymeric solid content of around 45 to 50 %.

Some important things need to be observed when handling and formulating water-borne coatings. For the sake of thinning of the paints deionized water needs to be used, since ions will have an adverse effect on the stability and the pH of the dispersions in the formulation.

For the cleaning of the spray and mixing equipment amine-containing water is recommended in order to avoid paint precipitations that would happen in case acidic cleansing conditions are used.

Additionally, stainless steel needs to be used for mixing devices, spray equipment and for the manufacturing of coatings. Due to the nature of water-borne paints – they consist of polymer droplets dispersed in water – their shear stability is limited compared to

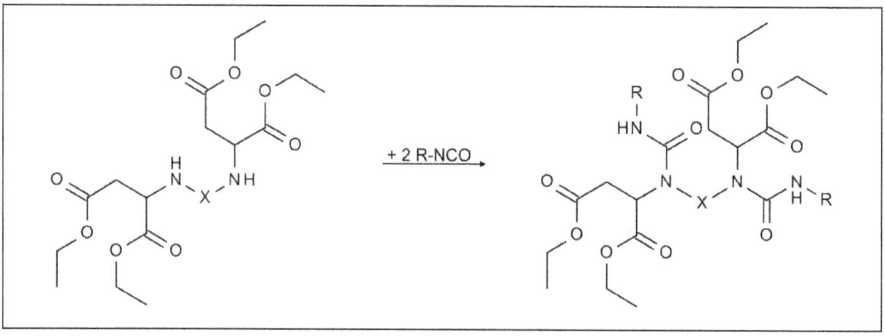

Figure 4.4: Aspartic acid esters polyisocyanate reaction; X = (cyclo)alkylene

solvent-borne systems. This fact is important when paints are strongly stirred or pumped through the lines of automated coatings processes such as in the automotive OEM industry. The storage stability of water-borne paints is limited to a range of around + 5 to 30 °C. At times biocide additives need to be added to water-borne systems in order to avoid deterioration by fungi.

Water-borne polyurethane coatings can also be one- or two-component systems. One-component water-borne polyurethane coatings are formulated by utilizing polyurethane dispersions (PUDs) (see Chapter 3). These paints based on PUDs are preferred for applications where highest levels of durability and elasticity of films are required. If such polyurethane dispersions are modified in their backbone with polycarbonate units, the durability of such films – especially against chemicals and UV light – can be even further increased.

One-component coatings based on polyurethane dispersions are typically used for the coating of wood (see Chapter 5.1), plastics (see Chapter 5.5) and textiles (see Chapter 5.7), and in new applications like in medical and cosmetics where high levels of elasticity are required (see Chapter 8).

Since polyurethane dispersions are easy to pigment, water-borne base coats for automotive applications (see Chapter 5.3) broadly utilize this technology. Often polyacrylate dispersions (PAC) are blended with PUD in order to achieve a good balance of hardness and cost (PAC) on one side and elasticity and pigment wetting (PUD) on the other side.

Thermally activated polyurethane crosslinkers that are chemically blocked are used when typical polyurethane coating properties are needed as a one-component system. Such formulations have no measurable pot life and can be applied with excellent results even after they have been stored for several months or years. They need to be cured by baking at temperatures between 100 to 180 °C. After the blocking agent has been split off, coatings with properties nearly comparable to those of two-component polyurethane coatings are obtained. Their main fields of application are automotive paints, coil coatings, and can coatings. The suitability of thermally activated polyurethane crosslinkers for a number of other applications is also being tested.

4.2 Aspects of one- and two-component coating technology

With the exception of film lamination, the formation of an organic coating usually involves a liquid phase. This is also true for powder coatings, since these are converted from the solid phase to a liquid phase through melting after application. The liquid coating material then solidifies on the substrate. Generally speaking, two drying mechanisms are involved in the formation of a solid paint film: physical drying with evaporation of the medium in which the coating is dissolved or dispersed (the thinner), and chemical curing, i.e.

film formation by means of a chemical reaction. Both mechanisms may overlap during the film formation process.

4.2.1 Physical drying

Resins which form a film solely by physical drying mechanisms are used in a number of applications. Such resins are dissolved in organic solvents and include, amongst others, polymers, polyurethanes and polyurethane ureas.

Above all, physically drying binders in the form of aqueous dispersions have come to dominate in formulations for do-it-yourself construction applications, so-called architectural paints. These one-component systems are easy to handle and, as a rule, fulfil the purpose for which they are intended, unless ambitious demands are made on their resistance properties. It should be pointed out that physically drying systems are not crosslinked – the film hardness results from associative interactions between the molecular chains. As a result, they can be re-solubilized (swollen/dissolved) by solvents and cleaning agents.

To improve the resistance properties in such systems, high mean molecular weights of up to 150,000 g/mol are the goal.[1] However, an infinite increase in the molecular mass is not feasible due to the associated increase in viscosity of the coating and the resulting application problems. The situation is different in the case of aqueous and non-aqueous dispersed two-phase systems. There, the limitations imposed by high molecular weights are applicable only to a limited extent.

4.2.2 Chemical curing

In case of reactive coatings, a distinction must be made between systems with one or two and more components.

One-component polyurethane systems can be divided into those that crosslink by baking and those that crosslink at room temperature. The latter exploit the reactivity of polyisocyanates with water in the form of atmospheric moisture and are known as moisture-curing polyurethane coatings. The drying rate of the paint film depends on the atmospheric humidity and on other factors like:

- temperature,
- reactivity of the NCO-groups,
- film thickness,
- hydrophilicity of the coating material,
- degree of branching (morphology) of the polymer,
- molecular weight of the polymer, and
- type and amount of any catalysts used.

Typical systems

In addition to one-component binders based on moisture-curing NCO-functional prepolymers, stable one-component systems can be formulated by combining polyisocyanates with polyols or polyamines. However, this requires the modification of the reactive groups of the individual components or of the overall system. The following coating systems of this basic type are currently in use:

- thermally activated polyurethane hardeners/polyols:
  - in dissolved form for baking coatings or
  - in solid form as powder coatings
- polyisocyanates/blocked amines that react with atmospheric moisture to yield free amines or react with isocyanates at elevated temperatures
- microencapsulated systems in which the co-reactants are separated by a diffusion barrier and the crosslinking reaction is initiated thermally.[3]

4.2.3 Air-drying coatings

One-component polyurethane coatings which cure at room temperature include those systems with unsaturated polyhydrocarbon chains which crosslink with oxygen. These are produced, for example, by the incorporation of suitable fatty acids or fatty alcohols into the polyurethane polymer. The addition of driers (catalysts) based on salts of cobalt or manganese (primary driers) and of e.g. magnesium, zinc, and calcium (auxiliary driers) allows these products to crosslink in air. Depending on the resin structure and the ambient conditions, different reactions will occur.[4] The crosslinking reaction is usually based on a radical mechanism with secondary reactions yielding ketones, aldehydes and other oxidation products.

In contrast to the fatty acid-modified polyurethanes described above, coatings that contain double bonds activated by carbonyl groups in the alpha position or allyl ether groups can be cured in seconds by exposure to high-energy radiation such as UV light (see Chapter 4.10).[5] The chemical precondition for this type of curing process is an adequately high density of double bonds in the one-component system and the addition of photoinitiators.

4.2.4 Dual-cure technology

Of particular interest are dual-cure systems that combine polyurethane and UV-crosslinking mechanisms (see Chapter 4.6). The advantages are, on the one hand, the cost-effective process control and, on the other hand, the fact that the coating cures chemically in those areas that are not covered by the radiation source. The quality of the resulting paint films is high and investigations are underway to optimize the suitability of these systems in coating metal, plastics and wood.[67]

4.2.5 Silyl-modified polyurethanes

Polyurethanes modified with trialkoxysilyl groups are a relatively new class of products that can also be processed as one-component systems.[6] These products are obtained by reacting NCO-functional prepolymers with special aminoalkoxysilanes. Catalysis with metal salts hydrolyzes the alkoxysilyl groups to form silanols. Crosslinking results from the subsequent silane polycondensation (see Chapter 7).

4.2.6 Application

The advantages of (moisture-curing) one-component coating systems are offset by several disadvantages including the complex formulation procedure and the need for carefully pre-drying the pigments.

In contrast to moisture-curing systems, pigmented two-component coatings can be formulated without pre-drying of the ingredients, provided they are added to the polyol component. A further advantage of two-component technology lies in the blister-free curing of the paint film at high film thickness. The pot life can be varied widely by the use of different catalysts.

Two-component polyurethane coatings are applied by a number of different methods and the pot life plays an important role. In the case of binder combinations based on aromatic polyisocyanates, rather than aliphatic polyisocyanates, the pot life is so short that, as a rule, they require two-component units specially designed for processing highly reactive two-component materials (see Figure 4.5).

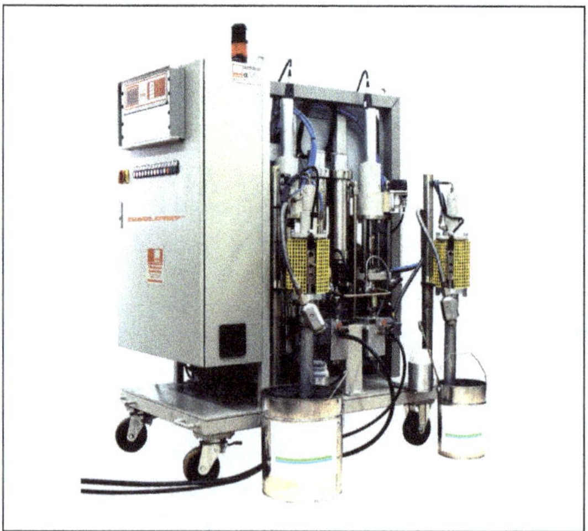

Figure 4.5: Two-component metering unit[7]

The reactivity of polyisocyanates with aliphatic amines is so high that even two-component units are not suitable for their application. Due to the fast crosslinking reaction, various components of the equipment (hoses, spray guns, etc.) become blocked as a result of polyurea formation. To facilitate a controlled crosslinking reaction, these amines must be suitably modified, e.g. blocked or sterically hindered (see Chapter 3.7).

4.2.7 Catalysis in polyurethane coatings

The reaction between an aliphatic polyisocyanate and most nucleophile moieties, e.g. alcohols, water, thiols and amines, is rather slow compared to aromatic polyisocyanate systems and to the drying speed needed in typical industrial coatings applications. Exceptions are the sterically non-hindered aliphatic amines, which react rather fast with aliphatic isocyanates. In this chapter we will focus on the formation of polyurethanes by the reaction of aliphatic isocyanates and alcohols.

A lot of research has taken place in the field and the respective catalysts that are suitable for catalyzing the isocyanate/alcohol reaction have to fulfill a range of different properties and deliver a wide performance range such as:
- high selectivity towards the urethane reaction vs. side reactions such as the formation of urea or other NCO-NCO reaction,
- high but controlled reactivity at low levels of catalyst,
- low toxicity,
- fit to changing legislation and sustainability,
- good availability and low additional cost for the formulation,
- color neutrality over time.

In general, two major classes of catalysts have been suggested:
- Lewis acidic metal-based catalysts, and
- amines.

The most broadly usable class of catalysts in polyurethane coatings have turned out to be metal compounds such as salts or carboxylates of lead, zinc, manganese, iron, zirconium, bismuth, aluminum, nickel, and others.[2, 58, 59]

Organotin catalysts have gained the most widespread practical importance within this group and particularly dibutyl tin dilaurate (DBTL) is very common in polyurethane coatings (see Figure 4.6).

DBTL offers several advantages when compared with other catalysts:
- it is broadly commercially available at reasonable costs
- it shows high activity with aromatic and aliphatic isocyanates and a high activity selectivity for the urethane bond formation
- it displays a high catalytic activity at very low concentrations
- it is soluble in and compatible with most common formulation ingredients

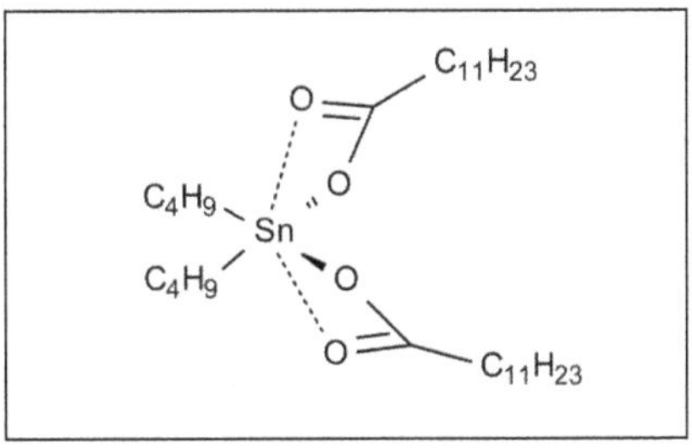

Figure 4.6: Schematic structure of dibutyl tin dilaurate (DBTL)

- it shows a reasonable stability against air, light and moisture
- it has a low influence on yellowing in coatings.

Nowadays, DBTL is used to catalyze several polyurethane related reactions, such as in two-component systems based on aliphatic or aromatic polyisocyanates, in one-component moisture curing polyurethane systems, for the activation of thermally activated polyurethane hardeners, for the curing of polyurethane powder coatings, for the synthesis of polyurethane dispersions (PUDs) and of polyurethane prepolymers.

The excellent catalytic performance of DBTL is based on its Lewis acid properties. It is well-known that complexation of the tin atom of the DBTL to the NCO group is one of the key steps in the catalysis of the urethane reaction. Figure 4.7 shows the reaction mechanism of the organotin-catalyzed reaction of a (poly)isocyanate and an alcohol (without

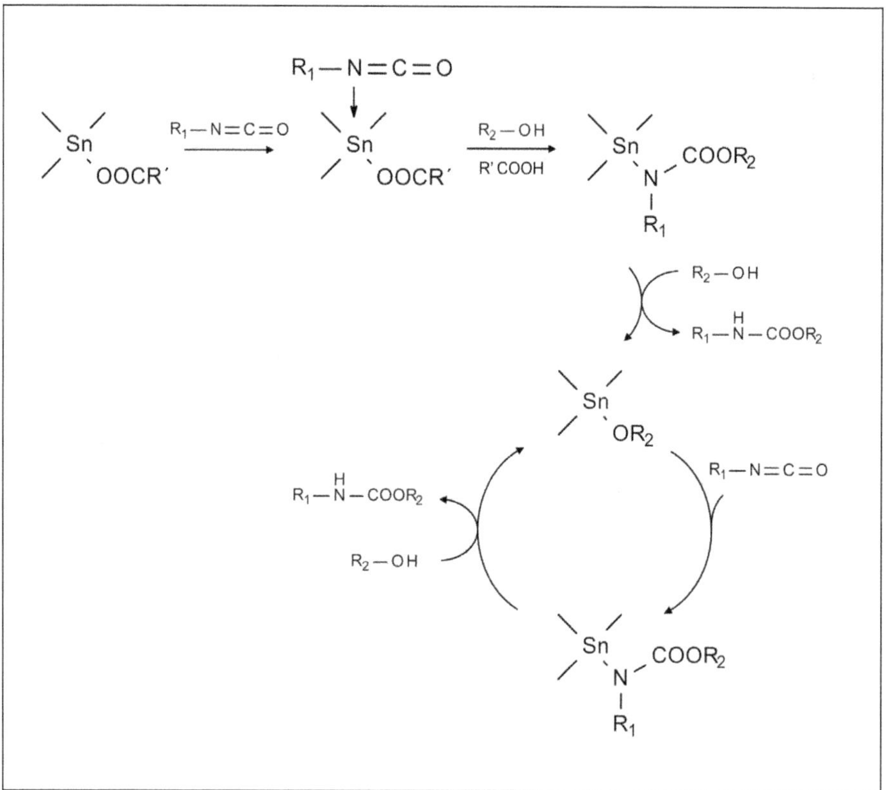

Figure 4.7: Schematic reaction mechanism of organotin compounds such as dibutyl tin dilaurate (DBTL) in the reaction of (poly)isocyanates with alcohols[45, 75]

considering possible side reactions), as originally proposed by *Bloodworth* and *Davies* and reported as such by *Delebecq* et al.[45, 72]

A special case of PU systems are two-component (2K) water-borne formulations: despite all attempts so far, the good performance of DBTL could not be reached in water-borne two-component polyurethane systems. Here DBTL (and other metal catalysts) lead to a dramatic decrease in pot life and to a significant deterioration of the optical properties of water-borne coatings such as gloss and haze. DBTL and metal catalysts do not distinguish between hydroxyl groups coming from polyols and those coming from water in the water-borne system. The excess of water will undoubtedly lead to the formation of polyurea caused by the reaction with polyisocyanate in the system and thus to reduced optical properties of the coating. However, this fact allows use of these kinds of catalysts very effectively in one-component moisture cure coatings. Nevertheless, to date no adequate (and selective) catalyst for water-borne 2K PU systems are known.

Due to regulatory pressure on dibutyl tin dilaurate (DBTL), there is a need to identify alternatives. Recently there has been an increasing discussion about toxicity of organotin compounds. In 2017 the European Commission decided to re-classify DBTL and other dibutyl tin-based substances as reproductively toxic in category repr. 1B H360FD, muta 2 H341, STOT RE 1 H372 (immune system).

This change in labelling led to an increased sensitivity along the value chain of PU coatings and adhesives producers, applicators and users to the presence of DBTL in formulations and thus to an intensified investigation for alternatives. Some brands, e.g. IKEA, have started voluntary actions to restrict or even ban tin-organic catalysts from their products.[60] Other brands, related to the textile industry have bundled their interests in reducing the use and impact of harmful substances such as tin-organic catalysts in the apparel and footwear supply chain in a group (AFIRM).[61]

Alternatives to dibutyl tin dilaurate (DBTL) with improved toxicological profile have been under investigation for years, but in spite of this increased research there has not been a breakthrough hit. Most of the proposed alternatives have already been known for a number of years. Some of them have been used in niche applications where the performance of DBTL or other organotin catalysts has not been sufficient to meet the specific technical requirements.
The following catalysts show some potential as DBTL replacement products[8–11]:
- dioctyl tin dilaurate,
- bismuth carboxylates,
- zinc carboxylates,
- chelates of zirconium and aluminum,
- tertiary amines such as DABCO (1,4-Diazabicyclo[2.2.2]octane), and
- 2,3-dimethyltetrahydro-pyrimidine or dimorpholinodiethylether

All of these catalysts have in common that they are able to replace DBTL only in certain specific applications or formulations but are not a universal replacement. In many cases the dosage of the alternatives has to be significantly increased compared to DBTL. Often handling and properties of the polyurethane system get adversely impacted i.e. the pot life of the formulation or the drying speed of the polyurethane system gets even reduced. If DBTL is used in the synthesis of prepolymers with well-defined structures, often the alternatives show far less selectivity yielding unwanted mixtures of different regiomers and by-products.

Another approach recently taken by the coatings and adhesives industry is to redesign (often with the help of modern modeling methods) typical polyurethane catalysts in such a way that pot life and drying speed of the corresponding systems can be decoupled and "curing on demand" can be achieved.

A smart example of such a two-component polyurethane system was launched by Ciba Inc. by the use of photo-latent metal catalysts, aliphatic polyisocyanates and polyols (polyacrylates). The photo-latent catalysts were designed in such a way, that they were almost inactive when mixed into the 2K PU system, yielding a "non-catalyzed" slow polyurethane reaction and a very long pot life. As soon as the PU system is applied onto a surface and light as external trigger is applied, the photo-latent catalyst rearranges and an almost instantaneous reaction of the aliphatic polyisocyanate and the polyol is initiated.[62]

More recently[63, 64] a thermo-latent aliphatic polyisocyanate hardener that is almost inactive at temperatures below 70 °C and starts to become active at around 80 °C has been developed. This system utilizes a thermo-latent, ring-type, catalyst that can be universally used in aliphatic two-component polyurethane systems. Compared to the previously described system of utilizing photo-latent metals the latter system has the advantage of being more practical, since most of coatings and adhesives applications are typically thermally and not photochemically cured.

DBTL continues at this time to be the catalyst of choice for aliphatic polyurethane systems. However, it is expected that by legal and voluntary constraints the use of DBTL will further decrease in the future and further attempts will be made to identify and replace DBTL by alternatives. Modern quantum chemistry methods will play a role in the search.

Another alternative is to identify productive polyurethane systems that do not require any catalysts or a new class of catalytic systems. One example is the above mentioned latent catalysts. Another method to increase the productivity of polyurethane systems can be the use of polyaspartic systems as referenced in other chapters and which do not require the use of DBTL or the switch to water-borne polyurethane UV systems that combine a high productivity, low VOC content and the absence of DBTL.

4.3 Solvent-borne and solvent-free systems

Solvent-borne one- and two-component polyurethane coatings account for the largest volumes applied. They are used in a broad spectrum of industrial applications, e.g. in transportation coating (automotive OEM and refinishing, coatings for aircraft, trucks, busses and railcars), in plastic, protective metal, textile, and in wood coating.

4.3.1 Classification

Depending on the solid content (low, medium or high), a distinction is made between low, medium and high-solid coatings (LS, MS and HS coatings), although there is no universal definition of the boundaries between these three types. As a rule, the point of reference is the solid content at spray viscosity. In order to reduce the organic solvent content and thus satisfy increasingly stringent environmental protection requirements, there is a clear trend towards high-solid coatings with a solid content of more than 60 %. These coatings require low viscosity (low molecular weight) co-reactants – polyisocyanates and polyols – if necessary, blended with reactive thinners.

4.3.2 Applications

Liquid solvent-free polyurethane coatings are used in the construction and corrosion protection sectors, and also as underbody sealing systems for vehicles. A particular requirement is blister-free coatings, even at high film thicknesses. This requirement can only be met by using solvent-free one- and two-component formulations, which restricts the choice of suitable polymers.

Solvent-free one-component polyurethane systems are based on moisture-curing aromatic or aliphatic prepolymers containing NCO groups. However, only relatively thin films can be produced as these systems tend to foam due to CO_2 generation. Other solvent-free one-component polyurethane formulations are based on thermally activated polyurethane prepolymers and may contain plasticizers. They are used as baking systems, e.g. as seam sealers for automotive applications.

The various systems cure at a wide range of temperatures. Moisture-curing polyurethane coatings harden at room temperature, whereas thermally activated crosslinkers may require baking at up to 500 °C of air temperature and very short process times, e.g. in the case of wire enameling.

Two-component polyurethane systems dry relatively quickly without exposure to heat due to their reactive NCO groups. However, these systems are often force-dried at up to

80 °C or baked at temperatures above 120 °C. Higher temperatures yield shorter curing times, resulting in faster production cycles and increased efficiency (equipment utilization). However, there are limits to these possibilities since, as a rule, temperatures of more than 200 °C and baking times of more than 30 minutes trigger thermal degradation reactions in polymers.

4.3.3 Quality characteristics of two component polyurethane coatings

Whether using solvent-borne or solvent-free two-component polyurethane coatings, the film quality obtained is largely independent of the drying temperature. Even at the lowest temperatures, an adequately long curing time results in a degree of crosslinking equivalent to that achieved under baking conditions. This fact was instrumental in the switch to this technology for the coating of large objects like aircrafts and for refinishing in automotive body shops, both segments where baking at high temperatures is not an option. Two-component polyurethane coatings have also made an important contribution towards the use of plastics in automotive production. For reasons including design (color matching with the body), they are used to finish add-on components such as fenders. They also provide long-term protection and prevent the plastic substrate from becoming brittle after exposure to weathering or from brittle fractures at low temperatures.

The curing conditions for two-component polyurethane coatings can be influenced with catalysts as required as discussed above (see Chapter 4.2). Particularly suitable for this purpose are metal salts, e.g. tin or zinc salts, and tertiary amines (chemical drying).

4.4 Water-borne systems

The significance and market share of water-borne products in the coating market as a whole has been growing steadily for several decades. This development is not only driven by ecological motives or legislation (VOC guidelines or better regulations) alone – there are also tangible economic reasons. Particularly in times of increasing crude oil prices and due to sustainability issues, the loss of significant amounts of valuable raw materials to the atmosphere in form of organic solvents must be avoided. Typically, there is the need to recover them by expensive processes or incinerate them, which requires capital investment for containments and incineration. However, in many industrial processes incineration units are used to generate thermal energy, which in turn is used for baking processes. In addition, the health risks associated with organic solvents, as well as the danger of fire and explosion, require a considerable expenditure in the area of safety technology. Modern environmentally friendly coatings therefore have a greatly reduced content of organic solvents (high-solid coatings, solvent-free or water-borne systems). In many cases, water-based coating systems are comparable with their solvent-based counterparts with regard to technical properties; often they even exceed them.

Water-borne coating systems based on polyurethanes are extensively used due to their high quality, and they continue to rapidly gain importance. Typical uses are automotive OEM, refinish and transportation finishes, wood, metal and plastic coatings, construction applications, as well as the coating of textiles, glass, and paper. In addition, water-borne systems have great potential in non-conventional coatings applications such as cosmetics and medical technology. A great deal of innovation can be expected in these fields.

Water-borne polyurethane systems are usually based on polyurethane dispersions that may, if desired, be formulated with further polyacrylate-, polyester-, or polycarbonate-based dispersions, with or without crosslinking agents. With regard to processing methods, they are essentially not different to their solvent-containing counterparts. Water-borne coatings are applied by conventional methods such as painting, spreading, curtain coating, dipping, or spraying.

After the application of dispersion-based coating systems, the water and other volatile components evaporate from the paint film. A number of different models have been proposed for the actual mechanism of film formation.[12] What is undisputed, however, is that good film formation only occurs, if there is a polymer chain exchange between adjacent particles as soon as these come into contact with each other (see Figure 4.8). Film formation is therefore largely governed by the polymer structure, the glass transition temperature of the soft segment phase, the type and amount of any coalescing agent or solvent used, the application conditions, and lastly the film thickness.[13]

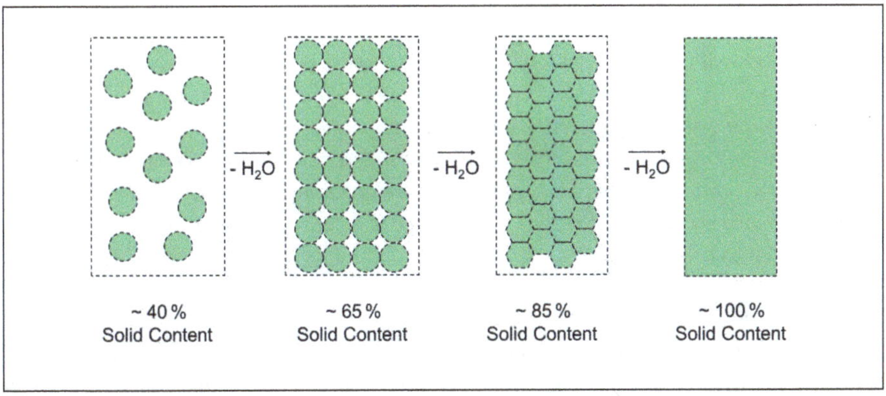

Figure 4.8: Film formation of a polyurethane dispersion

Despite their favorable properties, the use of high molecular weight physically drying polyurethane dispersions as binders often does not suffice to yield coatings that satisfy all requirements in terms of resistance and mechanical properties. This can only be achieved by post-crosslinking the dispersion in the film. In principle, such reactive coatings can be divided into one- and two-component systems.

4.4.1 Water-borne one-component polyurethane systems

Depending on the area of application, water-borne one-component polyurethane coatings can be customized in terms of crosslinking, curing and film properties. The most important polyurethane dispersions for water-borne one-component coatings include:

- Polyurethane dispersions for crosslinking with thermally activated polyurethane hardeners for baking coatings.
  These are OH- or NH-functional dispersions combined with raw materials that contain polyisocyanates with thermally reversible blocking. The blocked polyisocyanates suitable here can be used in their unmodified (i.e. hydrophobic) form. In this case, the resin dispersion (OH- or NH-terminated) must have a co-dispersing function. In contrast, hydrophilically-modified thermally activated polyurethane hardeners themselves form stable dispersions and are mixed into the resin dispersion.[14, 15] The blocked NCO group can also be bound directly to the OH- or NH-terminated polymer backbone. Such systems are designated self-crosslinking dispersions.[16]
- Polyurethane dispersions for melamine or epoxide crosslinking
  COOH- or OH-functional polyurethane dispersions combined with polyepoxides or alkoxymethyl melamines (melamine crosslinking).[17–19]
- Radiation-curing polyurethane dispersions
  Modification with hydroxyalkyl acrylates or allyl alcohols yields polyurethane dispersions for coatings that cure within seconds on exposure to high-energy radiation.[20]
- Polyurethane dispersions that dry by oxidation
  Polyurethane dispersions modified with unsaturated polyester units can be dried by reaction with atmospheric oxygen. These are generally fatty acid-modified polyurethane dispersions, the crosslinking of which is often accelerated by the addition of driers such as cobalt salts.[21]

Crosslinking via aziridines[22] and polycarbodiimides[23] are described as additional curing methods. Polyurethane dispersions that contain ketoester groups can be crosslinked with polyamines or polyhydrazides[24], and polyurethane dispersions that contain trialkoxysilyl groups, crosslink by condensation.[25] Occasionally, mixed crosslinking mechanisms are used, e.g. in the manufacture of baking systems. These combine an OH-functional dispersion with a thermally activated polyurethane hardener and a melamine crosslinker.

4.4.2 Water-borne two-component polyurethane systems

High-end water-borne two-component polyurethane coatings were first described in 1988.[26–29] Free polyisocyanates are emulsified in an aqueous hydroxy-functional polymer dispersion or solution. Dispersions of hydroxy-functional polyurethane, polyacrylate, polyester or urethane-modified polyester are normally used.

Key coating properties such as gloss and resistance depend directly on the formation of a homogeneous and dense polymer network. The main prerequisite for achieving this is to ensure that the hydroxy-functional polymer dispersion and the polyisocyanate are

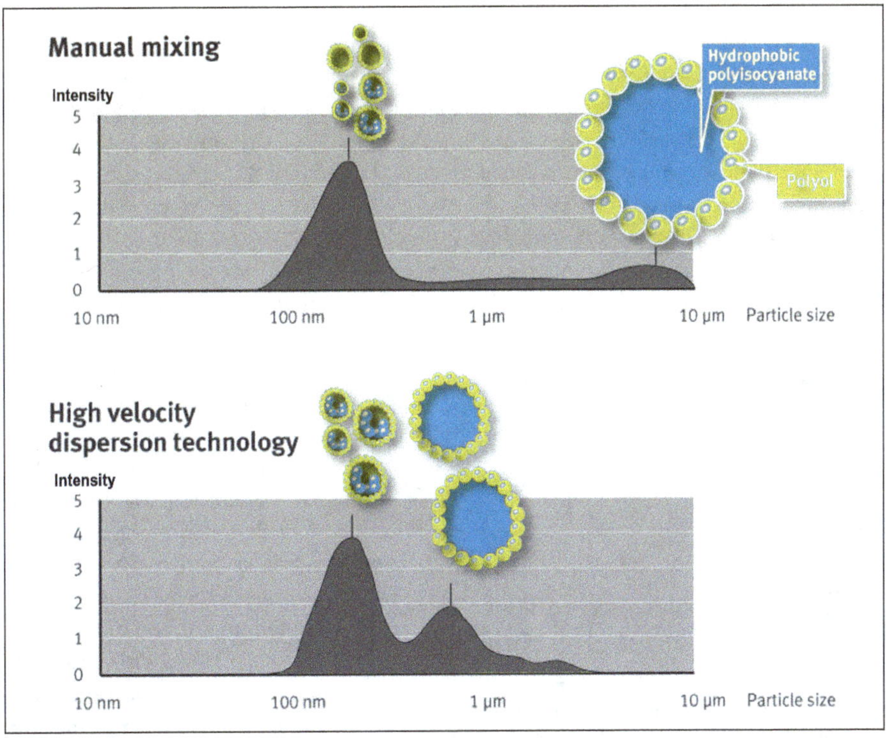

Figure 4.9: Emulsifying hydrophobic polyisocyanates

mixed as homogeneously as possible. The dispersion of the polyisocyanate in the aqueous phase is supported by three factors: reducing the interfacial tension, applying high shear forces during emulsification, and ensuring the low viscosity of the dispersed phase (the polyisocyanate component).[30]

The interfacial surface tension can be reduced by using a polyisocyanate with internal hydrophilic modification (see Chapter 3). External emulsifiers such as a suitable polyol will also produce the same effect.[31–33] However, excessive hydrophilic modification will have a negative impact on the coating properties. In order to fine-tune the hydrophilicity of the hardener portion polyisocyanate hardeners with low viscosity can also be used. These low viscous polyisocyanates however require the use of a dispersion technique in which high shear forces are applied (see Figure 4.9). Often a blend of both – a low viscous polyisocyanate combined with a polyisocyanate with internal hydrophilic modification is used. Modern internally modified polyisocyanates thus combine high functionality (highly branched coating systems) with low viscosity (good mixing). Anionically modified polyisocyanates exhibit the best property profile in this regard (see Chapter 3.6). Simply by stirring without any appreciable shear forces, these hydrophilic self-dispersing polyurethane crosslinkers spontaneously form a dispersion (see Figure 4.10).

After mixing the polyisocyanate with the aqueous polyol, a number of reactions take place in parallel during the pot life and curing (see Figure 4.11).

In addition to the desired polyurethane forming crosslinking reaction between the polyisocyanate and hydroxy groups of the polyol dispersion, the aliphatic NCO groups also react with water to form urea and carbon dioxide (see Figure 4.11). However, this reaction is relatively slow and in addition, increases the crosslinking density. The carboxylate groups of anionically stabilized polyol dispersions likewise react with polyisocyanates. However, this can only occur to any significant extent once the neutralizing agent has evaporated from the paint film. The consequence of the slow reaction between the NCO groups and water is that the application properties of the coating remain fairly constant throughout the pot life. With a judicious choice of components, pot lives of typically 4 in special cases up to 8 hours can be achieved. When a suitable binder is used, the carbon dioxide, that is formed as the film cures, does not lead to blistering, since it can diffuse out of the film at the time it is generated.[34, 35]

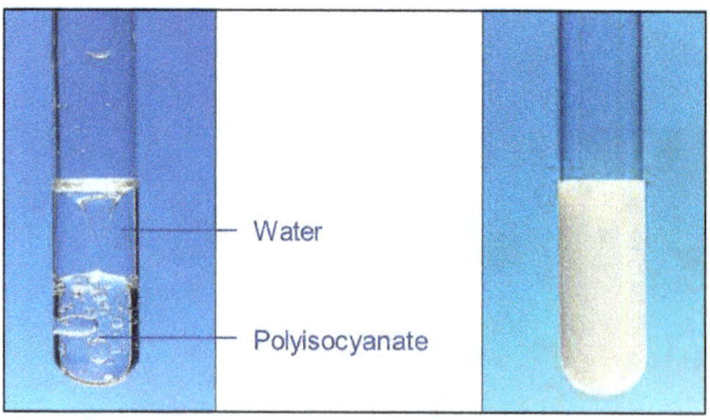

Figure 4.10: Phase formation of hydrophobic polyisocyanate and water (left); dispersion of hydrophilically-modified polyisocyanate in water (right)

Depending on the particular requirements, polyester-, polyacrylate- or polyurethane-polyol dispersions are used as a reaction partner for the polyisocyanates in water-borne two-component polyurethane coatings. Due to the tendency to hydrolyze in water, polyester dispersions are quite limited as a raw material for water-borne two-component polyurethane coatings. Often these polyesters are sold as solvent-borne or solvent-free raw material and dispersed right before paint is formulated. Polyacrylate polyol dispersions are used when high levels of properties in industrial coatings are needed. In general, their performance is rather similar to their solvent-borne counterparts. Water-borne two-component polyurethane coatings based on acrylates are used for wood, transportation and plastic coatings.

4.5 Process technology

Most polyurethane coating raw materials are used in reactive systems. Depending on whether they are one- or two-component systems, the various coatings require the use of specific application technologies. This does not apply to one-component polyurethane coatings based on thermally activated polyurethane hardeners that are non-reactive at room temperature. Since these only crosslinks above the cleavage temperature of the blocking agent, they can be applied by conventional techniques such as spraying, roller coating, curtain coating, and dipping.

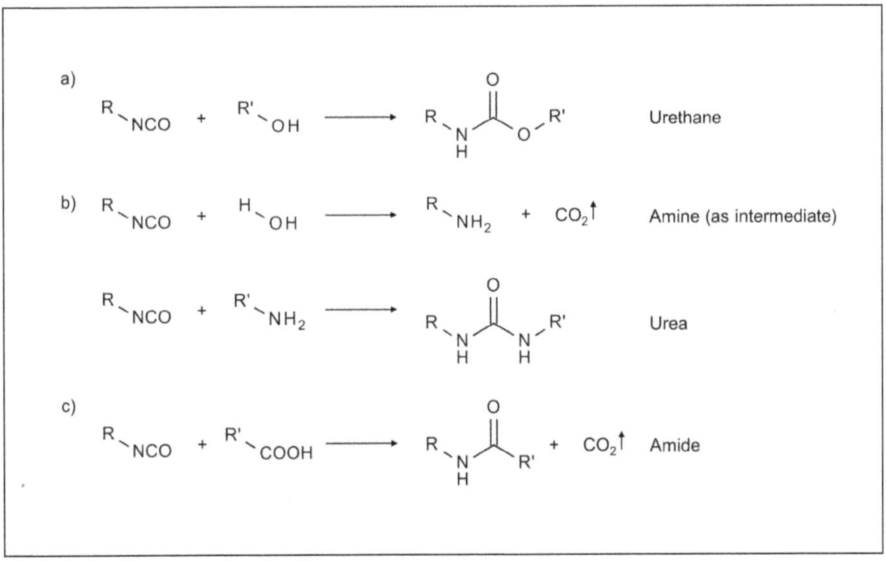

Figure 4.11: Possible parallel reactions in water-borne two-component polyurethane systems

4.5.1 Processing of one-component polyurethane coatings

One-component polyurethane coatings are reactive at room temperature, crosslink through reaction with ambient atmospheric humidity. To prevent uncontrolled crosslinking, a viscosity increases or thickening, e.g. during storage of the coating, the containers must be kept tightly sealed prior to application. Only solvents with a defined low water content of < 0.05 %, referred to as polyurethane-grade solvents, can be used. To improve the storage stability of pigmented coatings, water-scavenging chemicals are added to bind the residual moisture from the pigments. Suitable substances are e.g. "Additive TI" and "Additive OF" (Borchers) and "Desmodur" MT (Covestro).

4.5.2 Processing of two-component polyurethane coatings

The main limiting feature in application of all two-component polyurethane coatings is their restricted pot life, because the crosslinking reaction starts as soon as the components are mixed. This results in an increase in viscosity to a threshold viscosity at which the coating begins to gel (gel time).

The pot life is the period of time in which the coating can be applied without any detrimental effects on the quality. In practice, it ranges from minutes to several hours, depending on the formulation. The shorter the pot life, the more difficult the manual application will be, i.e. mixing of the components by hand and subsequent application. Generally speaking, two-component units and spray guns are used to apply highly reactive two-component coatings with a pot life of less than 30 minutes and for industrial finishing.[36]

Two-component spray application

There are two options for the industrial application of two-component coatings. In the first case, the material is homogenized either by hand or in a mixer unit and is then fed via a

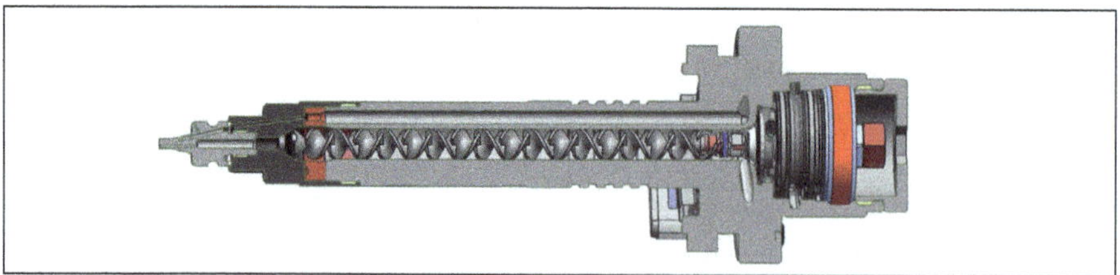

Figure 4.12: Static mixer for two-component coating systems for integration in an atomizer[42]

hose to the spray gun. The disadvantage is that there may be long hoses that require rinsing, resulting in a loss of coating material and solvents. It is very important to clean out the coating reservoir and the feed unit within the reaction time (pot life) of the components that have been mixed. The alternative is to use two-component spray guns with integrated or attached static ("Kenics") mixers (see Figure 4.12).

Of the variety of mixer units available, static mixers are the ones most frequently used in two-component application. [36] The mixing efficiency is essentially determined by the viscosities and the product affinities of the coating components. High viscosities or wide differences in viscosity require a correspondingly large number of screw elements in the mixer. To improve the feed properties, mixing and application, it is common to process high-viscosity materials in the "hot" state, e.g. at 40 to 60 °C. The supply tanks and feed hoses are therefore designed to be heatable. The noticeable limitation of this mode of operation is that the pot life for the two-component system is significantly reduced.

Under certain circumstances, static or dynamic (Ultra-"Turrax") mixers may not be able to achieve adequate homogenization of two-component coatings with critical miscibility. This is true in case of water-borne two-component polyurethane coatings, for example, in which non-hydrophilically-modified polyisocyanates are used as the hardener component. [37]

Depending on the viscosity of the two-component system used, a spray coater will use a spray gun with air atomization (high- and low-pressure processes) or, in the case of high viscosity, a hydraulic (airless) spray gun, i.e. techniques that are also commonly used for one-component coatings. In addition, an air-assisted hydraulic atomizer can also be used, in which a fast-flowing stream of coating is homogenized by air jets. [38]

Electrostatic application is primarily used in industrial coating environments. [39] An electric field is created between the spray gun on high voltage and the grounded workpiece. The coating material is charged in or by the spray gun. Due to the electric field and the attractive forces exerted by the workpiece, the coating droplets are drawn toward and deposited on the workpiece with minimal overspray. The most favorable conductivity range for coating materials is between 10^{-10} to 10^{-8} S/cm. Polyurethane coatings also fall within this range. The electrostatic coating processes can be divided into three categories, depending on the atomization and electrostatic charging mechanisms: the electrostatic compressed air and the hydraulic (airless) process, the electrostatic atomizing process, and the high-speed rotation process. [39]

Electrostatic spray guns atomize the coating material by the same mechanism as conventional spray guns, but the coating droplets receive an electrostatic charge as well. Nearly all coatings can be sprayed in this manner.

In the purely electrostatic atomization process, the coating is fed into a high voltage-conducting rotating bell (up to 3,000 rpm) or disk and is atomized by strong electrical forces. Highly conductive coating systems such as water-borne coatings cannot be applied by this process. By contrast, in the high-speed rotation process, atomization takes place through purely mechanical forces and is largely independent of the electrical properties of the coating. For example, electrostatic high-speed rotation bells with speeds between 15,000 to 70,000 rpm are used. The coating is expelled tangentially from the bell, which is under high voltage (90 kV) (see Figure 4.13). Such processes are used primarily in automatic coating units, particularly in the field of automotive finishing, and increasingly in general industrial metal coating applications. The application of a water-borne two-component polyurethane coating is described in Chapter 5.3. [41]

Two-component metering units

Two-component metering units meter the hardener and mill base components in the desired volume mixing ratio. Depending on the type of unit, the material may be mixed and conveyed as well as metered (see Figure 4.14).

The materials – mill base and hardener – are fed into the two-component metering unit from pressure pots or via paint circulation systems, for example. Two types of pumps can be used to meter the components: piston pumps or gear pumps. Piston pumps are robust and are also suitable for operation at high pressure to meter viscous materials. They have a dual effect: they convey the material on both, the upward and the downward

Figure 4.13: Example of a robotic rotating bell application in automotive OEM painting [42]

strokes. Due to the intermittent acceleration of the piston (minima at the turning points), pulsation may occur in the spray stream which, in the case of longer hoses, can be attenuated by bleeding the hose.

Because piston pumps operate by forced feed, the metering accuracy is ensured by the constant volume of the cylinder and the defined stroke length. However, because of the time and effort involved in cleaning, it is not possible to achieve a quick changeover to a different color. If several colors are to be used, parallel piston units are normally used. Suitable units are available for a variety of applications.

Gear pumps ensure pulsation-free application. In order not to restrict the metering accuracy (slippage), they are normally used in the low to medium pressure ranges (up to 50 bar). Variable air-driven and electric motors are used to drive the pumps. The feed

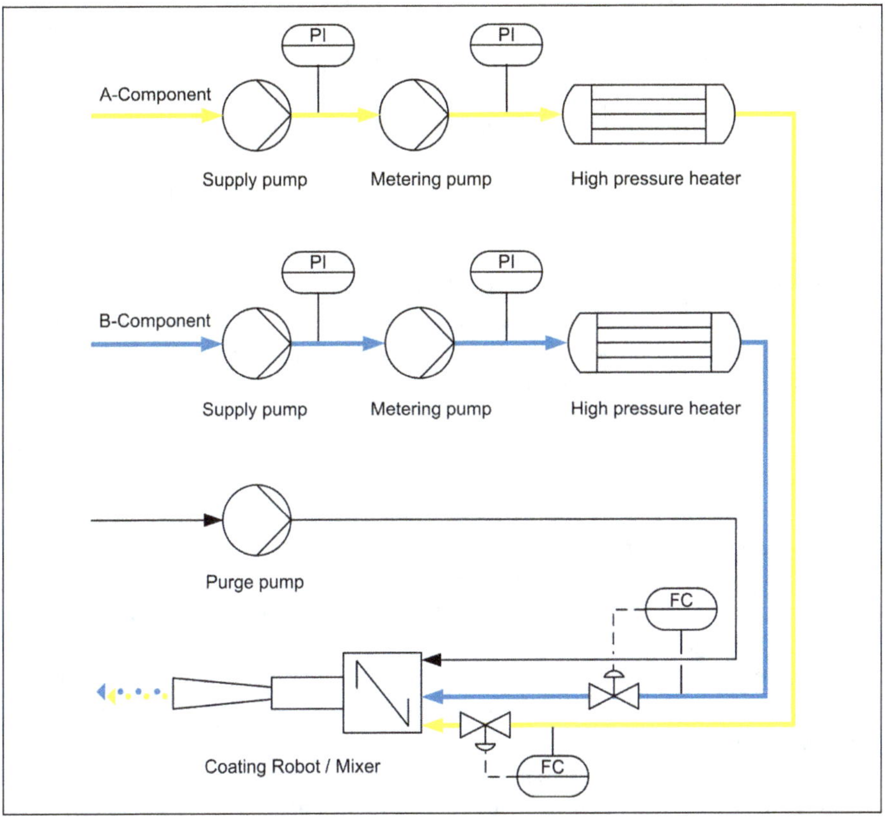

Figure 4.14: Schematic illustration of a two-component metering unit[7]

volume is governed by the chamber size and speed of the gear pump. Because of the small volumes involved (from 2 ml), flushing can be achieved quickly and efficiently, especially if the unit is equipped with flank flushing with alternating air and solvent intervals (less than 30 s). This allows rapid changeover to a differently colored material. When using flushable gear pumps, only one metering unit is required for the mill base.

In continuous coating operations, e.g. production line coating, electronically controlled and monitored two-component units are used.

A microprocessor control and regulation device compare the actual values against the specified metering ratios, and if necessary, automatically adjusts the amounts within a tolerance threshold of ± 5 %. Color changeover with automatic intermediate flushing takes place in the color changeover block.

General conditions for applying two-component polyurethane coatings

A number of important prerequisites must be met when applying two-component polyurethane coatings, so as to ensure optimal coating quality. These include careful observation of the pot life (see above), good mixing of the mill base and hardener, and observance of the tolerance threshold of ± 5 % for metering accuracy, in particular at the given mixing ratio of the two components. If operating within these thresholds, there is no reason to expect any deterioration in the film quality. In order to test the function of the unit, it is recommended that the mixing ratio of the two components, mill base and hardener, is checked regularly by measuring the volume flows of the materials.

One key requirement for the successful application of two-component polyurethane coatings is careful handling. This includes flushing the spray guns and other components of the application unit (hoses) during the pot life. A 1:1 mixture of butyl acetate/xylene is suitable as the flushing solvent when applying solvent-borne systems. When flushing the spray gun, the operator must avoid inhaling the solvent vapors that are emitted. When applying water-borne two-component polyurethane coating systems, a mixture of water, butylene glycol and dimethylethanolamine in a ratio of e.g. 89:10:1 is used as the flushing solvent on the mill base side (component A). On the hardener side (component B), butyl acetate is used to flush the unit.

If the unit contains hardener, exposure to water or moisture, e.g. via the flushing solvent or leaks in the unit, must be prevented. For this reason, the hardener tanks are overlaid with dry nitrogen or dry air. When the filling level falls as the hardener is used, more air must be fed in through a drying cartridge. Brief contact with atmospheric moisture, e.g. when transferring the product from one container to another, has no detrimental effect on the hardener.

Moving parts such as piston rods in the metering pump are protected from their surroundings by a fluid seal. Plasticizers such as "Mesamoll" are especially suitable as lubricants and sealants, since they prevent the formation of polyurea on the piston rods.

An overview about general new trends in painting and technology can be found here[57].

Safety precautions

Stainless steel is normally used to manufacture two-component units, pipes, and transport containers. Zinc-galvanized surfaces are entirely unsuitable due to the catalytic effect of the zinc. Plastics are generally unsuitable as container materials because of their permeability to water vapor. Multilayer plastic composites, which are considered impermeable in this respect, have not yet found use in this field. The hoses used to convey the hardener, e.g. made from nylon, must be emptied and flushed before long interruptions in operation, e.g. on weekends, to prevent gelling as the result of moisture diffusion. This does not apply to "Teflon" hoses, but their limited flexibility restricts the handling of the spray gun.

Compared to one-component coatings, the processing of two-component coatings is inherently more complex, and more expensive if two-component units are used. However, there are numerous case studies showing that, if the units are operated properly and maintained regularly, they can function for many years without breakdown.

Two-component technology is particularly strongly represented in automotive OEM finishing. Two-component polyurethane coatings are being utilized increasingly in the automotive industry and can be applied reliably when suitably maintained two-component units are used. The manual spray application of two-component polyurethane coatings has also proven successful and is employed worldwide, e.g. in car repair shops.

4.6 Crosslinking technologies based on the polyaddition reactions of polyisocyanates

In previous chapters, various crosslinking technologies have been discussed that are based on the high and versatile reactivity of (typically aliphatic) isocyanates towards active-hydrogen containing resins. The various addition reactions that monomeric (di)isocyanates can undergo with themselves or other NCO-functional species were mainly described as a convenient route to produce isocyanate-functional crosslinking oligomers (e.g. via trimerization, uretdione formation) or oligomer derivatives with non-free-isocyanate crosslinking reactivity (e.g. via carbodiimid formation). These structures are designed for being efficient crosslinkers for typical coatings or adhesive applications.

Interestingly, when looking at the way aromatic isocyanates are used in the polyurethane hard foam insulation industry, they are often used as main reactive building blocks, e.g. in the production of flame and temperature resistant so-called polyisocyanurate (PIR) foams. In this case the trimerization of isocyanates is even the dominating crosslinking mechanism.[68, 69]

In the last few years, the concept of using the isocyanate-isocyanate addition reactions of polyisocyanates as the primary crosslinking mechanism has started to gain interest in the coatings and adhesives technology arena as well. In Figure 4.15, two main examples of crosslinking mechanisms are shown that were recently developed for the coatings and adhesives industry based on the addition reactions of polyisocyanates.

This "new" crosslinking mechanism for aliphatic polyisocyanates can offer several benefits regarding the processing and application of coatings and adhesives compared to

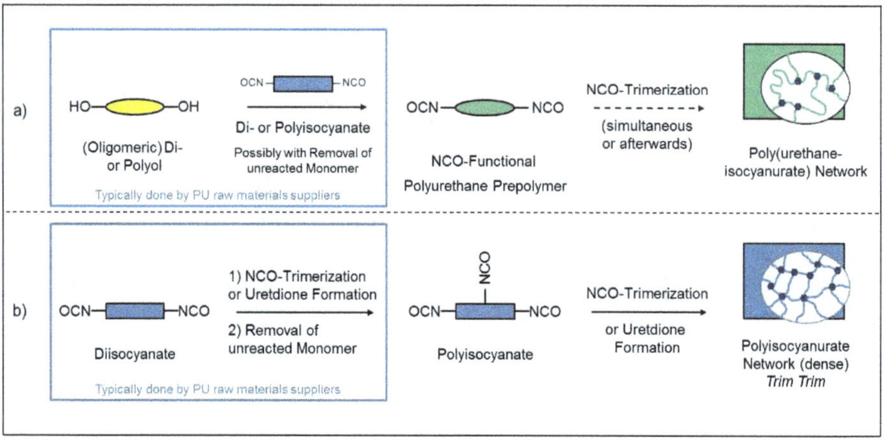

Figure 4.15: Two main processes of catalyzed polyaddition reactions of polyisocyanates: a) trimerization during/after polyurethanization, leading to a poly(urethane-isocyanurate) network; b) trimerization (or possibly uretdione-formation) of polyisocyanate precursors, leading to a typically dense polyisocyanurate network (trim trim)

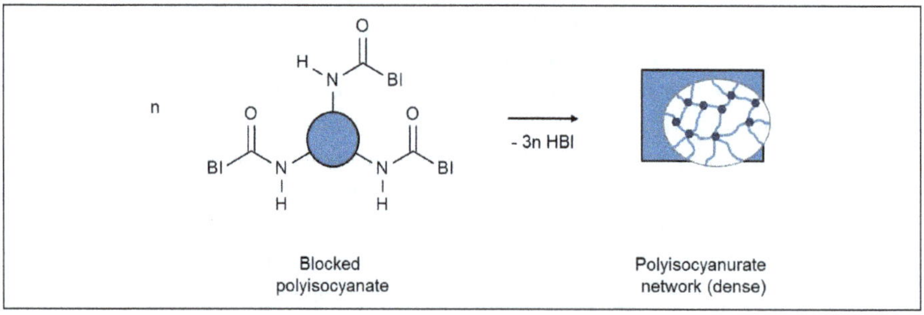

Figure 4.16. Thermally activated polyurethane hardeners as precursors for reactive crosslinking via trimerization.

BI = blocking agent

conventional polyurethane crosslinking mechanisms.[70, 71] The particular benefits of this type of crosslinking include:

- Wide applicability; i.e. All NCO-functional materials are suitable building blocks. The reactivity of the isocyanate in polyaddition reactions can be tuned in a large range by choice of catalyst and reaction conditions. Consequently, also high-solid low-viscosity systems can be formulated via the existing choices of available di- and polyisocyanates. Of particular interest is the fact that also thermally activated polyurethane hardeners can be used as a reactive crosslinking precursor, allowing for thermally activated reactive crosslinking formulations (see Chapter 3.5), as shown in Figure 4.16.
- The possibility to offer one-component crosslinking polyurethane formulations significantly improves the handling of these systems for paint and adhesive users. Since the crosslinking mechanism is based on a polyaddition reaction of NCO groups – given the use of suitable catalysts – no addition or release of chemicals is required prior to crosslinking. Also, multiple building blocks can be easily combined into such one-component systems to tailor the material properties. An example of the concept is shown by the specific tailoring of the glass-transition temperature of a densely crosslinked polyisocyanurate network by varying the ratio of IPDI-trimer and HDI-trimer, respectively, as shown in Figure 4.17.
- High index (i.e. high ratio of NCO to active-hydrogen species) polyurethanization reactions are possible. Since the excess of NCO groups is consumed by polyisocyanate addition reactions, successful crosslinking depends less on the precise stoichiometry

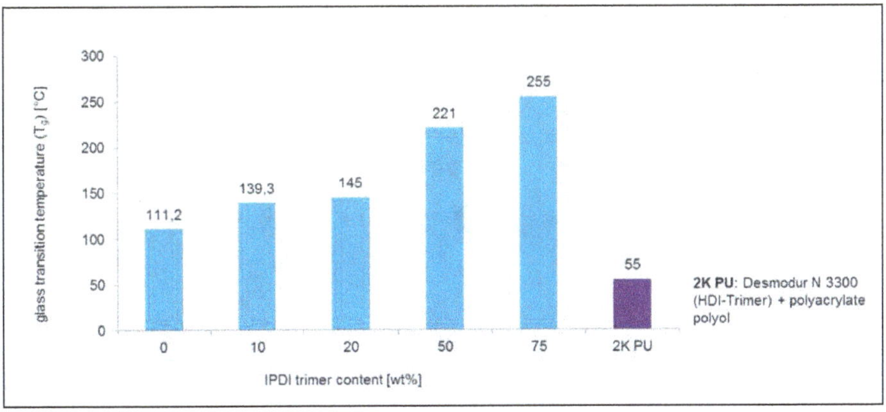

Figure 4.17: Glass-transition temperature (T_g) of a densely crosslinked polyisocyanurate trim trim network in dependence of the ratio of IPDI-trimer and HDI-trimer. As a comparison a typical two-component polyurethane coating based on HDI-trimer and a highly functional polyacrylate polyol is shown, which has a T_g of < 80 °C.

of reactive groups. Also, because NCO groups are a reactive site throughout the reaction, mixtures are more likely to stay homogeneous at every stage of the reaction. With the choice of the right catalyst(s), both co-continuous and step-wise reaction schemes are possible.

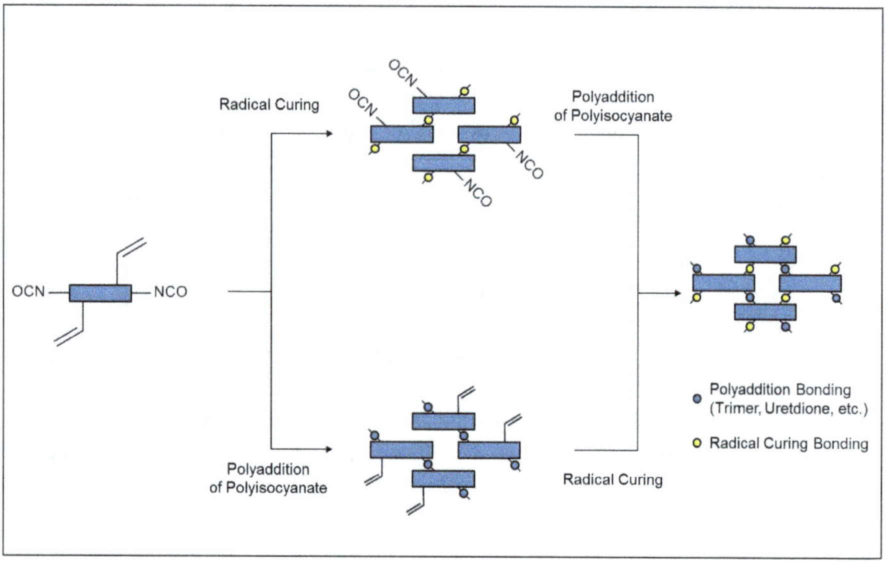

Figure 4.18: Dual curing polyurethane mixture based on radical-curing and polyaddition reactions of polyisocyanates

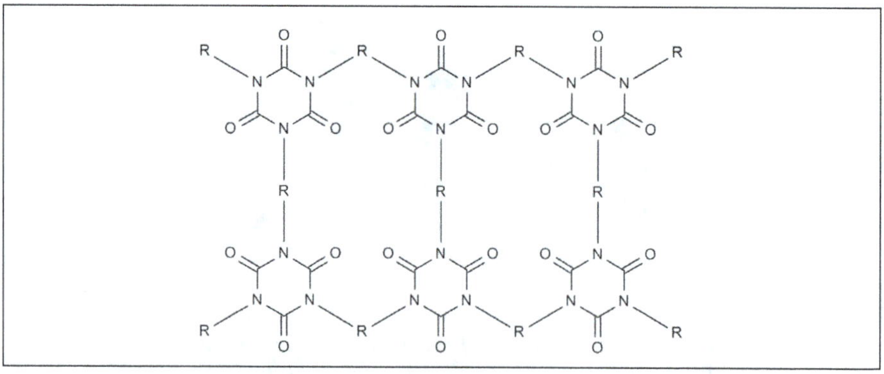

Figure 4.19: Idealized two-dimensionally simplified molecular structure of a dense polyisocyanurate network (trim trim)

Table 4.3: Chemical resistances of a composite material with ca. 80 % glass fiber content based on a densely crosslinked HDI-based polyisocyanurate (trim trim) and a reference high performance two-component polyurethane based on HDI trimer and glycerol. The chemical resistance is reported as positive, if the interlaminar shear strength (ILSS) change after 90 days at 23 °C in the respective medium is below 15 % compared to the original value (DIN EN ISO 14130).

Medium	H_2SO_4 (10 %)	HCl (10 %)	Gasoline	Motor oil	Ethanol	Acetone	Ethyl acetate	NaOH (10 %)
HDI-polyisocyanurate trim trim	+	+	+	+	+	-	-	+
Two-component polyurethane (HDI trimer and glycerol)	+	+	+	+	-	-	-	-

H_2SO_4 = Sulfuric acid; HCl = Hydrochloric acid; NaOH = Sodium hydroxide

- The addition reactions of NCO groups are suitable for the design of various new dual- or even triple-curing systems. The various curing steps can be initiated independently from each other; e.g. radical curing and polyisocyanate polyaddition, as shown in Figure 4.18, or radical curing, silane curing and polyisocyanate polyaddition.
- Suitable catalysts for isocyanate addition reactions (typically strong nucleophiles) are known and available and can be tailored to control the crosslinking reaction; for details, see Chapter 4.2 and references thereof. Transition-metal-free and amine-free catalysts are also available, such as common alkali-carboxylates and alkali-alcoholates.[72, 73]
- Crosslinking of a polyisocyanate via trimerization leads to the formation of chemically and mechanically highly stable crosslinked materials.[74, 75] Figure 4.19 shows an idealized two-dimensionally simplified molecular structure of a dense polyisocyanurate network (trim trim).

The material properties tend to improve with increasing isocyanurate-group density, as is shown in Table 4.3. The first application areas for this relatively new utilization of the polyaddition of aliphatic polyisocyanates are found in the composite[76] and 3D-printing[77, 78] industries.

In the coatings and adhesives industry, first application areas are in development for high-performance coatings, adhesives for the transportation and construction industry, as well as for the packaging industry. Drivers are the combination of excellent processing and product properties and the availability of a broad range of suitable (transition-)metal-free catalysts for the crosslinking reaction.

4.7 Powder coatings

The economic and ecological benefits of powder coatings have ensured the continued dynamic growth of this technology since inception in the 1940/50s with one of the first patents focused on fluidized bed applications registered by *Gemmer* in 1953.[79] Electrostatic powder spray was introduced in the mid-1960s and this shift allowed for the efficient application of powders 1) without preheating and 2) as thinner films, i.e. 30 to 200 µm without the high film builds common from fluidized bed applications (typically 200 to 1,000 µm).

Bate[80] reported that the global powder production in its infancy in 1962 was 8 tons and grew to 5,800 tons by 1972. Worldwide consumption in 1995 was approximately 450,000 tons, while demand in 2017 had reached 2,300 kilotons.[81, 82]
Steady growth is expected to continue, although overall annual growth rates have dropped below double digits and are expected to settle near annualized GDP growth rates. Asia Pacific, the Middle East and Africa are growing at higher rates than other regions. Figure 4.20 gives an overview about the powder market sorted by region.

Powder coating combined with electrostatic powder spray (EPS) became a cost-effective process that yields efficient and high quality decorative and protective surface coatings suitable for a wide range of applications. One of the main reasons for the dynamic growth of this technology is the fact that the coatings can be applied without any solvent emissions, resulting in environmentally friendly coatings. Additionally, the efficient utilization is a major benefit – transfer efficiency as high as 94 % on the first pass, coupled with the fact that overspray can be blended with virgin powder allows transfer efficiencies to approach 100 % material utilization without color changes.

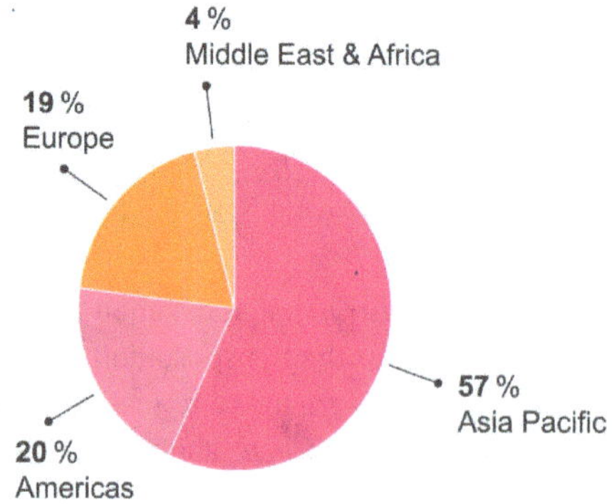

Figure 4.20: Global Market for Powder Coatings in 2017[82] sorted by region

Powder coating technology is important with respect to exceeding the tightened worldwide environmental legislation in the coatings industry and delivering quality finishes in a range of performance and economic levels, gloss levels, and protective capabilities (Solvent Emissions Directive, Decopaint Directive, VOC Restrictions).

4.7.1 Overview of powder coating technology

A powder coating is a finely ground blend of solid materials with latently reactive components in the case of thermoset powder coatings. Important parameters for material selection include material received, particulate size range, material glass transition temperature, melt viscosity, mutual material solubility, chemical stability in the melt state and chemical reactivity for powder curing, ease of pigment dispersion, pigment to binder ratio, reactive group stoichiometric balance, and additive loading levels.

Powder coatings are most often manufactured in the following steps [83]: weighing, pre-mixing, extrusion, grinding, sieving, and packaging. Extrusion is performed in single-screw, twin-screw, or planetary screw extruders. The extrusion process is linear plug flow with very little mixing outside of a given plug zone. Therefore, the pre-mix step is elevated in importance as it is required to develop material uniformity of size and composition near the scale of the extruder plug volume. Grinding may be performed using mechanical and/or air separation mills with the most common particle size distribution ranging between 20 to 120 µm. Applied powder thicknesses and smoothness values are related to a ratio of powder particle size (i.e. smaller particles provide thinner and smoother films and vice versa). Additionally, fines (smaller than 20 microns) and oversized particles (larger than 200 microns) create poor transfer efficiency.

Polymer-based powder coatings are applied by electrostatic powder spray (EPS) or fluidized bed applications. EPS has evolved to two major methods, corona charged powder application and tribo-electric spraying. The applied powder and substrate are then heated and cured almost simultaneously (very small time and temperature window in some chemistries for film formation, flow and levelling, and optimal adhesion development), before undergoing thermal curing to yield a vitrified thermoset film.

4.7.2 Application methods

Powder coatings are normally applied either by electrostatic spraying or fluidized bed. Electrostatic spray involves either the corona or tribo application process. [83] In the corona process, a voltage range between -50 and -100 kV is applied between the spray gun and the object to be coated. The corona is a poorly defined zone – usually just a few millimeters thick – of air ions (including some free electrons) which forms close to a high-voltage electrode as the result of the high field strengths that exist there.

New application systems automatically adjust between voltage/amperage to compensate for and manage overcharging, back-ionization, or Faraday cage related issues occurring during application.

During corona application the powder coating is forced pneumatically from the spray gun and charged during spray application (see Figure 4.21). Specifically, powder particles become charged from the corona zone created in the spray gun nozzle center, the powder follows the field lines to the ground object and adhere to the surface due to their electrostatic charge.

In the tribo application process, the electrostatic charge of the powder particles is produced by tribostatic friction in the spray gun most often through a slightly tortuous path with a "Teflon" tube. Tribo is short for triboelectric, meaning electric by contact or friction. This application process eliminates the need for a high voltage application. Additional application methods include bells and other variations on the same technology with higher degrees of automation and continuous production process robots are widely used for spraying.[84]

In the fluidized bed method, mainly metal objects are preheated and dipped into a fluidized bed of plastic powder. Although the fluidized bed process is lower in initial cost, it consistently yields thicker coatings and most often requires post-application thermal annealing. Technologies have been developed to allow for high speed powder coating applications using electrostatic cloud spray and electromagnetic brush methods to facilitate wire, flat stock, and coil coating application rates.[85]

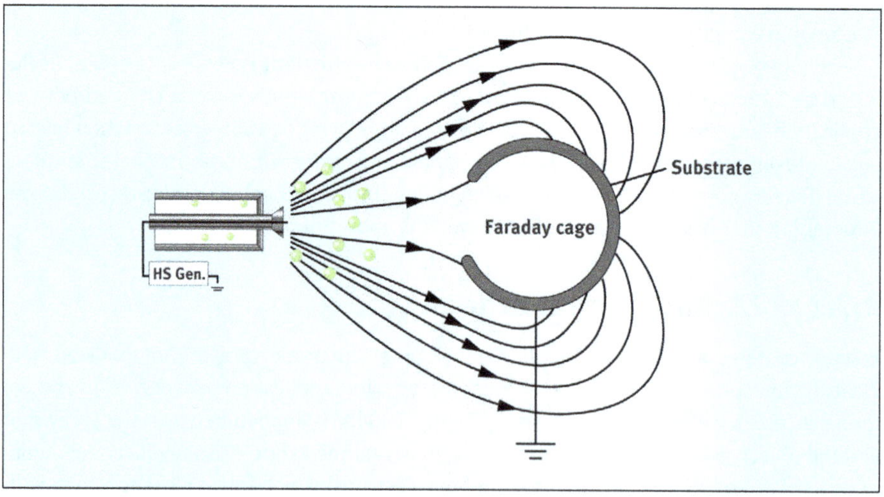

Figure 4.21: Powder coating application using the corona process

4.7.3 Application efficiency

The major advantage of the spray application for powder coatings compared to liquid coatings is the higher deposition efficiency (termed first pass transfer efficiency FPTE) and combined with the fact that the powder which is not deposited on the object – the overspray – can be collected and mixed with fresh powder without any great need for cleaning. This results in a high material yield of nearly 100 % which, together with the absence of solvent emissions and off-gas purification, minimal disposal costs and low operating costs, makes powder coating an interesting alternative compared to the application of liquid coatings, both from ecological and economic aspects.

4.7.4 Uses

Powder coatings are utilized in each metal-based coatings market and many other specialty areas. The major markets are generally as follows: general industrial, architectural, furniture, automotive OEM, domestic appliances, home comfort, and functional powder applications.

The use of powder coatings is still limited by certain required process conditions and infrastructure changes from conventional liquid systems including capital investments for liquid to powder conversions:
- relatively high baking temperatures are required (120 to 220 °C);
- regularly achieved uniform film thicknesses range between 30 to 70 μm and these are relatively thick compared to liquid coatings (5 to 30 μm) and advantageous in some cases but add weight and cost;
- in many cases, powder coatings still possess higher roughness or waviness than liquid coatings and results in an orange peel surface that also can impact resistance properties (i.e., the larger peak to trough topography can limit resistance performance at the level of the thinnest coating values).

In recent years, considerable effort has therefore been invested in overcoming these disadvantages and expanding the range of applications for powder coatings, e.g. to automotive finishing.[86] Although the films yielded by the coatings are already of a high quality, there is still the need to increase reliable control over thickness/thinness, gloss levels, and smoothness for competition with liquid coatings and overall cost reasons. The success of these efforts is shown by the fact that powder coatings, including polyurethane-based systems, are now used as high quality automotive and recreational vehicle primers, with blackout accessories and rock/chip resistance layers.

4.7.5 Polyurethane powder coatings

As described, polyurethane powder coatings usually consist of hydroxyl-bearing polyesters or acrylic copolymers cured with externally or internally blocked cycloaliphatic thermally activated polyurethane hardeners. Other polymers and crosslinkers are regularly used in specialty applications in combination with polyurethanes powder coatings. A special use of polyurethanes is in the hybrid technology development of UV-curable powder coatings. The photo-curable resins range from unsaturated polyesters to acrylated polyurethanes.

Powder crosslinkers

Practical powder coating crosslinkers are thermally activated polyurethane hardeners, either externally added or internally blocked; the NCO-functional group blocking is necessary as powder coatings are manufactured by extrusion and using temperatures up to 140 °C. Polyurethane-based powder coatings result in good film formation, excellent physical properties, and chemical storage stability for extended periods of time. The reactivity can be controlled by the type of polyisocyanate and blocking agent used (see also Chapter 3.5).

All commercial thermally activated polyurethane crosslinkers are amorphous in nature, i.e. these are not crystalline and blend well with available polyesters and acrylics without diminished storage stability. Today, multifunctional crosslinkers are primarily based upon either isophorone diisocyanate adducts or bis-(4-isocyanatocyclohexyl)methane diisocyanate adducts. These adducts are usually isocyanurate or trimethylol propane based-derivatives to deliver functionality higher than the base isocyanates, generally three or higher on average and have isocyanate equivalent weights ranging between 278 to 320 g/NCO.

Isophorone diisocyanate (IPDI) and bis-(4-isocyanatocyclohexyl)methane diisocyanate ($H_{12}MDI$) are the leading commercial building blocks for powder crosslinkers as these materials exhibit acceptable glass transition temperatures for storage stable powder coatings for thermally activated polyurethane hardeners. [87]

The primary blocking agent utilized in powder coatings is ε-caprolactam [88] and dominates the market of externally blocked crosslinkers used in polyurethane powder coatings, in regions which accept blocking agent emissions. ε-caprolactam is the prevailing blocking agent because of its thermal dissociation profile (sufficiently reactive and yet chemically and physically storage stable), cost, low impact of emission on films, and resulting quality of finish. Specifically, some of the ε-caprolactam remains in the paint film during the cure process and promotes the flow of the melt as caprolactam can be trapped as a low viscosity liquid within the film during the cure process. [89] Some of the caprolactam is emitted during cure, which explains the detectable VOC content of such powder coating systems, albeit often only a fraction of the theoretical limits, e.g. 39 % of theoretical. [90]

Some attempts to remedy and capture caprolactam during curing have been shown possible, e.g., chemical tethering of polysiloxane in ε-caprolactam blocked thermally activated hardeners may simultaneously result in ease of extrusion process due to decrease in viscosity during extrusion and improved polyurethane powder coatings properties.

Several alternative chemistries, architectures, and catalysts have been published for urethane powder coatings. The use of allophonate linkages with imidazole blocking agents was reported with some technical success. [91] Further modification by the incorporation of fluorinated polyol in ε-caprolactam blocked hardeners was shown to increase the polyurethane powder coating hydrophobicity. [92]

Pilch-Pitera reported another novel approach towards practical crosslinkers using trimethlyhexamethylene diisocyanate (TMDI) as a biuret adduct and blocked with ε-caprolactam. [93]

Internally blocked thermally activated polyurethane hardeners, mainly based upon isophorone diisocyanate are commercially available and designed using uretdione linkages as a method to advance the crosslinker molecular weight and arrive at an acceptable glass transition temperature to produce storage stable powder coatings (see Figures 4.22 and 4.23).

Uretdione crosslinkers are generally lower in functionality than the externally blocked analogues and therefore, often utilized with slightly higher functionality prepolymers. The commercially available materials are based upon isophorone diisocyanate *Spyrou*

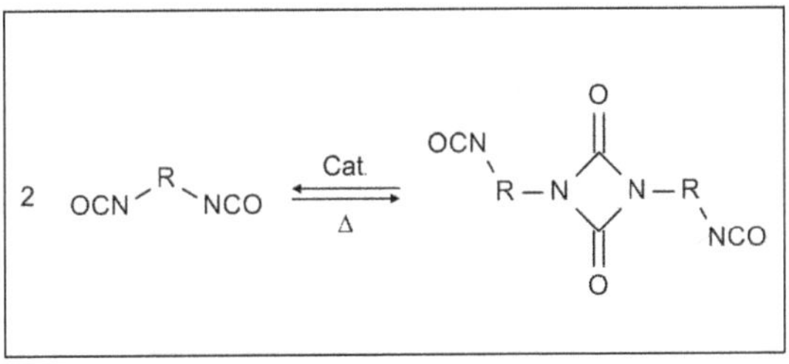

Figure 4.22: Dimerization of isocyanates

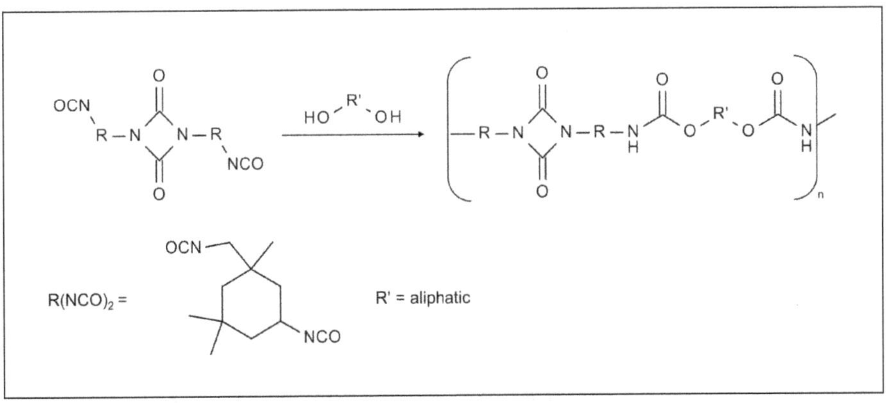

Figure 4.23: Structure of an IPDI uretdione powder coating crosslinker

and co-workers have identified 10 isomers via ^{13}C-NMR and correlated the different isomers with the uretdione generating steps and differences in reactivity from the isomers.[94]

During the baking process, the powder melts, urethane hardeners produce favorable rheological properties with sufficiently long open times for film formation and film substrate wetting. The polyuretdione hardeners containing no blocking agent represents an important step toward the emission-free or extremely low emission application of polyurethane powder coatings. Typical baking conditions for these second-generation polyurethane powder coatings range from 10 to 30 minutes at 180 to 220 °C peak metal object temperature.

Catalysis focused research has proven successful in further reducing the necessary cure profile as polyuretdione hardeners can now be crosslinked fully by baking for 10 minutes at 140 °C.[46, 95] There is no doubt that this development continues to yield additional application possibilities for powder coating systems. Uretdione powder hardeners are characterized by their very good pigmentability.[96] The outstanding optical properties such as gloss and levelling are also worth mentioning. These are a result of the combined low melt viscosity yielded by polyuretdiones and the absence of vitrification or cessation of flow until nearly full cure (greater film formation open time).[97]

Specialty catalysts (e.g. zinc acetylacetonate, tetraalkyl ammonium carboxylate) have been discovered that allow for reduced curing temperatures, e.g. from 180 °C at 30 minutes to 120 °C for 30 minutes to achieve similar performance with specific low melt viscosity polyesters (see Figure 4.24[47]). Tetraalkyl ammonium carboxylate provides a range of structures during a 140 °C cure profile with a commercial hydroxyl functional polyester (T_g = -57 °C, hydroxyl value 58 mg/KOH). The reported ranges are 5 to 20 % isocyanurate, 80 to 95 % urethane, and 0 to 5 % of allophanate. The weathering results presented using super-durable polyesters confirm that the catalyst and the reduced cure profile do not produce negative results with regards to gloss retention.[98]

There are some patent and peer reviewed literature references that describe interesting developments not yet commercialized (not known by the authors to be commercial now) and yet interesting research results. Several examples are described herein. *Chang et al.*[99]

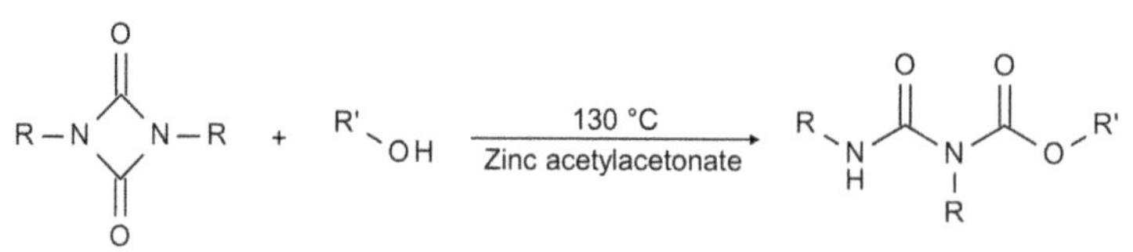

Figure 4.24: Low temperature crosslinking of a IPDI polyuretdione powder coating crosslinker with the formation of polyallophanate

describe a powder coating crosslinker based upon a blend of 1,6-hexamethylene diisocyanate: the 1,6-hexamethylene diisocyanate trimer (isocyanurate) blocked with 1,2,4-triazole. *Subramanian* and coworkers [100] presented new low temperature crosslinkers for powder coatings based upon combinations of triazole blocked IPDI polyurethane hardeners with good performance attainment after 160 °C for 30 minutes.

Several other thermally activated polyurethane hardeners were summarized in the same conference paper including 3,5-dimethyl pyrazole (DMP) that was shown to unblock at lower temperatures than caprolactam and triazole based crosslinkers. The experimental crosslinker comprised of $H_{12}MDI$ and DMP exhibited low temperature cure and excellent flow characteristics. Adjusting the prepolymer functionality with specific hydroxylated polyesters and an added catalyst provided good MEK resistance with oven curing at 145 °C for 30 minutes. A solid diethyl malonate (DEM) based polyurethane crosslinker was blended with a hydroxyl function prepolymer and cured at 130 °C via transesterification mechanisms. [100]

Powder resins

Hydroxy-functional polyesters based upon terephthalic acid and isophthalic acid with varying OH values and functionalities ranging between 2 to 5 have proven ideal as the coreactants for the polyurethane powder hardeners described above. Due to the variety of binders available on the market, it is now possible to produce polyurethane powder coatings with high gloss and dead matte films with good levelling and at any gloss level in between, [101] using various technology pathways for gloss control and each very dependent upon the formulation, raw materials, and processing.

New raw materials and polyesters based upon a variety of building block monomers have been studied and introduced. *Noordover* et al. published data on bio-based step-growth polymers for powder coatings synthesized from isosorbide and citric acid separately with a blend of other traditional monomers. According to the authors, the isosorbide-based polyesters exhibited similar weathering resistance as commercially common terephthalic acid-based polyesters when crosslinked using powder coating specific blocked isocyanates. [102]

In a typical industrial coating formulation, polyesters with a hydroxyl number of 30 to 50 (1870 to 1122 g/hydroxyl equivalent) are used. Often blended at 1:1 stoichiometric ratio for full crosslinking, these materials yield coatings with good mechanical resistance (reverse impact) and weather stability (Florida, 5° South: at least 50 % gloss retention after one year). Stoichiometric crosslinking of polyesters with high hydroxyl numbers yield coatings with excellent chemical resistance (hydroxyl values are available with 100 to 300 and these correlate with 561 to 187 g/hydroxyl equivalent). Polyesters with a hydroxyl number of 100 or more are suitable for use in the easy-to-clean coatings that are now increasingly in demand.

Polyesters comprised of solely or primarily isophthalic acid esters exhibit powder coatings with particularly good weather stability. These polyesters are often referred to as super durable polyesters (achieving test results Florida, 5° South: at a minimum of 50 % gloss retention after two years).[103] Hydroxyl-bearing copolymer based polyacrylates are also used singularly to a small extent and as blending resins with polyesters and polyacrylate copolymers for a variety of specialty results.

Photo-curable powder coatings have been an academic interest and potential solution for many of the limitations for traditional powder coatings since 1972. The commercial applications of UV powders are those dependent upon low temperature cure, high speed of cure and often high thermal loads are circumvented using IR for annealing. Many polymer backbones have been utilized from standard solid epoxies cured by cationic initiation to acrylate and vinyl ether modified polyesters, polyacrylates, epoxies and unsaturated polyesters. The most recent materials from commercial suppliers include urethane acrylate resins. Commercial applications for engineered wood most often target 105 to 120 °C annealing profiles and attempt to deliver optimum properties, e.g. scratch, abrasion, solvent, and stain resistance as in kitchen cabinet manufacturers specifications.[104]

4.7.6 Properties

A powder coating is not only required to remain chemically stable during prolonged storage, even at temperatures of 40 to 50 °C, but must also remain free-flowing, retain the original average particle size and particle-size distribution, and retain its application and post cure performance properties. This must be borne in mind when selecting the binder, crosslinker, additives, pigments, fillers, and storage conditions. As a rule, a polyurethane powder coating is formulated to have a glass transition temperature (T_g) > 50 °C. Due to their pronounced polar properties, polyurethane powder coatings are characterized by their particularly good tribo spray application properties (see above). Depending on the formulation and application, the coatings are cured for 10 to 25 minutes at temperatures between 150 to 210 °C. Table 4.4 describes a typical guide formulation for a blocking agent-free polyurethane powder coating. These formulations dominate European markets. The ε-caprolactam emissions are not considered acceptable in Europe and several other regions of the world. However, in the United States, ε-caprolactam-blocked systems are still used widely.

Ratliff and *Faecke* presented a summary of data supporting excellent corrosion resistance and weatherability from polyurethane based powder coatings. The study titled "Corrosion Resistance of Polyurethane Powder Coatings" included different substrates, varying powder curing agents comparing blocked and internally blocked urethanes, TGIC, polyester resin type and functional equivalent weight, film thicknesses, and cure profiles.[105]

A limited number of curing chemistries have been successful over the past 50+ years as a result of the complicated nature of powder coatings, i.e., each combination must be physically and chemically storage stable while retaining the final powder particle size for utility and facile application. Therefore, it is important to see differences in energy of activation and in turn relative differences in rate and degree of cure as coatings experts attempt to reduce cure time, temperature, and maintain performance. PU systems depending on their thermally activation chemistry show relative low energies of activation in comparison to other common powder coating crosslinking reactions. [106–112]

Table 4.4: Guide formulation for a polyurethane powder coating

Component	**Parts by weight**
Hydroxyl polyester [1)]	52.8
Polyuretdione hardener [2)]	9.2
Flow promoter [3)]	1.5
Catalyst [4)]	1.0
Antifoamer [5)]	0.5
Titanium dioxide [6)]	27.0
Barytes [7)]	8.0
	100.0

1) E.g. "Uralac" P 1480 (DSM), "Crylcoat" 2840-2 (Allnex), "Rucote" 102 (Stepan Company), "Albester" 3120 (Synthomer)
2) E.g. "Crelan" EF 403 (Covestro), "Vestagon" BF 1540 (Evonik)
3) E.g. "Modaflow" P III (Allnex), "Resiflow" PV 88 (Worlee Chemie)
4) E.g. "Addocat" 9594 (Lanxess)
5) E.g. Benzoin
6) E.g. "Kronos" 2160 (Kronos), "Tronox" R-KB-5 (Angene)
7) E.g. Sachtleben micro (Sachtleben/Huntsman)

Formulators should keep in mind that some crosslinkers are blended intimately during extrusion and others are intentionally fine powders that maintain heterogeneity and do not melt during extrusion. The latter provide intentional heterogeneity resulting in greater chemical storage stability and therefore improved physical storage stability. However, these systems require thermal homogenization during the high temperature cure process; making these systems complicated in design and handling.

4.7.7 Uses of polyurethane powder coatings

The main uses for polyurethane powder systems are as coatings for household appliances, bicycle frames, metal furniture, add-on automotive components, auto OEM primer-surfacers, toys, and sports equipment, as well as many architectural applications, e.g., window frames (see Figure 4.25, see page 120).

A comprehensive overview of the properties of PU based powder coatings versus different powder coating systems can be found here [114, 115].

A desired primary development goal exists in increasing the reactivity of polyurethane powder coatings. By reducing the minimum baking temperature to < 140°C many additional applications such as plastics finishing, and wood coating would be facilitated and provide large market opportunities.

4.8 DirectCoating/DirectSkinning

Reaction injection molding (RIM) is a technology widely used to prepare solid polyurethane plastic parts by injecting two component polyurethane systems into a mold. When the polyurethane reaction has sufficiently progressed, the mold is opened, and the part can be taken out. Since the late 1990s, RIM technology has also been used to apply a coating on top of a part in a mold by injecting a two-component polyurethane coating into the gap between mold and the part in the mold. First applied on wood veneer trim parts for automotive interiors that have been coated hitherto by a tedious process using unsaturated polyester coatings, the technology is being increasingly used on thermoplastic plastic parts and composite substrates.[116]

Several terms like overmolding or flooding are often used along with trademarked terms coined by equipment suppliers[117] to describe this technology. To distinguish between thin, compact layers and thicker foamed skin-like layers, the terms "DirectCoatings" (compact) and "DirectSkinning" (foamed) are employed for this RIM version of in-mold coating technology.

For "DirectCoating" and "DirectSkinning", where the liquid coating is cured in a closed mold, only solvent-free coating systems are used. This avoids the emission of solvents and the development of overspray, representing a clear ecological and economical advantage over conventional coating methods. Suitable coating systems are low-viscosity two-component polyurethane or polyurea systems with a solid content of 100 %, as well as radically curing one-component systems, e.g. acrylic based systems.[118] For polyurethanes, depending on the required property profile, polyacrylate, polyether, polyester or polycarbonate polyols can be combined with aliphatic or aromatic polyisocyanates to achieve a wide variety of performance properties. The coating components are usually preheated to reduce the viscosity. To avoid pot life problems with the highly reactive coating systems, RIM equipment has proven reliable for the homogeneous mixing of the coating components and injection of the coating. Curing times of less than 60 seconds can be

Figure 4.25: A window frame finished with a one-shot matte polyurethane powder coating[113]

achieved at temperatures of about 80 °C with suitably catalyzed polyurethane systems. Alternatively, polyurea systems, which combine aliphatic polyisocyanates with amine-terminated polyethers, can be used. Due to the higher reactivity of the amine group in comparison to the hydroxyl group, these systems react more quickly yielding better demolding characteristics. The high amount of urea groups, however, makes it harder to achieve high scratch resistance.

Shot weights between a few grams and several hundred grams have been demonstrated, allowing for small as well as large parts to be coated by RIM technology. In recent years, equipment manufacturers have developed 3-component-mixing-heads for RIM facilitating quick color change, thus further building on the value proposition of the "Direct-Coatings" technology.

The wide variety of polyurethanes makes it possible to control the properties of "DirectCoatings" over a broad range. Depending on requirements, hard coatings with good chemical resistance or soft coatings can be formulated which are very flexible even at low temperatures and yield more pleasant haptics.[119] The RIM method offers excellent

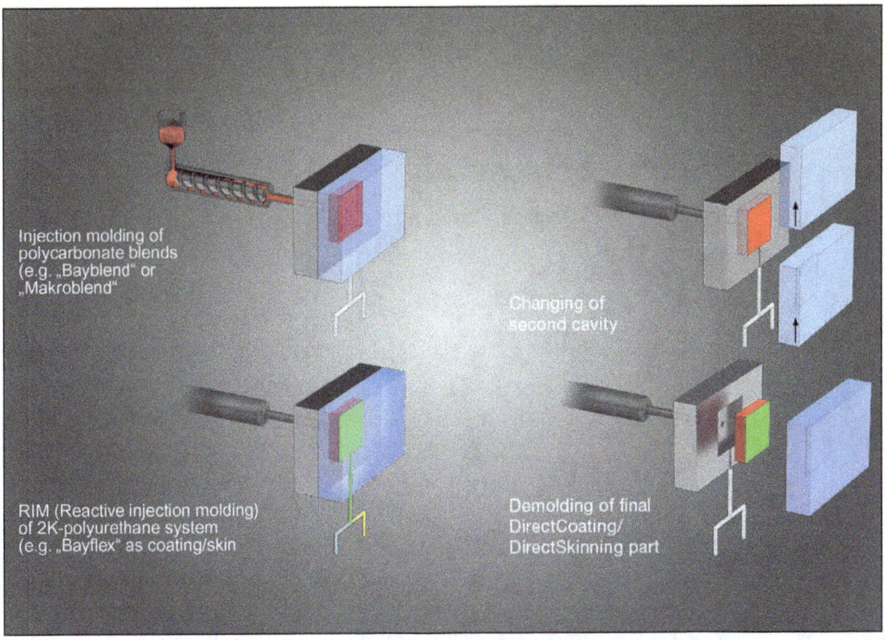

Figure 4.26: The two-cavity process: the component is produced by injection molding in the first cavity. It is then pushed into the second cavity, where it is coated with a two-component polyurethane system.

appearance in high gloss applications such as piano black or behind glass optics. Because surface structures are being imposed by the mold surface, matt and structured surfaces can easily be achieved without the use of matting agents as in conventionally applied coatings. High gloss, matt and surface structured appearances can be easily combined in one part allowing designs not possible with any other coating technology. Nevertheless, the method still has some limitations, which is why it has not yet prevailed to a larger extent on the market. The following aspects require special attention:

- mold design needs to be adjusted to the requirements of filling the form fast and allow for efficient venting of the air,
- sealing the tools from the exit of the (thin) fluid coating,
- creation of undercuts and shear edges,
- mold design that facilitates the de-molding, especially for high gloss areas at small angles to the de-molding direction, and
- production of thin coats (< 100 μm).

The advantages of "DirectCoatings" technology have led to increased interest in the market and have initiated intense development activity in order to overcome the technical challenges. Raw material and coating manufacturers, plastics processors and tool producers are involved in interdisciplinary projects.

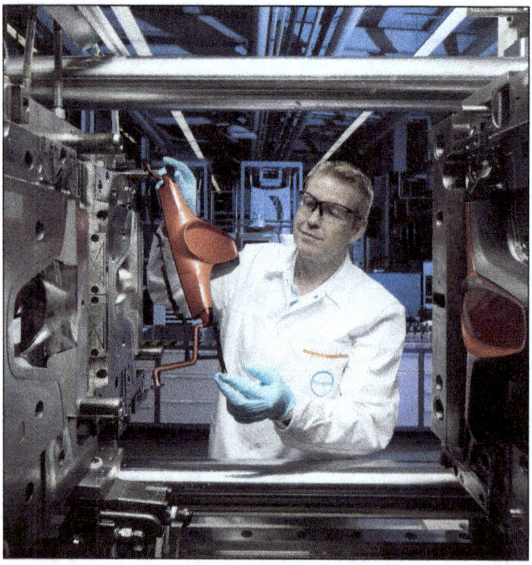

Figure 4.27: Two-cavity tool

Test and prototype tools are seeing increasing investment and a few automotive applications have been published. [120] Figure 4.26 and 4.27 describes the two-cavity process. The plastic component is manufactured by injection molding in the first cavity. It is then pushed into a second cavity where it is coated with a two-component polyurethane system. With optimal coordination of the processing properties and process timing it is possible to produce a plastic component in one cavity while a second component, molded before, can be coated at the same time in the second cavity, thereby enabling the highest possible productivity.

4.8.1 Outlook

It is expected that the "DirectCoating"/"DirectSkinning" method in combination with polyurethane coatings will become increasingly important in the production of coated plastic parts due to the economic, ecological and qualitative advantages. Combinations with labels or inserted functionalized films e.g. capacitive switches which are simply overmolded by the RIM coating offer interesting potential for the future.

4.9 Digital printing as disruptive application technology for coatings and adhesives

The use of digital printing technologies as an application method for coatings and adhesives is a new trend in the industry which is still in its infancy.

Drivers include the elimination of overspray (which offers significant savings due to reduced material waste), reduced cleaning and masking efforts, improved work place hygiene, individualization and process automation. One prominent example of how fast the shift to a digital process can happen is in the ceramic tile industry. [121] After solving the problem of handling unstable ceramic pigment inks the nearly complete shift from screen printing to digital inkjet printing took place in just a few years. Today digital printing is already well established for the ceramic tile and textile industry (see Figure 4.28).

Figure 4.28: "Colaris", example for a high-performance printing machine for carpet or textiles [122]

Other fast-growing applications of digital printing can be found in the (vinyl)-flooring and laminate flooring industry and for particle board decoration. [123]

All these applications are primarily about providing decoration and graphics. The function of classic coatings and adhesives besides providing color and decoration also include protective properties like scratch, weather and chemical resistance, adhesion, haptics etc. to name just a few.

Due to the high versatility of the raw material offerings, two-component polyurethane systems should be able to create multiple combinations on the substrate meeting specific requirements by digital printing of selected polyols (component A) and polyisocyanates (component B). The performance of the polyurethane print will be designed on a "targeted dot per dot" basis (see Figure 4.29).

In the meantime, many companies are working on extending the application of digitally applied coatings and adhesives to multiple applications in the automotive, aviation and furniture industry. [124, 125]

Ideally, in addition to precise application, the following named properties become available on a drop per drop basis opening up completely new ways to improve the performance and efficiency of coating and adhesive materials.

In order to establish digital application of high quality polyurethane coatings and adhesives in a broad range of uses further development is needed. State-of-the-art printing technologies focusing on providing high resolution at the lowest possible layer thickness have to be adjusted to allow for higher material throughput to satisfy the needs of e.g. coatings with protective functions at sufficient process speed.
Examples for emerging printing technologies include:

- new large nozzle piezo printhead designs without capillary systems; [126, 127]
- drop on demand printing technologies like valve jet enabling the printing of high viscous material, direct printing of coating formulations, coloration of towels and carpets; [128–130]
- printing technologies commonly used in 3D-printing, enabling the precise application of high viscous fluids or even thermoplastic materials. [40, 131]

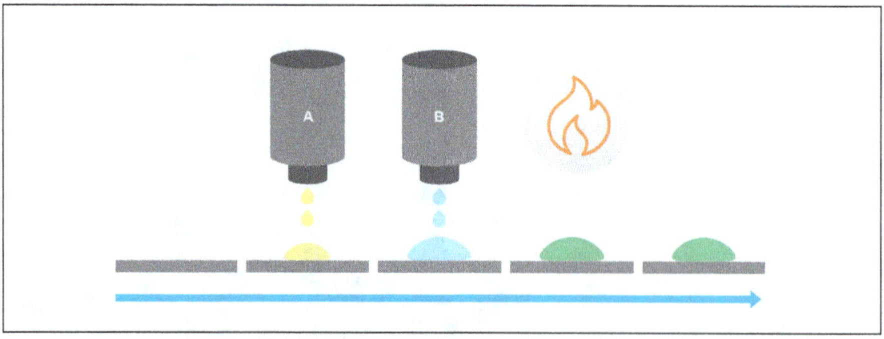

Figure 4.29: Multiple printheads printing of two-component polyurethane on the same spot on the substrate to achieve reactive mixing in situ

Even considering the steady advancement of computer and printing technology, the digital printing of industrial quality coatings and adhesives materials is still in early development. Most of the existing technologies are not suitable for handling the materials needed for coatings and adhesives applications, others are still in a prototype phase and still a lot of development has to be invested to establish first industrial production lines.

However, there are great opportunities for the development of new coatings and adhesives materials to be used successfully in such new application technologies.

4.10 Radiation curing

4.10.1 Technology and coating formulation

In coating technology, the term radiation curing usually describes the curing of coatings or printing inks with either accelerated electrons (electron beam curing, EBC), in the energy range 90 to 250 keV (higher if necessary) or ultraviolet light (UV curing), with wavelengths of 420 nm to approximately 240 nm (see Figure 4.30).

Radiation curing coating systems have shown above-average growth for many years, replacing processes that are less productive or less environmentally friendly [48]. Figure 4.31 and 4.32 show the radiation curing market split by chemistry type and regions.

Traditionally mercury and gallium doped medium and high pressure lamps have been used. However, with increasing power, energy efficiency and more choice in wavelengths, LED lamps are gaining share rapidly and will allow more complex lamp geometries to be built, thus allowing one to efficiently cure more complex parts. [43]

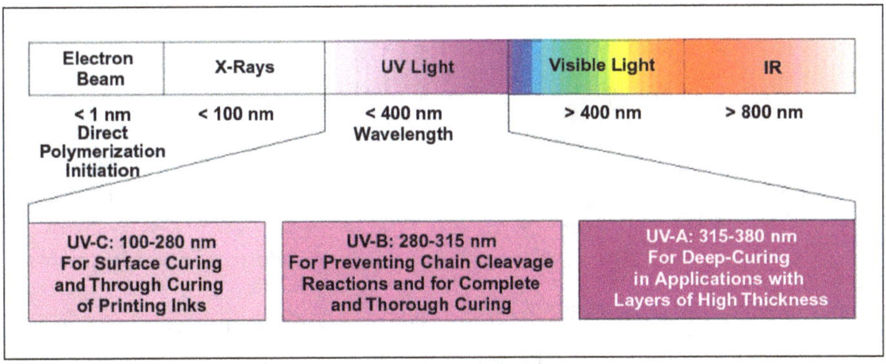

Figure 4.30: Radiation sources for curing coatings and printing inks

The vast majority of radiation curing coatings are based on acrylate chemistry and crosslinked via radical polymerization. For special applications (aliphatic) epoxy-based resins are used that crosslink via living cationic polymerization.

The majority of radiation curing coating formulations is solvent-free (100 % solids). In this case reactive thinners that are incorporated into the final solid by copolymerization take over the role of solvents in conventional coating formulations.

Besides 100 % solid systems, radiation curing water-borne systems are a steadily growing coating technology especially for application on wood and plastic substrates.

Radiation curing powder technology is a hybrid of conventional powder coatings with a radiation induced post-cure. At present, radiation curing powder coatings are still limited to special applications.

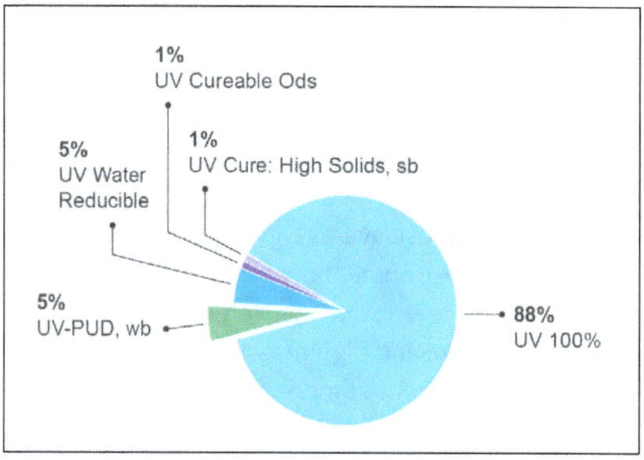

Figure 4.31: World radiation curing consumption of coating raw materials for industrial coatings by technology, total: ~ 530 kt

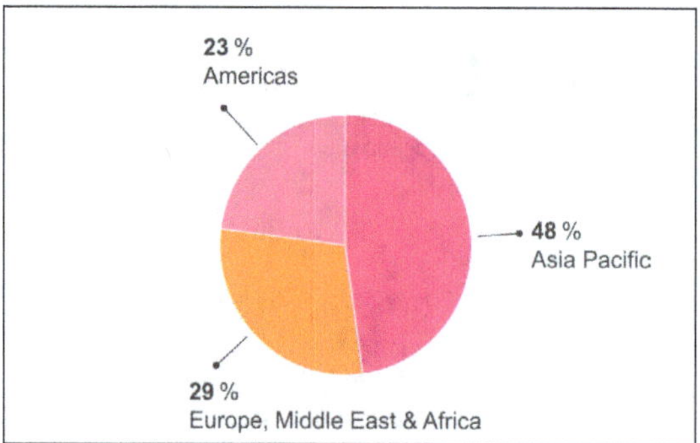

Figure 4.32: UV/EB market formulated product (kt) in 2016[48]

A typical radiation curing coating formulation is comprised of the following components:

- **Binder/oligomer:** Responsible for the desired film properties such as resistance to scratching, abrasion and weathering, etc.
- **Reactive thinner or water:** Reactive thinners are low-viscosity acrylates used primarily to lower the viscosity. The functionality and double bond density of the reactive thinners also influence the properties of the cured paint film
- **Photoinitiator:** Absorbs UV radiation, decomposes into radicals and initiates the radical chain polymerization. Photoinitiation is unnecessary in EBC
- **Additives/fillers/pigments:** E.g. for good flow, degassing, gloss and weather resistance, etc.

4.10.2 Binders for radiation curing

Binders are typically formed from reaction of functional acrylates with epoxy groups, polyesterpolyols or polyetherpolyols. Urethane acrylates are the most versatile of all the binder types and hold a share of approximately 15 %. They exceed the dynamic growth of UV/EB systems because they help to overcome drawbacks of UV/EB systems like brittleness and adhesion problems and they participate over average in new applications, e.g. digital printing, 3D printing, onsite application, automotive refinish, metal, including coil coating, textile and leather.

Urethane acrylates exhibit outstanding properties in comparison to epoxy, ester or ether acrylates in regard to a combination of:

- good mechanical properties (resistance to abrasion, scratching, impact, etc.)
- superior weather stability
- good adhesion

Urethane acrylates are used as binders in mono-cure and dual-cure systems, as well as in UV curing dispersions and powders.

4.10.3 Urethane acrylates for UV (mono-cure) and electron beam curing applications

Interior applications

The addition of hydroxyalkyl acrylates to diisocyanates and monomeric polyols yields oligomers with high-viscosity due to intermolecular urethane hydrogen bonding. Products of this type are most often commercially available diluted in reactive thinners. However, progressive amounts of reactive thinner, inevitably leads to deterioration in the desired coating properties because positive effects caused by the resin will be reduced due to the dominating polymer formed out of the reactive thinner.[49]

Figure 4.33: UV-curing of parquet flooring panels coated with urethane acrylate

Table 4.5: Typical formulation for UV-curing sealer for parquet

Component	**Parts by weight**
Resin: aromatic urethane acrylate 6000 mPa s	81.0
Reactive thinner: dipropylene glycol diacrylate	16.0
Photoinitiator: "Omnirad" 2022	3.0
	100.0
Technical properties	
Viscosity at 23 °C	~ 2000 mPas
Application/curing conditions	By roller, 1 UV Hg lamp (80 W/cm, 10 m/min, ~ 350 mJ/cm²)

Table 4.6: Typical formulation for UV curing glossy topcoat for parquet

Component	**Parts by weight**
Resin: aromatic urethane acrylate 6000 mPa s	45.2
Synergist: amine modified PET acrylate 800 mPa s	23.7
Reactive thinner: dipropylen glycol diacrylate	28.6
Photoinitiator: "Omnirad" 500	2.5
	100.0
Technical properties	
Viscosity at 23 °C	~ 500 mPa s
Application/curing conditions	By roller, 1 UV Hg lamp (80 W/cm, 5 m/min, ~ 700 mJ/cm²)

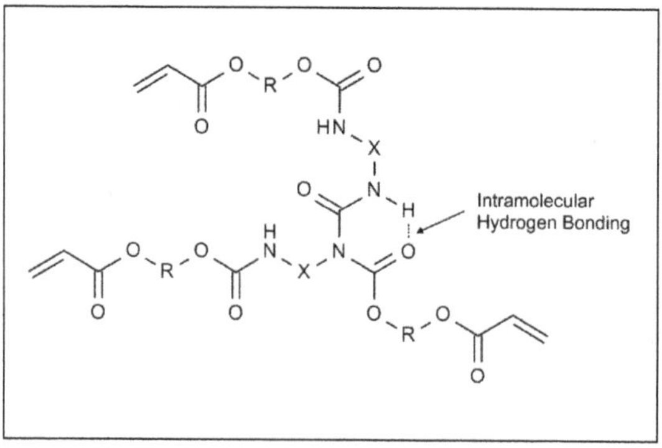

Figure 4.34: Structure of an aliphatic allophanate urethane acrylate with reduced viscosity (50 Pa s/23 °C/100 % supply form) compared to an isocyanurate type (150 Pa s/23 °C/100 % supply form),

X = hexamethylene group, R = e.g. C_2H_4

Second generation urethane acrylates, are produced from low molecular weight polyester acrylates and diisocyanates via polyaddition. By combining these two technologies, resins with high curing performance and a viscosity lower than 10,000 mPa s (23 °C) can be manufactured. Requiring only small amounts of reactive thinner, their property profiles can be largely transferred to application-ready formulations. Applications include floor, furniture and plastic coatings (see Figure 4.33). Basic formulations are shown in Tables 4.5 and 4.6.

Exterior applications

Exterior applications include clear coats for automotive add-on parts as well as coatings for construction materials like facade tiles. Weather-stable coatings must not only be resistant to yellowing and hydrolysis, but also to thermal and photochemical induced degradation processes. In the radiation curing segment, these requirements are best served by aliphatic urethane acrylates, and in particular, the reaction products of hydroxyethyl(propyl) acrylate and HDI-based polyisocyanates. This type of product has been extensively described in the literature.[51]

HDI-based urethane acrylates containing allophanate structures (see Figure 4.34) are especially effective.[50, 52, 53] A high density of allophanate and urethane structures combined with a comparable low-viscosity and a well-balanced functionality and double-bond density, results in superior weather stability and mechanical properties. This product class is used in a wide range of plastic coating applications (see Figure 4.35) and provides specific properties in formulations for metallic substrates.

4.10.4 Water-borne UV-curing polyurethane coatings

Water-borne radiation curing coatings combine the advantages of water-borne coatings with the high crosslinking density and crosslinking speed of radiation curing coatings. Water-borne UV curing coatings are thus characterized by:

- decoupling of molecular weight and viscosity,
- ease and versatility of application (spraying, rolling, brushing),
- easier down-glossing,
- good adhesion,

Figure 4.35: Weather and scratch resistant coating for automobile head-lamp lenses with urethane acrylates

- low/no shrinkage during radiation curing, and
- high flexibility.

Furthermore, water-borne radiation curing coatings are environmentally friendly due to their fast curing profile and low VOC benefits which influences its hazard categorization and resulting labeling.

Many radiation-curing water-borne coating formulations are based on polyurethane dispersions. The cured paint films perform similar to conventional two-component polyurethane coatings. In special cases radiation curing water-borne formulations can be combined with hydrophilic isocyanate based crosslinking technology or oxidative curing technology to achieve a further enhanced property profile.

Similar raw materials are used to achieve much the same end result as that yielded by two-component polyurethane coatings. However, very different routes are followed (see Figure 4.36). In a UV curing coating, in contrast to a two-component coating, the time-determining polyol/isocyanate reaction already takes place during manufacture of the polyurethane dispersion. All that is needed for the coating to crosslink on the substrate is a few seconds of exposure to UV light. It is obvious that these systems allow a considerably faster process speed.

Urethane chemistry offers the option of manufacturing products with specific properties. Polyester and epoxy acrylates with inherent hydroxyl functionality can be used. By variation of the saturated and unsaturated components, choice of diisocyanate, and variation

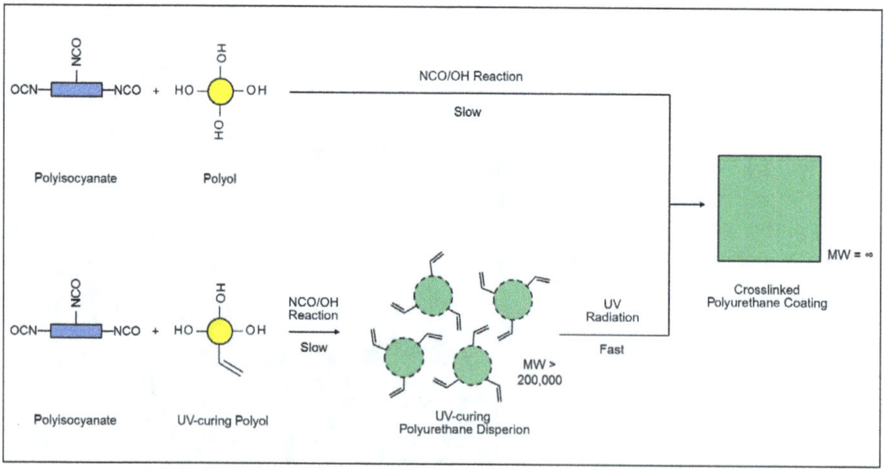

Figure 4.36: Crosslinking pathways for polyurethane coatings: comparison of a two-component polyurethane coating and a UV curing water-borne one-component polyurethane coating

of the diols and hydrophilic modification, many different property profiles can be created, for example:

- physical drying, (dispersions yielding tack-free films after evaporation of the water),
- UV reactivity,
- hardness/elasticity,
- weather stability, etc.

High molecular weight urethane acrylate dispersions (see Figure 4.37) are typically manufactured by the acetone process (see Chapter 3.8.1).[54]

Radiation curing polyurethane dispersions are most commonly applied via spray coatings, e.g. on wood and plastics substrates. Roller coating application technologies are used for radiation curing polyurethane dispersions on PVC and vinyl substrates and in printing as inks. Typical use fields for radiation curing polyurethane dispersions are in the furniture and wood flooring industry, PVC/Vinyl flooring industry as well as plastic coatings in automotive and electronics.

4.10.5 Urethane acrylates for UV powder applications

Reducing the thermal stress on the products being coated, shorter processing times and the decoupling of flow behavior and crosslinking kinetics are important development goals in the area of powder coating technology. Introducing the radiation curing method for the crosslinking and curing of powder coatings, these goals can largely be realized. UV curing powder binders must meet contradictory requirements of perfect grind ability and storage stability on the one hand, and low processing temperatures on the other. Combining polyurethane chemistry with radiation curing technology for use in reactive powder coatings offers significant advantages over conventional powder coating technologies with regard to balancing these difficult to combine requirements.[55]

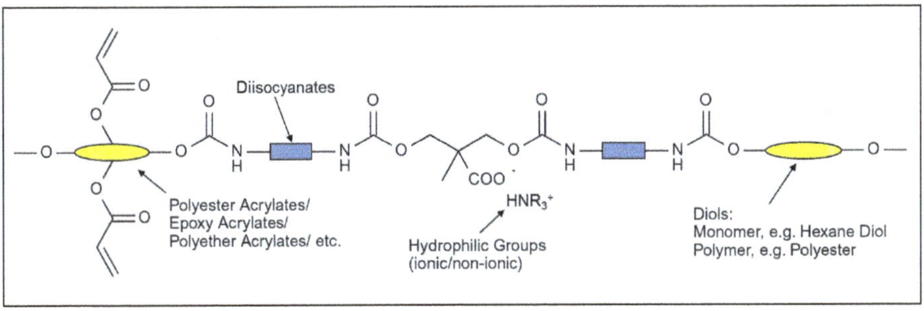

Figure 4.37: Structure of a UV curing urethane acrylate dispersion

Excellent results can be achieved using high T_g amorphous linear urethane acrylates as main base material. They are easily grindable and yield UV curing powders with good storage stability and excellent film properties. Until now, UV powder coating technology has been used primarily in finishing MDF furniture components.

4.10.6 Dual-cure technology

Dual-cure technology generally refers to a combination of two chemically different crosslinking mechanisms. Isocyanato-urethane acrylates are used in coating systems that are crosslinked via photochemically or thermally initiated radical polymerization and by isocyanate/polyol reaction.[56] Dual-cure systems are commonly crosslinked first by UV radiation and then via an isocyanate/polyol reaction to form the final polyurethane network (see Figure 4.38).

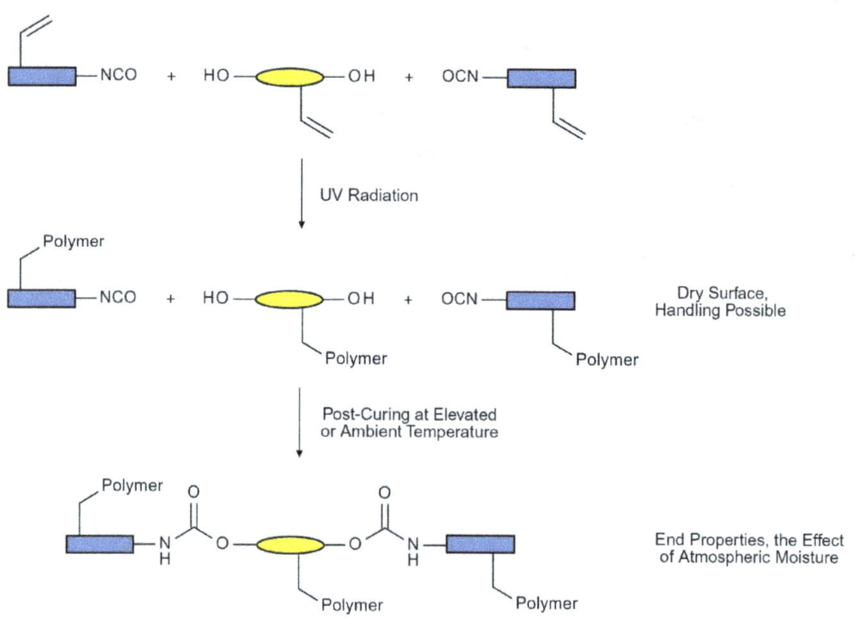

Figure 4.38: Dual-cure mechanism with initial UV-curing

In comparison to coatings that are cured only by exposure to radiation (mono-cure), dual-cure technology offers the following advantages:
- improvement of the adhesion due to the reaction of the isocyanate group with the substrate;
- fewer geometrical limitations because the NCO/OH reaction takes place even in those areas hidden from the radiation source;
- better through-curing of highly pigmented UV coatings, via an NCO/OH reaction in deep layers that radiation cannot reach easily.

The reversal of the dual-cure process with upstream thermal drying, as shown in Figure 4.39, opens completely new possibilities in coating technology. Formable films are films that are pre-coated but thermoplastic, enabling them to conform to the geometry of the

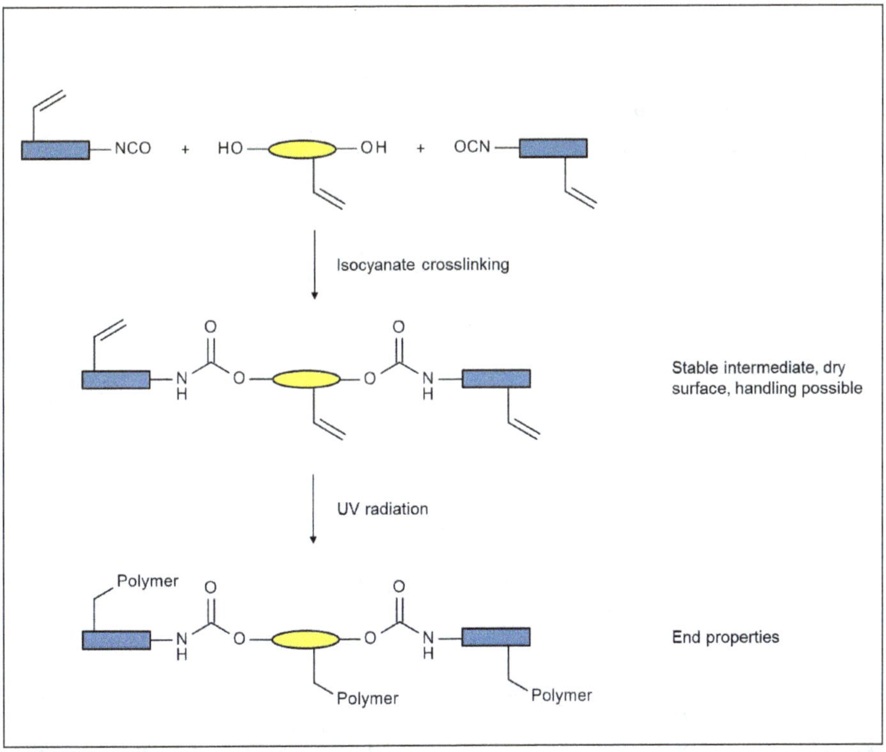

Figure 4.39: Dual-cure mechanism with final UV-curing

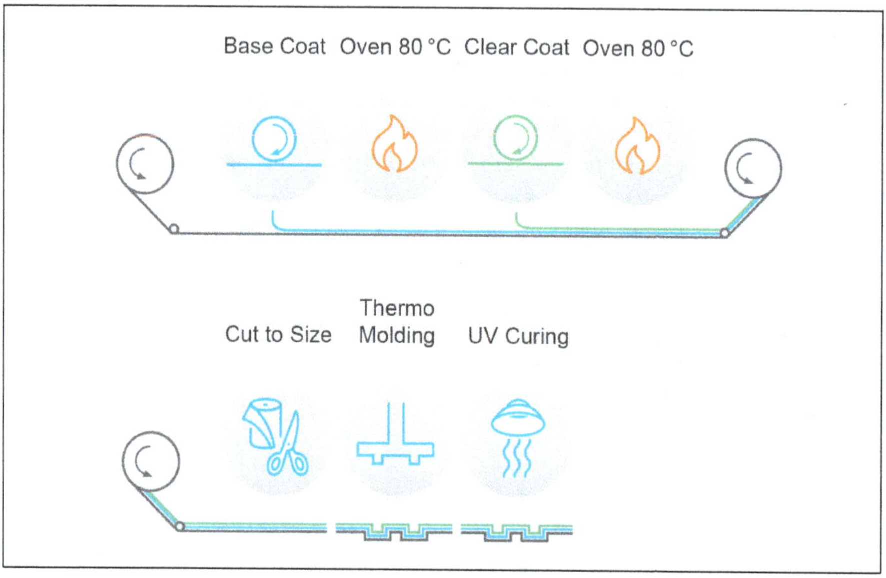

Figure 4.40: Manufacturing concept for coated post-formable films

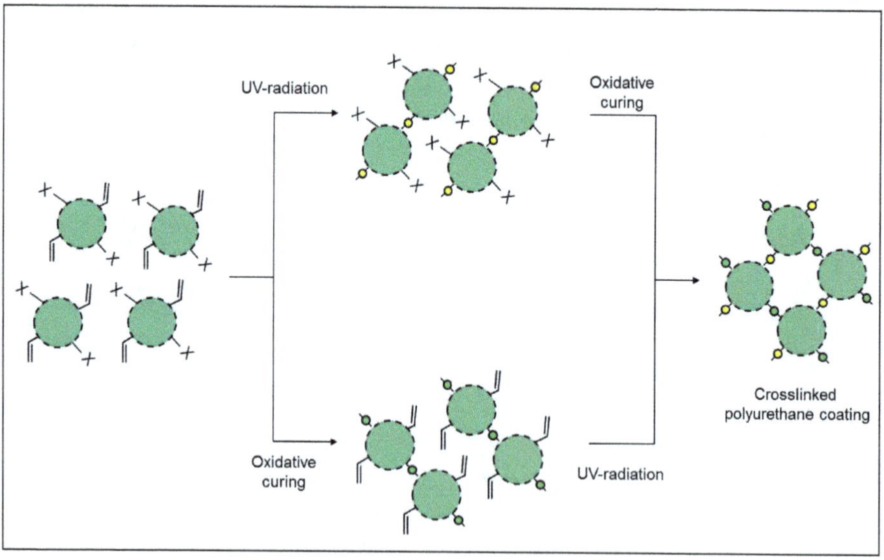

Figure 4.41: Schematic cross-linking systems on a self-crosslinking UV curing polyurethane dispersion. X = self-crosslinking units, which react by oxidative curing; // = double bonds that react under UV light.

respective workpiece. The end properties desired can be obtained via UV post-curing (see Figure 4.39 and 4.40).

An oxidative self-crosslinking UV-curable water-borne dispersion has recently been launched.[44] This new polyurethane solution introduces an additional crosslinking mechanism to the UV-curable one to increase robustness of the system (see Figure 4.41). In addition to faster water release and reduced energy consumption this two-step crosslinking has the advantages of improving productivity by additional shadow curing and post UV curing thermoformable options.

4.10.7 Outlook

Radiation curing is on the way to overcome previous limitations. As a VOC-free technology, the basic environmental requirement has been fulfilled; further development of lamps and photoinitiators, especially for LED, should allow UV technology to coat more complex geometries. Polyurethanes are the most important chemical tool for new UV/EB applications. In addition to traditional urethane acrylates, PU technology is employed in UV water-borne resins, powders and dual-cure systems and offers – comparable to conventional curing technologies – high versatility combined with high performance together with the high productivity.

4.11 References

[1] D. Dieterich in: Houben-Weyl, Methoden der Organischen Chemie, Bd. E20, H. Bartl, J. Falbe (Hrsg.). Georg Thieme Verlag, Stuttgart 1987, p. 1641 ff

[2] Q. Bell, Raw Materials and their usage, in: Solventborne urethane resin, Vol. 1: Surface Coatings, Chapman and Hall, New York 1993, p. 153ff

[3] D. Dieterich in: Ullmann`s Encyclopedia of Industrial Chemistry, Vol. A 21. VCH Verlagsgesellschaft mbH, Weinheim 1992, p. 684f

[4] W. Kern, H. Willersinn, Angew. Chem. 67 (19/20) (1955) p. 573

[5] R. Schwalm, UV Coatings: Basics, Recent Developments and New Applications. – 1st Ed. – Amsterdam u.a.: Elsevier, 2007. XI, ISBN 978-0-444-52979-4

[6] EP 596 360 (1992) Bayer AG; F. Gross, C. Jaboc: Farbe und Lack, 9, 2017, 18; M. Hallack, T. Unkelhäußer, H. Görlitzer, C. King, R. Lomölder, PPCJ, 10, (2014), p. 24–25; M. Groenewolt, "Silan-Technologie als Basis für zukünftige Klarlackgenerationen", Karosserielackierung intensiv, Sindelfingen 2016

[7] 2 Komponenten Maschinenbau GmbH (2KM)

[8] a) L. Thiele, Isocyanatreaktionen und Katalyse in der Polyurethanchemie, Fortschrittsbericht, Acta Polymerica 30 (1979) p. 323–342; b) A. L. Silva & J. C. Bordado (2004) Recent Developments in Polyurethane Catalysis: Catalytic Mechanisms Review, Catalysis Reviews, 46:1, p. 31–51; c) F.-Z Belmokaddem, J. Dagonneau, J. Lhomme, R. Blanc, A. Garduno-Alva, C. Maliverney, A. Baceiredo, E. Maerten, E. Fleury, F. Méchin Designed Monomers and Polymers, 2016, 19:4, p. 347–360, D. H. Sardon, A. C. Engler, J. M. W. Chan,

J. M. Garcia, D. J. Coady, A. Pascual, D. Mecerreyes, G. O. Jones, J. E. Rice, H. W. Horn, J. American Chemical Society, 2013, 135 (43), 16235
[9] L. Thiele, R. Becker, Catalytic Mechanism of Polyurethane Formation, Adv. Ureth. Sci. Tech. 12 (1993) p. 59–85
[10] D. Dieterich in: Houben-Weyl, Methoden der Organischen Chemie, Bd. E20, H. Bartl, J. Falbe (Hrsg.), Georg Thieme Verlag, Stuttgart 1987, p. 1604 ff.
[11] P. F. Haas, K. Uhlig in: Kunststoff-Handbuch, G. W. Becker, D. Braun (Hrsg.), 3. Auflage, Band 7: Polyurethane, Carl Hanser Verlag, München, Wien 1993
[12] R. M. Rynders, C. R. Hegedus, A. G. Gilicinski, Characterization of Particle Coalescence in Waterborne Coatings, Journal of Coatings Technology 67 (845) (1995) p. 59
[13] Bechara, Formulating with polyurethane dispersions, European Coatings Journal 4 (1998) 236
[14] DE 3 313 237 (1983) Bayer AG
[15] T. A. Potter, J. W. Rosthauser, H. G. Schmelzer, 11th Intern. Conf. Org. Coat. Sci. Technol., Athen Greece, July, 8–12 1985, Preprints p. 331
[16] DE 3 231 062 (1982) Bayer AG
[17] US 4 190 567 (1976) Kao Soap Co
[18] EP 285 898 (1987) Bayer AG
[19] EP 436 941 (1990) Hoechst AG
[20] DE 4 012 339 (1990) Bayer AG
[21] DE 3 219 413 (1982) Vianova Kunstharze AG
[22] US 5 133 997 (1991) Union Carbide Marble Care Inc; R. R. Roesler, K. Danielmeier, Progress in Organic Coatings, 2004, 50, 1–27
[23] DE 3 523 856 (1985) Bayer AG
[24] EP 442 652 (1990) Imperial Chemical Industries
[25] DE 3 737 245 (1987) Bayer AG
[26] EP 358 979 (1988) Bayer AG
[27] L. Kahl, M. Bock, H. J. Laas, B. Klinksiek, Wässrige 2K-PUR-Klarlacke für die Autoerstlackierung, Teil 1, Industrie-Lackierbetrieb 63 (12) (1995) 424; Teil II, ebd. 64 (2) (1996) p. 63
[28] M. Bock, J. Petzoldt, Aqueous Polyurethane Coating Systems for Plastics, Modern Paint and Coatings, (2) (1996) p. 22
[29] M. Sonntag, E. Jürgens, Wässrige 2K-PUR-Lacke für die Autoreparatur- und Serienlackierung, J. Oberflächentechnik (6) 12, 1997
[30] B. Klinksiek, B. Koglin, Verfahren zur Herstellung von Salben und Cremes, GVC Tagung Köln (1991), Vortrag 11
[31] EP 540 985 (1991) Bayer AG
[32] M. Shaffer, H. Bui, Rheology of 2-Component Waterborne Coatings, Waterborne, High Solids and Powder Coatings Symposium, New Orleans 1998
[33] W. Kubitza, Wässrige Zweikomponenten-Polyurethan-Reaktivsysteme, Farbe und Lack 97 (3) (1991) p. 210
[34] M. S. Gebhardt, L. E. Scriven, Reduction of air bubbles in spray-applied coatings, Journal of Coatings Technology 66 (830) (1994) p. 27
[35] W. Hovestadt, E. Jürgens, K. L. Noble, J. Probst, Applikationssicherheit wässriger 2K-Polyurethanlacke, Fatipec-Kongress, Interlaken 1998
[36] D. Ondratschek, O. Tiedje in besser lackieren!, Jahrbuch 2017, Vincentz Network, Hannover 2016, p. 404–410
[37] W. Hovestadt, B. Klinksiek, M. Melchiors, Farbe und Lack 106 (7) (2000) p. 40
[38] D. Ondratschek in besser lackieren!, Jahrbuch 2017, Vincentz Network, Hannover 2016, p. 383–389
[39] W. Kleber, U. Strohbeck in besser lackieren!, Jahrbuch 2017, Vincentz Network, Hannover 2016, p. 390–403
[40] C. K. Chua, K. F. Leong, C. S. Lim, World Scientific (2003) 124
[41] M. Mechtel, M. Melchiors, JOT 9 (2003) p. 42
[42] Dürr Systems AG
[43] Radiation-Cured Coatings: UV Today, LED Tomorrow, CoatingsTech, Vol. 14, No. 4, April 2017
[44] E. Tejada Rosales et al, European Coating Journal, 11, 2017
[45] A. J. Bloodworth and A. G. Davies, Proceedings of the Chemical Society, 264, 1963

[46] U. Freudenberg, H.-U. Meier-Westhues, H. J. Laas, Creative Advances in Coating Technology, 4th Nürnberg Congress, April, 7–9 1997, Paper 39

[47] F. Lehmann; M. Gedan-Smolka; D. Lehmann, Farbe & Lack, (2000), 106 (12), p. 62–70 M. Gedan-Smolka; F. Lehmann; D. Lehmann, Proceedings of the International Waterborne, High-Solids, and Powder Coatings Symposium, (2001), p. 405–419

[48] Nick Gruber, RadTech Europe 2017, Conference Proceedings

[49] M. Müller, European Coatings Conference 2004, Conference proceedings pp.11–20

[50] W. Fischer, H. Kuczewski, D. Rappen, J. Weikard 5, 2005, Conference proceedings

[51] e.g. EP-A 668 330 Bayer AG, EP-A 1 002 818 Bayer AG

[52] T. Fäcke, W. Fischer, D. Rappen, F. Richter, J. Weikard, RadTech Europe 2003, Conference proceedings, pp 219–224

[53] WO 2005/092942 Bayer MaterialScience AG

[54] EP-B 753 531 Bayer MaterialScience AG

[55] T. Fäcke, 3rd Int. Exhibition and Congress for Powder Coating Technology, 2002, Congress Sessions

[56] W. Fischer, J. Weikard, U. Meier-Westhues European Coatings Conference 2001, Conference Proceedings, pp 277–296

[57] a) D. Ondratscheck, O. Tiedje, besser lackieren!, Jahrbuch 2017 b) K.-H. Haas, G. Tovar Hrsg. Angewandte Nanotechnologie, Fraunhofer Verlag 2018, 37 ff. c) U. Hoffmann, M. Hilt, besser lackieren!, Jahrbuch 2017, Vincentz Network, Hannover 2016, Moderne Funktionsbeschichtungen, p. 158–173

[58] High Polymers Volume XVI, POLYURETHANES Chemistry and Technology Part 1. Chemistry J. H. Sauders; K. C. Frisch, p. 212, 1962

[59] (a) D. Guhl, European Coatings Journal 7–8 (2008), p. 36–40 (b) de Lima, Viviane; Pelissoli, Nicole da Silva; Dullius, Jeane; Ligabue, Rosane; Einloft, Sandra, Journal of Applied Polymer Science (2010), 115 (3), p. 1797–1802

[60] Ikea specification IOS MAT 066

[61] www.afirm-group.com

[62] A. Carroy, T. Hintermann, G. Baudin, D. Bauer, P. Contich, K. Dietliker, M. Faller, R. Kohli Steck, C. Lordelot, K. Misteli, Progress in Organic Coatings 68 (2010) p. 37–41

[63] Covestro, proprietary investigation 2010 ff

[64] F. U. Richter, M. Grahl, J. Weikard, Proceedings of the European Coatings Congress 2015

[65] A. Prabhakar, D. K. Chattopadhyay, B. Jagadeeh, K. V. S. N. Raju, J. Appl. Polym. Sci. 43 (2005) p. 196–1209

[66] H. Sardon, L. Irusta, M. J. Fernadez-Berridi, Progress in Organic Coatings 66 (2009) p. 291–295

[67] a) W. Fischer; J. Weikard; and H-U. Meier-Westhues, U (2001), "Dual Cure; Combination of Superior Properties", RadTech Report, November/December Issue b) W. Fischer, H. Kuczewski, D. Rappen, J. Weikard; RadTech Europe 2005, Conference Proceedings c) J. Weikard, Dual Cure: Neue Möglichkeiten mit UV-härtenden Lacksystemen, Conference Proceedings Esslinger Oberflächenseminar: UV-härtbare Systeme für Kunststoff, 2004

[68] K. Dusek, Networks from telechelic polymers: theory and application to polyurethanes, 289 in: E. J. Goethals, Telechelic polymers: synthesis and applications, CRC Press, Boca Raton, Florida, 1989

[69] J. Budinski-Simendić et al., Mater. Sci. Forum 518, p. 399–404, 2006

[70] U. Scherf et al., ACS Macro Lett. 4, p. 1268–1272, 2015

[71] T. Endo et al., J. Polym. Sci. Part A: Polym Chem. 51, p. 2631–2637, 2013

[72] E. Delebecq et al., Chem Rev. 113, p. 80–118, 2013

[73] H. Sardon et al., Macromolecules 48, p. 3153–3165, 2015

[74] T. Flipsen, PhD Thesis, University of Groningen, 2000

[75] P. Fabbri et al., Polymer 55, p. 488–494, 2014

[76] www.jeccomposites.com/events/past-events/jec-world-2017/innovation-awards/winners-profiles
[77] D. Achten et al., Thermosetting Resins, 25–27 September 2018, Berlin, Germany, 2018
[78] US20180162981 (2018) Covestro Deutschland AG
[79] T. A. Misev 1991 "Powder Coating Chemistry and Technology" John Wiley and Sons
[80] D. A. Bate, The Science of Powder Coatings Chemistry, Formulation and Application Volume 1, Selective Industrial Training Limited London, United
[81] Powder Coatings Europe 2001–2010: A Global Comparison: Europe, Americas and Asia IRFAB Chemical Consultants Scrl, Brussels, published 2002
[82] "Irfab" Global Industrial Coatings Markets (GICM), PRA World Ltd (2017, 2018)
[83] Pieter Gillies de Lange, Powder Coatings, Chemistry and Technology, Vincentz Network, Hannover 2004
[84] Zhongyan Du, Shaoguo Wen, Jihu Wang, Changle Yin, Dayang Yu, Jian Luo, Journal of Materials Science and Chemical Engineering, 2016, 4, p. 54–59
[85] T. A. Misev, R. van der Linde/Progress in Organic Coatings 34 (1998) p.160–168
[86] Various authors, 1. Internationale Konferenz: Das Pulverbeschichten von Automobil-Karosserien, June 1998, Berlin
[87] Performance Advantages of H_{12}MDI Based Urethane Crosslinkers in Powder Coatings", C. Sullivan, J. Weinberg, Powder Coating 2002 in Indianapolis 23–24 Sep 2002 Powder Coating Institute
[88] H.-U. Meier-Westhues, M. Bock, W. Schultz, Farbe und Lack 99 (1) (1993) p. 9
[89] M. Bock, H.-U. Meier-Westhues, 19th Intern. Conference in Org. Coatings, Science and Technology, Athens 1993
[90] J. M. Loutz, W. Demartequ, and D. Vandervorst, 1994 Elsevier Science, Characterization and Control of Odours and VOC in the Process Industries, Table IV, "Comparison concerning VOC emissions between different powder systems during the curing"
[91] B. Pilch-Pitera, Progress in Organic Coatings 77 (2014) p. 1653–1662
[92] B. Pilch-Pitera, Łukasz Byczyński, Bogdan Myśliwiec, Progress in Organic Coatings 113 (2017) p. 82–89
[93] B. Pilch-Pitera, M. Kedzierski, E. Olejnik, Szczepan Zapotocyzny, Structure and properties of polyurethane-based powder clear coating systems modified with hydrotalcites, Progress in Organic Coatings 95 (2016) p. 120–126
[94] E. Spyrou et al./Progress in Organic Coatings 48 (2003) p. 201–206
[95] M. Grahl, Württemberger Lacksymposien 2006, „Pulverlacke für Niedertemperaturhärtung – Möglichkeiten und Perspektiven für PUR-Pulverlacke", FH Esslingen, March 2006
[96] P. Thometzek, H. Basten, K. Köhler, H.-U. Meier-Westhues, Farbe und Lack 104 (2) (1998) p. 22
[97] H.-U. Meier-Westhues, P. Thometzek, H. J. Laas, Farbe und Lack 103 (4) (1997) p. 140
[98] E. Spyrou et.al., Highly Reactive Blocking Agent-Free Polyurethane Powder Coatings, 58, November 2005, www.pcimag.com
[99] US 5804646 (1998)
[100] R. Subramanian et al., "Recent Advances in Low Temperature Cure Crosslinkers for Polyurethane Powder Coatings" presented at the 81st annual Meeting of the FSCT in Philadelphia, 13–14 Nov 2003
[101] D. Brand, S. Molitor-Limberg, DFO-Tagung June 1998, Boppard, Vol 37
[102] B. A. J. Noordover A. Heise, P. Malanowksi, D. Senatore, M. Mak, L. Molhoek R. Duchateau, C. E. Koning, R. A.T.M. van Benthem, Progress in Organic Coatings 65 (2009) p. 187–196
[103] R. Destexhe, Pulverlackkongress 1996, Hamburg, DFO Congress Paper, 33
[104] V. Maurin, C. Croutxé-Barghorn, X. Allonas, Progress in Organic Coatings 73 (2012) p. 250– 256.
[105] K. Ratliff, J. Eastman and T. Faecke, Corrosion Resistance of Polyurethane Powder Coatings, International Waterborne, High-

Solids, and Powder Coatings Symposium, New Orleans, 22–24 Feb 2006
[106] N. Tudorachi and F. Mustata, Arabian Journal of Chemistry, In Press, 2018
[107] P. M. Stefani, S. M. Moschiar and M. I. Aranguren, Journal of Applied Polymer Science, 79, 1771–1779, 2001
[108] F. Hernández-Sánchez, H. Vázquez-Torres, Journal of Polymer Science: Part A: Polymer Chemistry, 28, 1579–1592, 1990
[109] A. N. Mauri and C. C. Riccardi, Journal of Applied Polymer Science, 85, p. 2342–2349, 2002
[110] E. D. Dumain, T. Agawa, S. Goel, A. Toman and A. S. Tse, Journal of Coatings Technology, 71, 69–75, 1999
[111] T. I. Kadurina, V. A. Prokopenko and S. I. Omelchenko, Polymer, 33, 3858–3864, 1992
[112] Robert E. Odom, Kinetics of epoxide-carboxylic acid reaction, Report No RK-TR-63-20, Defense Documentation Center for Scientific and Technical Information, Cameron Station, Alexandria, Virginia, 1963
[113] Decoral System srl, Arcole, Italy
[114] G. Merfeld et al., Development of Low Temperature Curing, 120 °C, Durable, Corrosion Protection Powder Coatings for Temperature Sensitive Substrates, Originally Presented at the 31st Annual International Waterborne, High-Solids, and Powder Coatings Symposium, February 18–20, 2004, in New Orleans, LA., JCT Research, Vol. 2, No. 8, October 2005
[115] C. Goemans, J. Oberflächentechnik (2) (1998) 32
[116] W. Nowotny, Conference Proceedings, PUR-Forum, Polyurethan Surfaces, Kunststoffzentrum Leipzig, Germany, 05/21/2014
[117] KrausMaffei Technologies GmbH introduced "ColorForm"; Engel Austria GmbH and Hennecke GmbH introduced „clearmelt". Taken from respective company homepages on 03/14/2018
[118] S. Schröbel, A. Schmidt, T. Hörl, M. Eichlseder, "Scratch-Resistant in One Step", Kunststoffe 2010/01, p. 52–55
[119] M. Glawe, J. Helmig, H. Mundstock, J. Weikard, "DirectCoating: Creating Designed High Value Surfaces on Automotive Plastics by an Innovative Process", Conference Proceedings, European Coatings Congress, Nuremberg, Germany, 2017
[120] T. Elfroth, "Serien direkt im Werkzeug lackieren", besser lackieren!, online journal, published 08/28/2017, taken from http://www.besserlackieren.de/Industrielle-Anwendungen/Sonstige-Beschichtungen/Serien-direkt-im-Werkzeug-lackieren on 03/14/2018
[121] Z. Pan, Y. Wang, H. Huang, Z. Ling, Y. Dai, S. Ke, Ceramics International (2015), 41 (10_Part_A), 2515–12528
[122] www.zimmer-kufstein.com
[123] www.print-und-produktion.de/nachrichten/messen-veranstaltungen/news/datum/2017/11/13/jupiter-digital-printing-lines-10-jahre-industrieller-digitaldruck-von-hymmen/
[124] US 2017203580 A (WO 17003277 A2)
[125] www.airbus.com/content/dam/corporate-topics/financial-and-company-information/Ralph-Maurer-Direct-Printing-Technology-AIRBUS-INNOVATION-DAYS-2016.pdf
[126] US 2010039480 A, www.ttp.com/, Vista Technology, The Technology Partnership
[127] www.ttp.com/printing, Vista Technology, The Technology Partnership
[128] US 2017361346 A, www.epainters.eu/, "EPjet" Printing Technology, Epainters
[129] www.epainters.eu/english/technologies/epjet-technology-concept/
[130] www.carpet.namatec.com
[131] US 5121329

5 Polyurethane coatings

5.1 Wood coating

The global wood coating market size is estimated to be 2,500 kilotons[1] Polyurethane (PU)-based coatings systems have long been and continue to be a leading technology for the formulation of wood and furniture coatings. The applications range from the coating of wood for outdoor use, e.g. windows and doors, to wooden materials for furniture and parquet. Due to their unique combination of chemical and mechanical resistance, PU coatings are most widely used wherever high durability of surfaces and long-term appearance is a prerequisite: kitchen cabinet, office furniture, high quality living room furniture and the finishing of parquet or hardwood flooring. On-site applied professional decoration of wooden doors is an important end use market for PU coatings in China.

The many choices of resin structure and the associated wide range of properties predestine polyurethane systems for meeting substrate-specific requirements in the wood segment. Furthermore, they can be applied in many ways and cured at ambient and even temperatures, while still offering fast cure that allows for high productivity in manufacturing of wooden goods.

In the global markets of wood and furniture coatings several technologies are used, depending on performance requirements and application setup. In addition to solvent- and water-borne polyurethane coatings, industries are using acid curing melamine crosslinking systems, nitrocellulose coatings, unsaturated polyesters and acrylics as well as water-borne polyacrylic-based coatings. Polyurethanes have been the dominant technology in this industry for decades, as they represent the best overall profile in terms of ease and efficiency of application as well as performance and cost. Figure 5.1 and 5.2 show

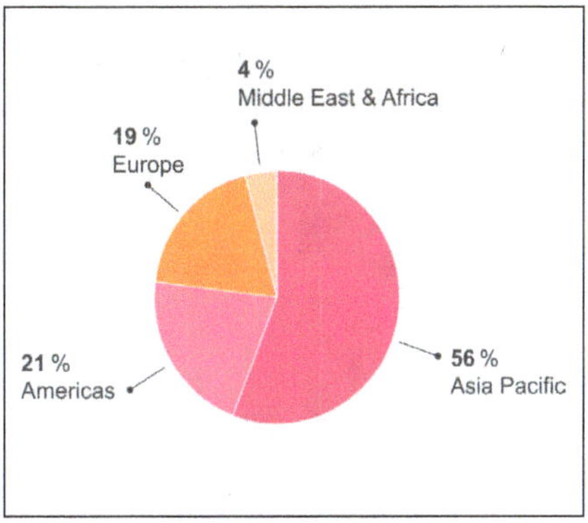

Figure 5.1: Coatings for industrial wood/furniture coating, world consumption by region, 2017; total: approximately 2.5 million tons[1]

the consumption of wood and furniture coating worldwide.[1] Solvent-borne coatings are still by far the preferred choice in global furniture industry 2017. China is the largest national market for wood coatings consumption in the world.[1]

Polyurethanes have been the leading technology for wood and furniture coatings for a long time. In 2017, they represent 32 % of global wood coatings production. However, water-borne and UV-curing systems represent fast growing alternatives. A particularly fast growth has been observed since 2017 in China. The Chinese furniture as well as joinery industries are moving towards low emission water-borne systems at high conversion rates.

5.1.1 Industry needs: requirements of wood and furniture coatings

Productivity

Due to the large number of individual components involved in modern furniture production high levels of productivity are needed that is fast drying and sandability of the coating. This is all the more important since wood and furniture finishes are in most cases multi-coat systems, which consist of at least a primer and a topcoat. Achieving the earliest possible sandability of the primer and/or primer surfacer is important for achieving rapid stackability of the topcoated parts.

Substrate-specific requirements

Woods differ in their color, grain, pore structure, density and the type and content of specific constituents. The various wood-based materials such as chipboard, hardboard,

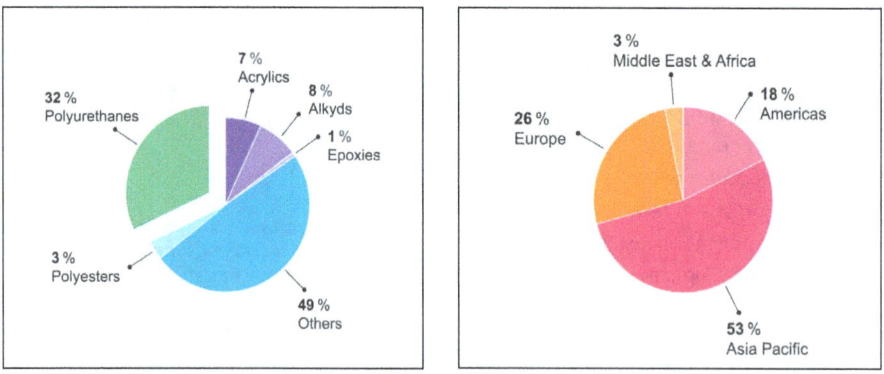

Figure 5.2: Coatings for industrial wood/furniture coating, world consumption by chemical system, 2017; total: approximately 2.5 million tons[1]

medium-density fiberboard (MDF) and plywood, which are also available with veneer or laminated film finishes, differ widely in their surface structures. Exact knowledge of a substrate allows selection of a polyurethane (PU) system which ensures optimum coating results. In order to achieve special surface effects and qualities, various coatings can be combined in the overall coating system. In contrast to alternative systems like nitrocellulose coatings, unsaturated polyesters or UV-curing systems, polyurethane coatings usually exhibit good compatibility, over-coatability and intercoat adhesion. This holds true for both two-component (2K) solvent- and water-borne PU formulations.

Subsegment-specific requirements and standards

Depending on their subsequent use, coated wood and furniture components have to comply with a variety of requirements. For example, the coating on a dining table is expected to have much better resistance properties than the coating used on bedroom furniture. Coatings for window frames have to satisfy different requirements compared to parquet sealers (see Figure 5.3). Generally recognized and widely used performance profiles for furniture coatings are contained in national and industry standards, such as:

- IKEA IOS MAT 066 (Global IKEA standard);
- EN 12 720 (European standard);
- GB/T-3324-2017 (Chinese standard for wooden furniture);
- US standards KCMA (kitchen cabinet) and BIFMA (office furniture).

Among the key requirements specified by these standards for furniture surfaces is resistance to household cleaning agents, scratch and abrasion. Additionally, European standard EN 927 specifically describes the requirements of wood coatings for outdoor applications. Besides surface properties, also composition of the coatings formulations and their respective safety in use and environmental impact are widely covered by national and industry standards.[2]

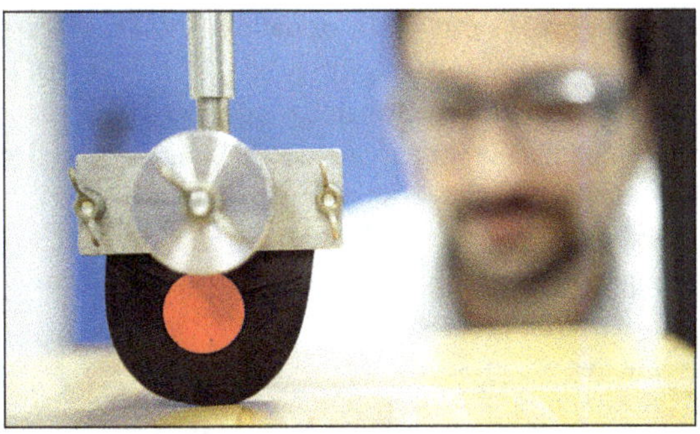

Figure 5.3: Testing the mechanical resistance of parquet coatings under actual conditions

Application requirements

Depending on requirements, in particular, on the shape of the substrates to be coated, there are a variety of application processes that can be exploited. Conventional spraying techniques (compressed air, airless, airmix) are widely used, as are the corresponding electrostatic variants to reduce overspray. Curtain coating has proven ideal for large flat components, such as doors. Flow and roller coating techniques are also used. The coatings are typically dried at room temperature (RT); in order to accelerate drying coatings can be force-dried at temperatures up to 70 °C in flat, vertical or rack dryers. Similar equipment can be used for solvent- and water-borne systems; however, special drying and application equipment for water-borne systems is used in many lines to maximize productivity.

Environmental aspects

All of the mentioned application technologies can be used for solvent-borne, UV-curing and water-borne one- and two-component polyurethane systems. The significance of water-borne systems is growing as they satisfy the stringent requirements of emission reduction, in many regions of the world. In addition to legislation that limits emissions into the atmosphere or within buildings (Europe, Asia, Americas), environmental labels such as Ecolabel, "Blauer Engel" (Europe), Ten-ring (China) are also becoming influential in the marketplace. Of particular interest in the US is the fact that polyurethanes enable the formulation of LEED (Leadership in Energy and Environmental Design) certified coatings.

Polyurethane-based systems in wood and furniture coating

Ever since polyurethane-based coatings formulations have been available, the use in wood and furniture coatings was an important end use market for this technology. The following chapters summarize advantages and limitations of all PU-based technologies, from urethane modified alkyd resins to the most recent water-borne coating technology.

5.1.2 Urethane-modified oil and alkyd resin coatings

Remarkable property variations can be achieved by using urethane modification of oils or long-oil alkyds which dry by oxidation. Those resins are synthesized by modifying OH-functional unsaturated alkyds with diisocyanates like TDI or IPDI. The products are adjusted to be free of reactive isocyanate groups and contain double bonds for air drying. This synergistic modification will improve properties of oils and long-oil alkyds in a smart way, such as

- solubility in mineral spirit;
- storage stability;
- pigment wetting;
- faster drying;
- greater hardness;
- better elasticity;
- lower abrasion;
- resistance to water and chemicals, especially alkaline liquids.

Urethane-modified oils and urethane alkyd resins are used to formulate varnishes and clear coats for wood in indoor and outdoor applications, e.g. wall and ceiling paneling, window frames and doors. These products are also used for hardwood floor coatings, clear coats for the carpentry business and in premium grade coatings for the do-it-yourself (DIY) sector.

Urethane-modified oils and resins are also suitable for the formulation of pigmented coatings. These may be high gloss or matte and have outstanding mechanical strength and high resistance to cleaning agents. This chemistry can also be used to formulate primers or fillers with early sandability, referred to as fast-sanding primers, which are normally applied by brushing or rolling.

Moisture-curing coatings

In the wood coating industry, one-component polyurethane systems are used as moisture-curing formulations. For reactivity reasons, mostly TDI prepolymers are used to ensure fast drying. These systems are used as sealers for wood and parquet floors. The cured paint films are characterized by outstanding properties, especially high abrasion resistance. Although they combine high performance with unmatched ease-of-use, solvent-based moisture coatings are being substituted by water-borne polyurethane dispersions in many regions of the world, due to the lower content of organic solvents in these modern systems.

5.1.3 Two-component polyurethane coatings

Solvent-borne 2K PU formulations: polyisocyanates for wood and furniture coatings

The principal requirements which must be satisfied by polyisocyanates for wood and furniture coatings are:

- very low monomeric diisocyanante content, preferably below 0.5 % or even 0.1 %,
- excellent compatibility with the polyol of a given formulation, and
- stability also at low concentrations.

A wide range of polyisocyanates with different structures and qualities enables selection of the right substance or combination to satisfy the requirements made of wood and furniture coatings. TDI-based aromatic polyisocyanates are mainly used for wood coatings. For non-yellowing coatings, HDI-based polyisocyanates are required, usually isocyanurates or biurets. However, non-yellowing is a prerequisite, e.g. in kitchen cabinet coatings, demanding highest stability of color over long periods of time. Lower requirements, e.g. in living room furniture, can be fulfilled by combinations of aliphatic- and aromatic crosslink-

Table 5.1: Comparison of coating properties yielded by various polyurethane crosslinkers

	HDI iso-cyanurate	HDI biuret	TDI adduct	HDI/TDI iso-cyanurate	TDI iso-cyanurate, low viscosity	TDI iso-cyanurate, high viscosity	Trend
Viscosity	——————→						Increasing
Reactivity	——————→						Increasing
Sandability	——————→						Increasing
Pot life	←——————						Longer
Flexibility	←——————						Increasing
Yellowing	←——————						Decreasing
Compatibility with							
NC*	←——————						Increasing
CAB**	←——————						Increasing
High viscosity alkyd resin	←——————						Increasing

*NC = nitrocellulose **CAB = cellulose acetobutyrate

ers or copolymers, like "Desmodur" HL. A qualitative comparison of properties yielded by the various crosslinkers is shown in Table 5.1.

Solvent-borne 2K PU formulations: crosslinker composition and crosslinking reaction

A typical aromatic polyisocyanate crosslinker is a blend of several TDI polyisocyanates. Due to their good overall properties (e.g. compatibility, elasticity), TDI adducts e.g. "Desmodur" L (see Chapter 4.1) are in most cases the main component. Performance can be fine-tuned by adding TDI isocyanurates such as "Desmodur" IL 1451. The latter will speed up drying at the expense of pot life, elasticity and compatibility. The mixing ratios of these two polyisocyanates in the hardener solution vary between 25:75 for primers to 60:40 for topcoats and depend in any case on the polyol used.

To facilitate mixing of the two-components, whole-number mixing ratios between component A and the polyisocyanate hardeners (component B) are preferred. For this, the hardener is thinned to a solid content of 30 to 50 %. The mixing ratio of components A to B (i.e. polyol to polyisocyanate crosslinker) varies between 1:1 and 4:1, depending on the hydroxyl content and the required crosslinking degree. In general, NCO/OH around 1 is used. The application viscosity is set using suitable thinners, mainly mixtures of esters and ketones. Aromatic solvents like xylene and toluene are less preferred, but are still in use, particularly in Asia. The resultant coatings have a solid content of 25 to 38 %.

The use of TDI polyisocyanate hardeners and the addition of cellulose nitrate to the polyol (see below) lead to a certain discoloration under the influence of UV light and temperature. However, this behavior is tolerated as the wood substrate often is dark or dark-stained. In addition, furniture is usually not exposed to direct sunlight, but only to sunlight filtered through windows. Yellowing of the coating is accepted far more readily than fading of the wood. If the yellowing is unattractive, e.g. on pale woods, the pure TDI hardener combination can be replaced with aliphatic/aromatic hardeners such as "Desmodur" HL or pure aliphatic hardeners like "Desmodur" N 3390 (e.g. in kitchen cabinet coatings).
The advantageous properties of these coatings are:

- reasonable productivity,
- outstanding grain wetting and aesthetic aspect, and
- chemical and mechanical resistance.

Above all, 2K PU coatings are economical and highly versatile, which is why they are widely used to coat chairs, tables, furniture for living rooms, bedrooms and dining rooms, frames and battens.

Solvent-borne 2K PU formulations: alkyd resins and polyesters as polyols

Combinations of hydroxy-functional alkyd resins (alkyd polyols) with aromatic polyisocyanates have been a standard in the market of wood and furniture coatings. The fundamental technical developments date back to the 1960s and 1970s, with the successful introduction of these products in Italy. These systems are therefore being referred to as "Italian Style".

Short-oil, highly condensed alkyd resins with an oil content of less than 45 % are the basis for manufacturing these systems. They are normally modified with soya, peanut, olive or coconut oil fatty acids. Mixtures of these fatty acids with castor oil or its fatty acid are often used to improve pot life and resistance properties.

The solid content of the resins in their supply form vary between 50 to 75 %. Due to the relatively high molar masses, a solid content of 50 to 55 % is preferred. The hydroxyl content is between 2.5 to 4.0 %. In order to shorten the drying time and increase the solid content, small amounts of high viscosity cellulose nitrate are often added. This also improves the pore-filling properties of the coating.

Solvent-borne 2K PU formulations: acrylic resin combinations

Polyurethane systems based on hydroxyl-functional polyacrylates crosslinked with aliphatic crosslinkers are widely used. In Europe and China, the main end use is for furniture coatings, especially lightfast coatings on kitchen cabinet surfaces. In Eastern Europe and the Middle East internal and external joinery is also an important market for such combinations.

To speed up drying, either cellulose acetobutyrate (CAB) or nitrocellulose (NC) are often added to the component A. Most of the binders are pure acrylates but less expensive variants can contain styrene. However, these products have poorer compatibility with CAB or NC. The hydroxyl content of the polyacrylic resins varies between 1 to 3 %, calculated on the solid resin.

However, the high solvent content of up to 75 % restricts the future importance of these systems in the face of more environmentally friendly alternatives such as water-borne and UV-curing coatings.

In contrast to the coating of metals, high-solid solvent-borne industrial wood coating systems have not been widely used for satisfying emission requirements. The main problem is the slower curing speed of high-solid systems which can be compensated only to a small extent by a higher curing temperature.

5.1.4 Water-borne polyurethane coatings

Water-borne polyurethane systems have been growing strongly in the last two decades in wood coatings and in the furniture industry. Depending on the required performance, water-borne systems can be tailored as 1K, 2K and, in addition, UV-curing formulations.

Just like their solvent-borne counterparts, water-borne polyurethane coatings offer many options for favorably influencing the technical properties of wood coatings. It was back in 1975 that urethane-modified, medium-oil, water-reducible alkyd resins were first used in a significant way to formulate wood varnishes.

Since the mid-1970s, polyurethane dispersions have been used increasingly in the formulation of wood coatings. The reasons for this are not only due to the environmental impact of the reduced solvent requirement but also economic drivers due to no longer needed thermal waste-gas incineration in furniture coatings lines. In addition, water-borne systems are used as trade and DIY coatings for the finishing of hardwood floors. In Europe and North America, water-borne coatings have taken a major share in these segments.

Water-borne one-component polyurethane coatings

One-component polyurethane dispersions (1K PUD) have achieved considerable market importance. These dispersions, usually based on the aliphatic diisocyanates IPDI and/or H_{12}MDI, are characterized by their high level of toughness and elasticity, resistance to chemicals and weather stability. They are primarily used to modify polyacrylate dispersions, to improve mechanical properties, such as scratch and abrasion resistance. Such combinations comply with the requirements of international quality standards. They are used in the industrial coating of wooden window frames and in furniture coatings. Polyurethane/polyacrylate hybrid dispersions based on aliphatic or aromatic isocyanates are being used increasingly for parquet coating. In addition, pure aliphatic polyurethane dispersions are often used, in particular for sealing light woods such as maple or beech. Properties such as abrasion and scratch resistance can be improved further by the addition of nanoscale fillers, usually based on inorganic materials or waxes.

Coatings with a low or no co-solvent content are requested by numerous national emission standards.[2] They can easily be formulated using polyurethane or polyurethane/polyacrylate dispersions produced by the acetone process (see Chapter 3.8). Of particular benefit is the low solvent content of the ready-to-use coatings, leading to low odor and indoor emissions for example in parquet-, hardwood floor- and DIY coatings.

Water-borne two-component polyurethane coatings

Water-borne two-component polyurethane coatings were first used in the wood and furniture industry as pigmented systems for coating kitchen and office furniture. This is the natural technology of choice combining high PU performance with lower environmental impact. Hydroxyl-bearing acrylates and hydroxyl-functional polyurethane dispersions (or combinations) are crosslinked with hydrophilically-modified aliphatic polyisocyanates (see Chapter 3.6). The use of flexibilizing HDI-based hardeners makes it possible to use polyols with very high glass transition temperatures (T_g) to achieve the necessary short drying times.

The wide range of raw materials available, both hydrophilic and low-viscous polyisocyanates, as well as water-borne polyols, are at a far more advanced stage of development than the first-generation products. Polyisocyanates with improved mixing and/or greatly enhanced resistance properties are available. The use of secondary acrylic OH-functional dispersions allows the formulation of water-borne wood and furniture coatings that have the typical look and feel of solvent-borne polyurethane coatings.[3] A sample formulation is shown in Table 5.2.

For faster drying systems, primary acrylic dispersions would be the polyol of choice because of their high molecular weight. Chemical and physical drying of water-borne 2K systems can also be improved by use high T_g Polyisocyanates-based on IPDI and high functional HDI crosslinkers. In addition, over-crosslinking is often recommended.

Water-borne two-component polyurethane systems are also used for trade-applied hardwood floor coatings. A water-borne two-component polyurethane formulation ideally fulfils the requirement for a highly durable and low odor system. Compared to coatings based on non-chemically crosslinked one-component systems such as PU- and acrylic dispersions, the water-borne formulations display greatly improved resistance to chemicals and enhanced mechanical strength (see Figure 5.4).[4, 5]

Table 5.2: Formulation and technical data of a glossy water-borne two-component furniture coating

	Parts by weight
Component A	
"Bayhydrol" A 2651, (Covestro)	68.8
"Surfynol" AD01 Surfactant, (Evonik)	1.8
"Tego" Flow 425, (Evonik)	0.1
Water	11.2
"Optiflo" TVS-VF, (Byk)	0.5
Component B	
"Bayhydur" XP 2655, (Covestro)	14.2
MPA (methoxypropylacetate)	3.4
	100.00

Pine, primarily Nordic pine, is most commonly used in the manufacture of window frames. Particularly at the knot-holes left by a tree's many branches, discoloration may occur as the result of phenolic constituents leaching from the wood and adhesion may be poor. Highly crosslinked water-borne two-component polyurethane systems based on acrylic resins have proven ideal as barrier coatings in industrial coatings of window frames.

5.1.5 Radiation-curing PU-based technologies

When increased productivity is required, radiation curing technologies are the best choice. This technology is mainly limited to flat surfaces and initial inversion needed (UV/EB line), so this is a technology that complements 2K technologies, for high-end segments.

Generally, two radiation curing methods are used for these products. Of these, UV curing has prevailed[6] over electron beam curing for wood and furniture coating on account of the much lower capital investment required.

Figure 5.4: Water-borne two-component polyurethane coatings are highly resistant to staining liquids such as red wine

UV-curable systems are used for industrial flat interior furniture including wooden doors, tables, cupboards, shelves, and industrial parquet coatings as well as paper coatings and UV-curable printing inks. The recent invention of portable UV curing machines is allowing the application of UV technology also for on-site parquet. On-site UV-cured coating applications are especially interesting in commercial environments, high-end residential jobs, and any quick turn-around situations where furniture is moving in as soon as the floors are cured.

In the last 20 years, the highest growth rates have been achieved with radiation-curing unsaturated acrylates, alongside water-borne coatings. In 2017, production of wood and furniture coatings based on unsaturated acrylates was about 375 kilotons worldwide[1] or particularly high-quality applications such as the coating of highly abrasion-resistant ready-to-lay parquet flooring, coatings based on urethane acrylates are used (see Figure 5.5). These coatings are characterized by their outstanding resistance to abrasion, toughness combined with elasticity and, as a rule, resistance to yellowing.

The viscosity of supplied urethane acrylates is set using reactive thinners such as di- or tripropylene glycol diacrylate. Products free of reactive thinners are also available. The application viscosity is also set using reactive thinners. The photoinitiators needed to trigger the curing reaction and all other additives required must be tailored to the formulation and application conditions (see Chapter 4.10).

Since the mid-1990s, the industrial wood and furniture coating market has been showing considerable interest in UV-curing dispersions. This is not only due to the fact that these systems allow reduction of solvent emissions to almost zero, but also to a long list of unique advantages linked to process efficiency and sustainability.

Water-borne UV-curing coatings can be applied using all conventional methods (curtain coating, spraying, rolling) and they allow cost-effective manufacturing due to the fast crosslinking reaction. When changing an existing line from solvent-borne systems to water-borne UV, capital investment can be relatively low, as compared to e.g. a conversion towards non-aqueous UV systems which usually requires full exchange of application and drying equipment. A further benefit is found in the outstanding property profile of the coatings. They are resistant to abrasion, scratching and chemicals and comply with international standards.[7] In short, application and properties are virtually identical to those of solvent-borne two-component polyurethane coatings but are coupled with the extremely low emissions.

Figure 5.5: UV-curing urethane acrylate coatings ensure long term resistance, also under tough conditions

Recently self-crosslinking UV-curable dispersions have been developed.[8] These systems show an improved chemical resistance in combination with critical UV-absorbing pigments, shadow curing including on 3-dimensional objects and have the advantage of improving water release, which is also an important pain-point of the UV water-borne technology.

5.1.6 Outlook

In today's (year 2018) wood and furniture industries, polyurethane coatings and in particular solvent-borne polyurethane systems are widely used throughout the world. The increasingly high market share of water-borne polyurethane coatings – in Europe and China driven mainly by the strict solvent emission legislation – has resulted in the availability of an enormous variety of environmentally friendly solutions. This, in turn, has driven the development of water-borne polyurethane systems, resulting in a large number of solutions that are of interest from a holistic view on sustainability. UV-curing water-borne coatings especially are seeing strong growth worldwide, due to their high productivity and efficiency (reduced overspray). Since the early 2010's there is a clear trend towards low emission (water-borne, UV curing) PU systems in China, using the proven coating systems from Europe. However, the trend towards a new generation of coatings systems particularly designed towards the needs of Asian markets can be clearly seen. Besides new market places, innovative new applications are also rising globally, e.g. urethane acrylate-based coatings for on-site applied and UV cured on residential and commercial hardwood and PVC floors.[10]

5.2 Metal coating

The trend towards enhanced surface protection and an attractive appearance has resulted in the widespread use of polyurethane coating raw materials in industrial coating applications. This is the case for solvent-borne, water-borne and powder coating systems (for powder coating systems see Chapter 4.7). Polyurethane coatings have become particularly well established in general industrial coatings and are also used in coil and can coating.

5.2.1 General industrial coating

The term general industrial coating is used to describe the industrial OEM application of coatings on all metal substrates, excluding coils and cans (see Chapters 5.2.2 and 5.2.3), protective coatings (5.2.4), automotive OEM (5.3), refinish and transportation coatings (5.4). Typical substrates are untreated, zinc-galvanized or phosphated steel and chromated aluminum.

Economy, quality and environmental compatibility govern the requirements for these coatings. An industrial coating is considered economical if it can be applied easily and efficiently, dries rapidly, and thus quickly achieves the performance necessary for assembly and packaging. Other criteria include early achievement of water resistance, good anti-corrosion properties, high abrasion resistance and good optical properties.

Depending on the application, one- and two-component polyurethane coatings are available for drying at ambient and higher temperatures (e.g. bake applications). Reactive two-component polyurethane coatings are used primarily for high-quality industrial goods whose size and structure make the use of one-component bake coatings impossible.

Industrially coated goods are now found in virtually all segments of the consumer and capital goods industries. The largest users of industrial coatings are within the market segments that include: agricultural-, construction- and earth moving-equipment, mechanical engineering and household appliances. At present solvent-borne and powder coatings dominate these markets while water-borne systems still have – with the exception of Europe – a relatively low share (see Figures 5.6 and 5.7).

With European legislators seeking to reduce solvent emissions from coating processes by means of new legislation, such as the Solvent Emissions Directive (SED) and the Decopaint Directive (DPD), there is a significant trend toward the use of more environmentally friendly coating systems, particularly in the field of general industrial coating. Powder coatings, high-solid coatings, water-borne coatings, and radiation-curing systems are becoming increasingly important. Since quality requirements continue to increase, above-average growth of polyurethane coatings, especially water-borne polyurethane coatings, can be anticipated.

Technical aspects of polyurethane coatings

At present, – beside powder coatings – liquid polyester and acrylic/melamine, alkyd and polyurethane coatings are the standard systems in many general industrial coating applications. Polyurethanes first entered into this field by the modification of alkyd resins

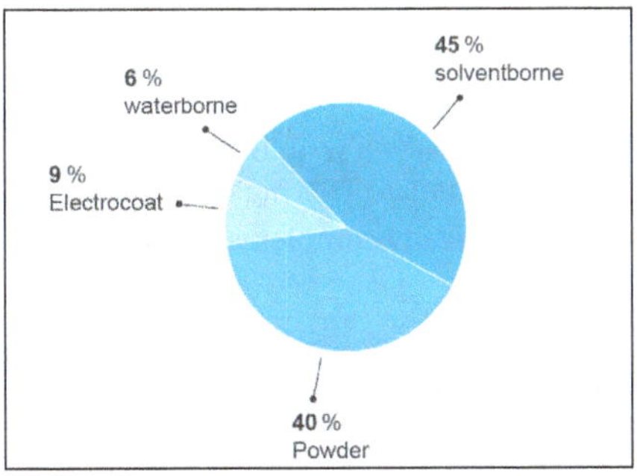

Figure 5.6: Worldwide use of coating systems in general industrial coating by technology, 2017, total: 5,470 kilotons[1]

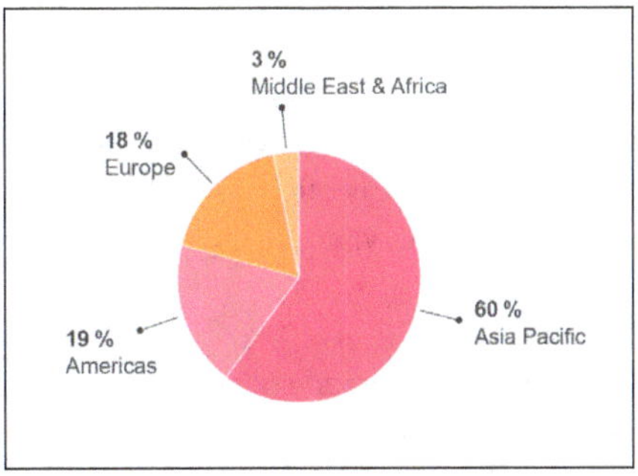

Figure 5.7: Worldwide use of coating systems in general industrial coating by region, 2017, total: 5,474 kilotons[1]

with polyisocyanates. On account of their good property profile and their application efficiency, two-component polyurethane coatings have now become firm favorites for protecting high-quality industrial goods.

Variants

Owing to the wide variability of their components (polyisocyanates and polyols), polyurethanes coatings may be used in many applications. Combinations of aromatic polyisocyanates, such as TDI adducts and trimers, with polyols are particularly suitable for use as primers and one-coat finishes. These coatings dry quickly, and the resulting paint films have good resistance to chemicals. However, the coatings crosslinked with aromatic isocyanates show a distinct tendency to yellowing, which limits their use in light-colored topcoats. Aliphatic polyisocyanates are more versatile and can be used to formulate primers, one-coat finishes and topcoats. These coatings have good resistance to weathering, chemicals and yellowing. HDI-based polyisocyanates flexibilize the paint films, whereas IPDI polyisocyanates exhibit rapid physical drying.

The properties of the paint films are governed largely by the polyol component. The resin type and hydroxyl content influence the gloss, adhesion and resistance properties of such coatings.

Primers are usually epoxy resin systems or two-component polyurethane formulations (mainly polyester-based). These provide good corrosion protection combined with good adhesion – especially at low hydroxyl contents. Polyacrylates with a hydroxyl content of 1 to 3 % are often used in the formulation of primers. If better weather stability is required, products with a hydroxyl content of about 4 % must be used, thereby yielding a higher crosslinking density. A hydroxyl content of more than 5 % produces good chemical resistance up to easy-to-clean properties.

Solvent-borne two-component polyurethane coatings normally have a solid content of 55 to 60 % on application. It is possible to formulate so called high-solid coatings with a solid content of 70 to 75 % using low viscosity aliphatic polyisocyanates combined with suitable polyols. Even higher solid content can be achieved if all or part of the binder in the formulation is a low viscosity amino-functional product. Mainly polyaspartic acid esters are particularly suitable, but also aldimines and ketimines. The high reactivity and very low product viscosities of amino-functional binders ideally complement each other.

Properties

The requirements made of coatings used for general industrial applications vary as widely as the range of possible applications. In case of two-component polyurethane coatings, it has been found that systems which are suitable for automotive refinish can also be used in many general industrial applications. They meet many of the requirements specified. Table 5.3 shows the formulation of a typical white coating for industrial applications.

Table 5.3: Basic formulation of a solvent-borne two-component white polyurethane coating for industrial applications (for alternative catalysts see Chapter 4.2)

		Parts by weight
Component A		
Polyol	e.g. "Desmophen" 650, (Covestro)	24.07
Flow promoter	e.g. "Baysilone" Paint Additive OL 17, 10 %	0.30
Defoamer	e.g. "Byk" 141, 4 %, (Byk Additives)	0.30
Light stabilizer	e.g. "Tinuvin" 292, 50 %, (Bodo Möller Chemie)	0.60
Catalyst	e.g. "Desmorapid" PP, 10%, (Covestro)	0.90
Thickener	e.g. "Bentone" 38, 10%, (Elementis Specialties)	2.08
Anti-settling agent	e.g. "Aerosil" R 972, (Evonik)	0.15
Pigment	e.g. Titanium dioxide	23.78
Solvent	e.g. Methoxy propyl acetate/solvent naphtha (1:1)	12.61
Dispersing agent	e.g. "Disperbyk" 162, (Byk Additives)	1.88
Component B		
Polyisocyanate	e.g. "Desmodur" N 75, (Covestro)	19.18
Solvent	e.g. methoxy propyl acetate/solvent naphtha (1:1)	14.15
		100.00

The main prerequisites for a uniform appearance and property profile of the paint films are homogeneous mixing of the components and a suitable application process. Depending on the specification, coatings of this type can be dried under conditions ranging from room temperature (RT) to baking at 180 °C for up to 30 min. Table 5.4 (see page 156) compares two industrial coatings, one based on polyurethane and the other on an alkyd resin. It clearly shows the characteristic advantages of polyurethane coatings in terms of hardness, resistance to solvents and weather stability.

Applications

Large objects such as construction and agricultural machinery are normally coated with air-drying systems (see Figure 5.8, see page 157). Smaller objects such as wheels and window frame profiles can be coated with a baking system in an online process, e.g. at curing conditions of 15 to 20 minutes at 130 °C.

Air-drying, solvent-borne one-component polyurethane coatings, usually applied in thick coats, provide resistance to abrasion and absorb noise. They are formulated on the basis of moisture-curing, aromatic or aliphatic polyisocyanates.

Solvent-borne one-component polyurethane baking coatings also have specific advantages. These combinations of thermally activated polyisocyanates with polyester and polyacrylate polyols are preferred to polyester/melamine coatings, above all because of their superior

Table 5.4: Comparison of a solvent-borne two-component polyurethane industrial coating and an alkyd resin industrial coating

	Two-component polyurethane coating	**One-component alkyd coating**
Solid content at spray viscosity (25 s/DIN cup 4 mm) [%]	63–65	44–46
Pot life	8 h	3–6 Months
Drying [h] - dust-dry - tack-free	 ~ 1.0 ~ 5.0–6.0	 ~ 1.5 ~ 4.5–5.0
Gloss 20°	85–90	80–85
Pendulum hardness after 7 days at RT [s]	110–120	25–30
Elasticity, Erichsen indentation [mm]	6–7	6–7
Solvent resistance after 7 days at RT*	1	5
Weather stability*	1	4

*scale from 1 to 5: 1 = very good; 4 = satisfactory; 5 = poor

mechanical and chemical resistance properties. This has led to the addition of thermally activated polyisocyanates to polyester/melamine coatings to improve their resistance properties or film strength. Such coatings are applied by spraying and are crosslinked at the temperatures needed to split off the blocking agent. Examples of applications for one-component polyurethane baking coatings include household appliances and parts for bicycles and cars.

Water-borne polyurethane coatings

In order to effectively reduce solvent emissions during the coating process, low-solvent and solvent-free binders have been under development since the beginning of the 1990s for industrial coating applications. The driving force here is the constant tightening of environmental legislation in Europe and North America. This trend has started in Asia as well. Following initial success with water-reducible alkyd and polyester coatings, which have the disadvantage of a relatively high co-solvent content, water-borne polyurethane systems, both air-drying and baking, now hold a significant market share.

Water-borne one-component polyurethane coatings

Compared with other air-drying coatings, air-drying one-component coatings based on polyurethane dispersions are characterized by their high mechanical resistance properties such as abrasion resistance. The reason lies in their chemical structure comprising urethane and urea groups. This is why they are also used as modifiers in other water-borne coating systems, e.g. in combination with acrylic dispersions.

Table 5.5: Properties of an air-drying water-borne one-component polyurethane coating in comparison with a solvent-borne one-component alkyd resin coating

	PES/PUD (dried oxidatively)	**Solvent-borne one-component alkyd resin coating**
Solid content at spray viscosity (25 s/DIN cup 4 mm) [%]	50–55	44–46
Drying at RT [h]	4–6	4–6
Gloss 20°/60°	~ 85/90	~ 80/85
Pendulum hardness after 7 days at RT [s]	70–90	30–40
Water resistance*	1	2
Weather stability*	2	4

*scale from 1–5: 1 = very good; 2 = good; 4 = satisfactory

The film properties can be improved further by intramolecular or intermolecular crosslinking. This is achieved in, for example, polyester/alkyd polyurethane dispersions by oxidative drying under the catalytic influence of metal salts. These dispersions yield coatings with particularly good resistance to hydrolysis, good pigment wetting properties and high gloss (see Table 5.5). Key applications for such coatings are in the production of automotive accessories and components for agricultural and construction machinery.
Water-borne one-component polyurethane baking systems are already widely used in industrial coatings.

Hydrophilic or hydrophobic thermally activated polyisocyanates are used as crosslinkers; polyesters, acrylates or epoxy resins are used as the hydroxy-functional polyols. The emulsifying effect of the polyol dispersions makes it possible to use hydrophobic polyisocyanates, e.g. in electrodeposition coating processes.

Water-borne one-component polyurethane systems can be applied continuously or discontinuously by spraying, dipping (including electrodeposition), flow coating or roller coating.

Figure 5.8: Polyurethane coating for extreme conditions[196]

Table 5.6: Basic composition of a water-borne two-component polyurethane (white) topcoat

Component A		**Parts by weight**
Mill base		
Polyol dispersion	e.g. "Bayhydrol" A 2809, (Covestro)	55.71
Dispersion aid	e.g. "Surfynol" 104 BC, (Evonik)	0.85
Dispersion aid	e.g. "Borchi Gen" SN 95, (Borchers)	2.73
Pigment	e.g. "Sachtleben" R-KB-4, (Sachtleben)	34.13
Flow promoter	e.g. "Borchi Gol" LA 200, (Borchers)	0.11
Let-down		
Water		5.35
Thickener	e.g. "Borchi Gel" PW 25, (Borchers)	0.17
		100.00
Component B		
Polyisocyanate	e.g. "Desmodur" N 3900, (Covestro)	80.00
Organic solvent	e.g. Butoxyl	20.00
		100.00
Mixing ratio		
Component A		100.00
Component B		17.00
Water		9.60

The necessary application rate and the substrate dimensions are the main factors in selecting the application technique.

The paint films yielded by air-drying, water-borne one-component polyurethane systems can already be handled a few hours after application. However, they do not achieve their end properties until about two weeks later. Force-drying at 60 to 80 °C accelerates this reaction. In the case of polyurethane baking systems, the crosslinking temperature and time are dependent on the blocking agent. Baking usually takes approximately 20 to 30 minutes at a temperature of 150 to 180 °C.

Water-borne two-component polyurethane coatings

The goal of development in the field of water-borne coatings was to reproduce the good properties of solvent-borne two-component polyurethane systems in water-borne systems. As described, the mixing and dispersion stability of the hydrophilically-modified polyisocyanates is of particular importance in water-borne formulations. To further improve the general resistance properties, low viscosity hydrophobic polyisocyanates can be used instead or in part. By combining hydrophobic and hydrophilic polyisocyanates, the results can be optimized with respect to the mixing of the hardener and the resultant film performance. To compensate for the side reaction between the polyisocyanate and water, it is usual to work with a surplus of polyisocyanate crosslinker, equivalent to a molar ratio (NCO/OH) of up to 1.5:1. The pot life of water-borne two-component polyurethane coating formulations is limited. Depending on the resin type, it varies between 3 to 5 h. Individual factors which govern the pot life are the functional group content, the stoichiometry and the pH of the ready-to-use coating. Table 5.6 shows a basic formulation of a water-borne two-component polyurethane topcoat.

After thorough mixing of the components, water-borne two-component polyurethane coatings are applied using standard spraying techniques.

Water-borne two-component polyurethane topcoats yield a high standard of quality, which corresponds to or even exceeds that of solvent-borne polyurethane coatings. Of particular note is the outstanding weather stability achieved using polyacrylate polyols (e.g. Florida weathering: around 90 % gloss retention after two years), as well as good optical properties, chemical resistance and hardness. In recent years, binders have been developed which are characterized by, e.g. particularly fast drying and a comparatively long pot life. Others make it possible to produce certain effects, e.g. textured coatings.

Thanks to these advantages, water-borne two-component polyurethane topcoats have a broad spectrum of use in industrial coatings. Examples of their application include construction and agricultural machinery, vehicle chassis and components, wall elements, doors and window frames, industrial machinery, containers, shelf units and steel furniture.

5.2.2 Coil coating

Coil coating is a continuous process used to apply an organic coating on steel or aluminum coils.[11] This process includes the cleaning and chemical pre-treatment of the metal surface

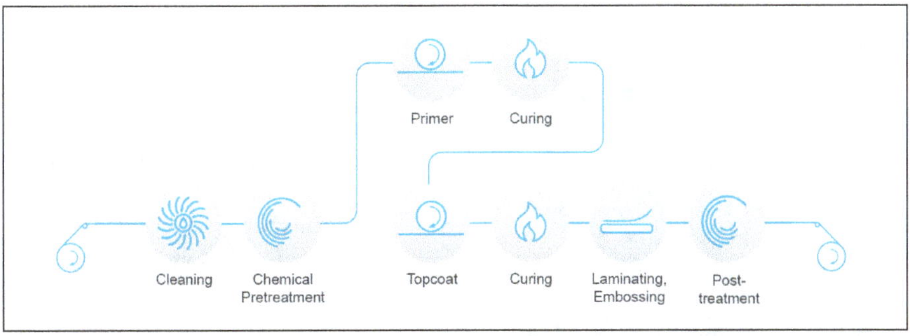

Figure 5.9: Flow chart and characteristics of the coil coating process[11]

Figure 5.10: Coated coils[197]

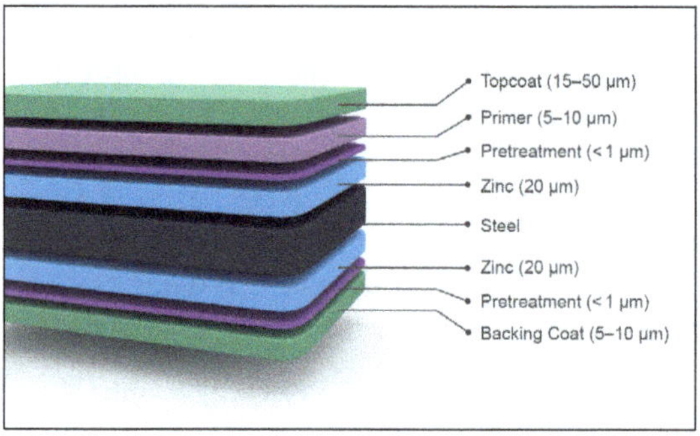

Figure 5.11: Schematic representation of coil-coated metal[11]

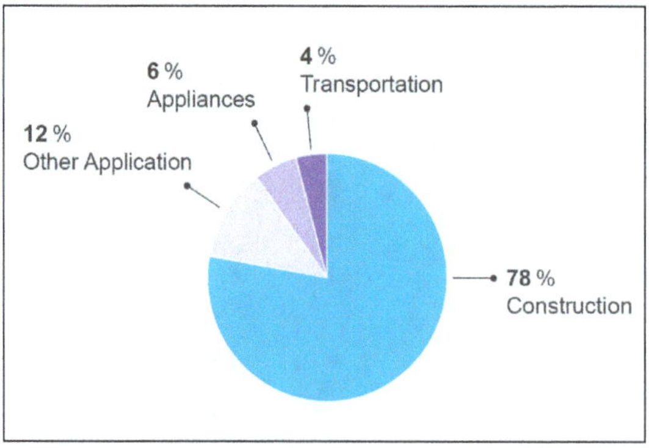

Figure 5.12: Global use of coil coated metal by market segments, 2017, total: approximately 970 kilotons[1]

followed by the single or multiple roller application of liquid coatings on either one side or both sides of the substrate. Heat is used for film formation. The process can also be used to apply powder coatings or to laminate the coils with plastic film (see Figures 5.9 and 5.10).

Because a coil coating unit is a closed system that is equipped with thermal waste gas incineration, it is both economical and environmentally friendly. Working by the motto: "Finish first, fabricate later", coil coating yields a material with a pre-coated surface (see Figure 5.11), which can be supplied to the metal-processing industry for cutting, shaping and processing as required.

Market evaluation

The introduction of this pre-coated material in manufacturing processes represented an important change. Subsequent processing was simplified because post-manufacture coating was no longer necessary. The outsourcing of production processes in many industrial sectors has given a significant boost to coil coating in recent years.

In 2017 approximately 970 kilotons[1] of coating materials were applied by the coil coating process worldwide. Figure 5.12

shows the consumption by market segments, Figure 5.13 the consumption by regions and Figure 5.14 the consumption by chemical systems.

The global annual average growth rate for coil coating was approximately 1 % over the last five years and a slightly higher growth is expected for the next years. Most of this volume is accounted for by solvent-borne coating systems. Today the majority of coil coating lines burn the solvents in an afterburning step and use the generated heat for operating the ovens. The use of water-borne systems does not allow for waste gas incineration and the resulting extraction of heat for the ovens. Additionally, the problems associated with applying powder coatings to a fast-moving metal coil have not yet been resolved. The use of radiation-curing coatings has been under investigation for some time, but as yet has not found significant industrial application.

Coating systems

Depending on requirements different chemical systems are used for primers, topcoats and backing coats. Table 5.7 shows typical properties of the most important chemical systems.

Polyester/melamine systems, which account for the largest share of the coating materials in terms of volume, are characterized by their good resistance and shaping properties. They are used for primers, topcoats and backing coats.

Polyurethane systems are characterized by an excellent balance between hardness and flexibility, outstanding resistance to chemicals and good weather stability. They are preferred for applications requiring very good shaping properties (suitability for deep drawing) and resistance properties. They are used for primers, topcoats and backing coats.

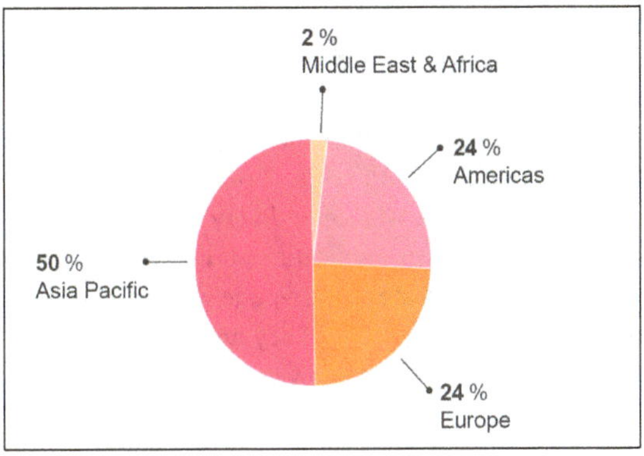

Figure 5.13: Global coil coating consumption by regions, 2017, total: approximately 970 kilotons[1]

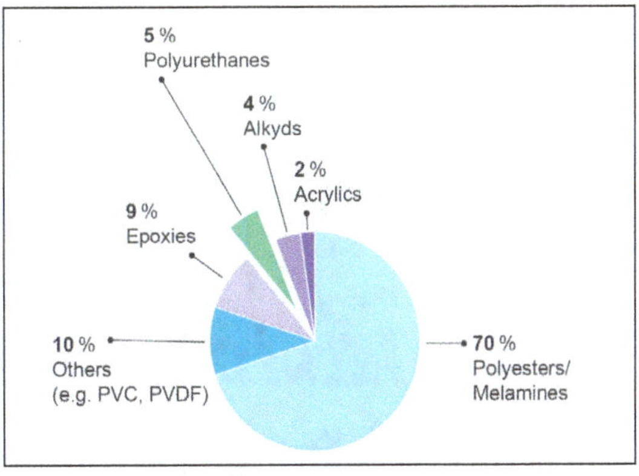

Figure 5.14: Global coil coating consumption by chemical systems, 2017, total: approximately 970 kilotons[1]

Table 5.7: Comparison of coil coating systems[12]

Chemical systems	Thickness range [µm]	Balance of hardness and flexibility	Chemical resistance	Weather stability
Polyesters/melamines	15–30	+	+	+
Polyurethanes	15–50	++	++	+
PVC plastisols	100–200	+	++	+
PVDF	20–25	+	++	++
Epoxies	5–15	–	++	–

++ = excellent, + = good, - = poor

Polyvinylchloride (PVC) plastisols are used exclusively in the formulation of topcoats. They can be applied in very high film thicknesses, yielding good corrosion resistance in aggressive atmospheres and very good shaping properties. However, these materials are sensitive to strong UV radiation and a high thermal load.

Polyvinylidene difluoride (PVDF) systems are also used exclusively for topcoats. They display high UV resistance and very low chalking. They are characterized by the longest durability in outdoor applications but attendant on higher costs.

On account of their inadequate weather stability, epoxy systems can only be used for backing coats and primers. They are characterized by good adhesion on metal and good corrosion resistance but show certain disadvantages in terms of flexibility.

Polyurethane systems

The standard polyurethane systems used in coil coating are combinations of thermally activated polyurethane crosslinkers and hydroxy-functional polymers, usually oil-free

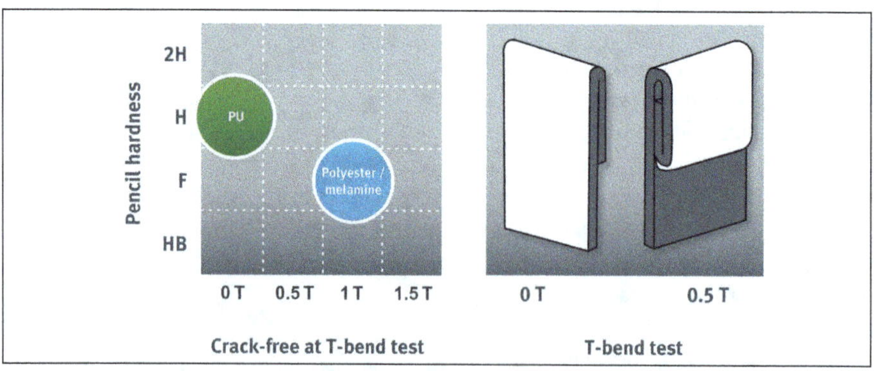

Figure 5.15: Shaping properties of polyurethane coil coatings[12]

Table 5.8: Properties of polyurethane coil coatings[12]

Properties	**Especially important for**
Corrosion resistance	All applications (primers)
Over-coatability	All applications (universal primers)
Balance between hardness and flexibility, low thermal cracking	All applications (primers, topcoats)
High-build coatings	Construction and automotive (primers, topcoats)
Chemical resistance (hydrolysis, chemicals, spotting)	Appliances and general engineering (topcoats)
Weather stability	Construction (topcoats)
Abrasion resistance	Construction (topcoats)
Printability	Automotive (topcoats)
Foam adhesion, bondability	All applications (backing coatings)

saturated polyesters. The ratio of binder to hardener is usually chosen to yield an equivalent ratio of hydroxyl-bearing and thermally activated groups.

Depending on the application, the coatings also contain fillers, pigments and additives. Polyurethane systems formulated in this way are stable at room temperature. They cure after application on the substrate and on exposure to the necessary baking temperature (peak metal temperature, PMT) by formation of a high molecular polyurethane network.

It is a special characteristic of the polyurethane systems that raw material and coating manufacturers can formulate tailor-made coating systems by specific choice of the assembly components. This module principle permits coating materials with a wide property spectrum that is unsurpassed.

The choice of thermally activated polyurethane crosslinkers affects the reactivity of the coating system and the properties of the resulting polyurethane paint film, such as a balance between hardness and flexibility and resistance properties. The reactivity of the thermally activated polyurethane crosslinkers depends mainly on the blocking agent, but can be increased by the use of a suitable catalyst, which lowers the baking temperature.

While aromatic types are mainly used for the production of backing coats and primers due to their low resistance to light, the lightfast aliphatic types can also be used for topcoats. Products based on linear aliphatic diisocyanates, such as HDI, are characterized mainly by their low viscosity and yield paint films with high resistance properties and very high elasticity. Products based on cycloaliphatic diisocyanates, such as IPDI or $H_{12}MDI$, range from highly viscous to solid and yield tough but flexible coatings with particularly good resistance to chemicals and abrasion.

Hydroxy-functional polyesters are available in a broad range. The typical hydroxyl contents are reflected in OH numbers between 5 to 100 and the molecular weight is between 3,000 to 30,000 g/mol. The ratio of aliphatic and aromatic components can also be varied within wide boundaries. The excellent balance of hardness and flexibility in polyurethane coatings is also demonstrated in Figure 5.15.

The superior flexibility of polyurethane coatings compared to polyester/melamine systems can be attributed to the chemical structure of the hardener used. Thermally activated polyurethane crosslinkers have a relatively uniform functionality of about three and yield a largely homogeneous network structure while forming hard and soft segments. In the case of melamine hardeners, a self-crosslinking reaction takes place in addition to the desired reaction with the hydroxy-functional polyester, as a result of which clusters develop which have an adverse effect on the flexibility (see Figure 5.16).

Table 5.8 shows typical properties of polyurethane coatings and the resulting fields of application. Polyester/melamine or epoxy systems are normally used for backing coats. Polyurethane systems are preferred for the production of sandwich panels for the construction and household appliance industries, because they form an outstanding bond with rigid polyurethane foams.

Polyurethane primers are used where high requirements are made in terms of flexibility, adhesion and corrosion protection and they are suitable as so-called universal primers.

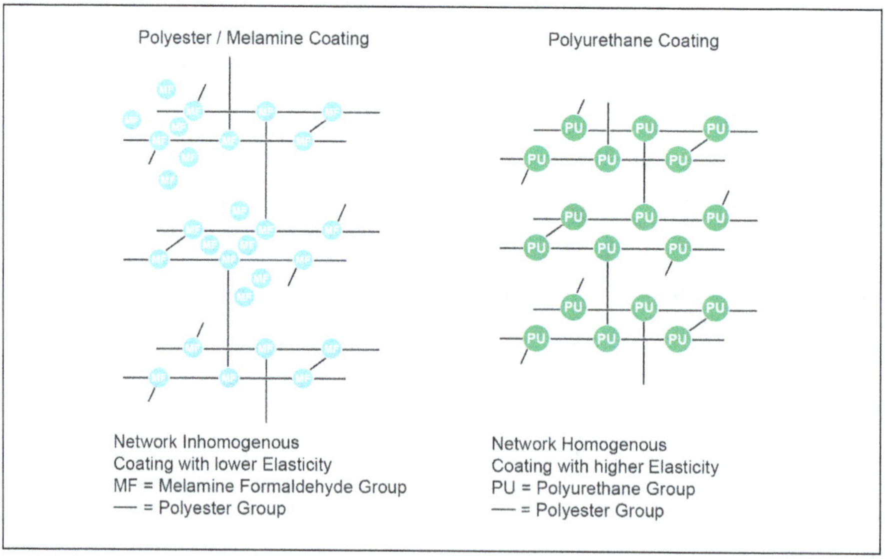

Figure 5.16: Idealized network structure of polyurethane and polyester/melamine coil coatings[12]

These coating materials can be applied to nearly all substrates (cold-rolled steel, electrolytically and hot-dip galvanized steel, galfan, galvalume, aluminum) and can be overcoated with a variety of topcoats based on different chemistries (polyurethane, polyester/melamine, PVDF systems). They are widely used in construction, transportation and the household appliances industry.

Special combinations of thermally activated polyurethane crosslinkers and hydroxy-functional epoxy resins are used in automotive engineering for the formulation of weldable anti-corrosion primers[13]. The outstanding corrosion protection properties eliminate the need for the labor-intensive application of cavity sealing systems.

At present, the main fields of application for polyurethane topcoats are in construction (cladding and roofing), but also in the production of sun protection systems, window frame profiles, garage doors (see Figure 5.17), the automotive industry, e.g. trucks, mobile homes, traffic signs, and the household appliance industry, e.g. white goods, lamps and electronic equipment.

Figure 5.17: Typical applications of polyurethane-coated coils[12]

Important criteria for the use of these coatings are their excellent flexibility, outstanding weather stability and resistance to chemicals, as well as special properties such as over-coatability and printability. High-build polyurethane coatings are being used increasingly as substitutes for PVC plastisols in the production of cladding and roofing. Combination of a polyurethane-based primer and topcoat can yield a total dry film thickness of about 80 to 100 µm. The coatings can also be modified with polyamide powder to optimize the abrasion resistance.

Polyurethane systems have gained increasing importance as primers, topcoats and backing coats in the coil coating industry. Their decorative, protective and other functional properties enable their use in demanding applications in the metal-processing industry.

Outlook

One of the key challenges will still be the reduction of energy in the coil coating process which also means cost reduction. Other challenges, e.g. digitalization or requests for more sustainability, could mean in the future, that material selection will be based more and more on fit for purpose rather than just a simple cost reduction of the coatings process.

5.2.3 Can coating

Can coating is the term used for the coating of metal packaging, cans and tubes or the sheet metal panels and coils used for their manufacture. The substrates used are cold-

Figure 5.18: Schematic representation of the coating sequence on a three-piece can

rolled packaging sheet metals such as electroplated steel (white plate), chromium plated steel (ECCS) and aluminum ranging in thickness from 0.1 to 0.5 mm.

One-component baking systems are the preferred coatings and are applied by roller. Interior coatings for two-piece beverage cans are applied by spraying.[14] These systems are baked for 10 to 12 minutes at temperatures ranging from 180 to 200 °C. Continuous high-performance plants are used for the process. Interior can coatings (golden varnishes) differ from the base coats (stamping paints) and clear coats (silver varnishes) for exterior can coatings (see Figure 5.18). The dry film thicknesses are in the range of 1 to 10 µm.

Sheet metal packaging is used in various forms for receiving, transporting and storing various fillings, preferably foodstuffs (food, luxury food, animal food) but also chemical products. Much of this packaging is produced from pre-coated sheets using cutting, soldering, welding, stamping and other shaping processes, such as deep and stretch drawing. Coil coated metal is processed to a lesser extent. Packaging produced by this resource-friendly process can be recycled after use.

Metal packaging has been divided into the following categories corresponding to its use[9]:

- beer and beverage,
- food cans,
- aerosol cans,
- general line,

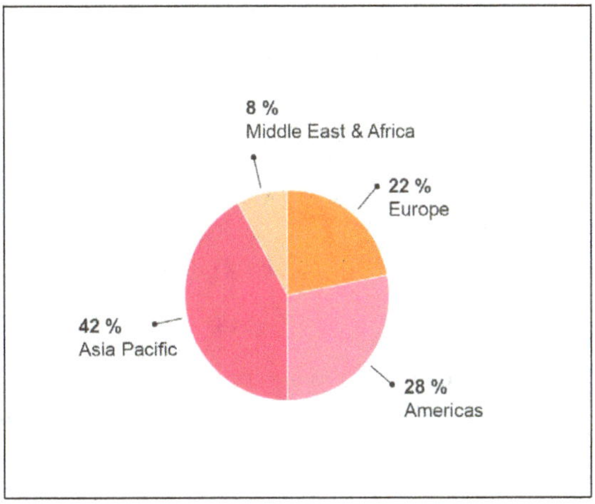

Figure 5.19: Global can coating consumption by regions, 2017, total: ~ 900 kt[1]

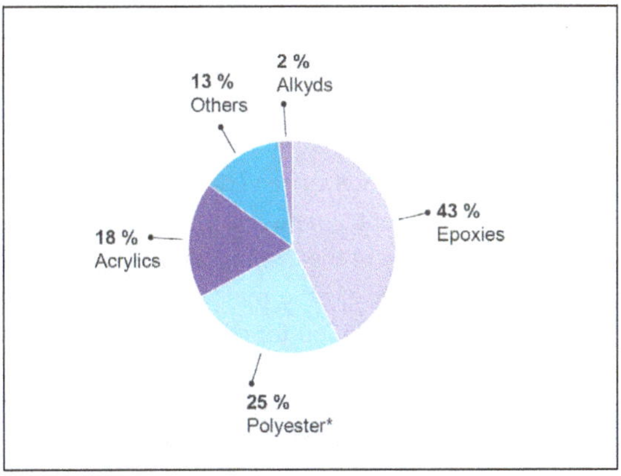

Figure 5.20: Global can coating consumption by chemical systems, 2017, total: ~ 900 kt[1]

* *in combination with amino resins and/or thermally activated polyurthane hardeners*

- crowns, caps and closures,
- can ends,
- collapsible tubes.

Market evaluation

In 2017 approximately 900 kilotons[1] of coatings material were used for can coating applications worldwide. Figure 5.19 shows the consumption by regions and Figure 5.20 the consumption by chemical systems.

The global annual average growth rate for can coating was approximately 3 % over the last five years and a comparable growth is expected for the coming years.

Corresponding to the variety of applications, a wide range of coating materials is found, the main functions of which are protective and decorative. In 2017 solvent-borne and water-borne coating systems had a share of approximately 50 %. Powder coatings and radiation-curing systems played a minor role[1].

Coating systems

Interior can coatings (golden varnishes) have the function of creating an impenetrable barrier between the metal substrate and the product contained by the can. As well as general coating properties such as adhesion, elasticity, hardness and resistance of the paint film to metal shaping processes, typical preservation properties are also required. These are resistance to acidic foodstuffs and food additives, even when subjected to heat sterilization, pasteurization or hot filling, as well as non-toxicity and tastelessness. Non-toxicity in respect of foodstuffs is governed by EU legislation (Commission Regulation (EU) 10/2011), the regulations of the U.S. Food and Drug Administration (21 CFR 175.300) and the National Health Commission (NHC) standard GB 4806.10 in China.

In the case of chemical products, a large number of special resistance properties are required. These include resistance to emulsion paints or to the propellants used in aerosol cans. Formerly most interior can coatings are based on combinations of epoxy and phenolic resins. Due to country regulations and requirements of the food industry now also alternative systems are in the market, e.g. systems based on polyester resins in combination with melamine resins and thermally activated polyurethane hardeners. This is one example of a starting formulation for a polyurethane interior can coating.

Base coats (stamping paints) are used as exterior can coatings. These are pigmented systems which, in addition to the general coating properties already mentioned, must display good hiding power, optimal levelling, good gloss, stackability, resistance to yellowing and perfect printability, even with UV-curing printing inks. Initially, modified alkyd resins and epoxy resins were used to formulate stamping paints. These were then replaced by polyester and polyacrylate resins, which are now being substituted by polyurethane systems to satisfy growing coating quality requirements. Polyurethane systems have the deep draw-

Table 5.9: Basic formulation of an interior can coating based on polyester resins in combination with melamine resins and thermally activated polyurethane hardener

Component	Parts by weight
"Desmophen" T1775, (Covestro), 75 % in solvent naphta 100	38.66
SFC 112, (SI Group), 65 % supply form	17.84
"Cymel" 1123, (Allnex), 98 % supply form	14.79
"Desmodur" BL 2078/2, (Covestro), 60 % in solvent naphta 100	4.83
"Byk" 310, (Byk Additives), 25 % supply form	0.10
1-Methoxy propyl acetate-2	23.78
	100.00

ing properties required in particular for processing two-piece beverage cans and satisfy sterilization resistance requirements.

Clear coats, which give a metallic silver effect on white plate and are therefore also called silver varnishes, provide the optical finish to an exterior coating. They are expected to protect complex printed designs from scratching and other damage and to yield a brilliant appearance with a strong promotional effect. Epoxy resins and combinations of polyester and amino resins were initially used for clear coats as well but are now being substituted increasingly by polyurethane systems. Figure 5.21 shows examples of application of polyurethane systems for base coats, clear coats and interior can coatings.

Polyurethane systems

Solvent-borne polyurethane systems for can coating applications consist of thermally activated polyurethane crosslinkers and hydroxy-functional polymers. HDI-based products are preferred for applications which require particularly high elasticity and deep drawing properties, while products based on IPDI or $H_{12}MDI$ are used for applications which require special resistance to chemicals and abrasion.

Figure 5.21: Applications of polyurethane systems for interior can coating

Figure 5.22: Offshore wind turbine towers use polyurethane topcoats above the water line[198]

Figure 5.23: Polyaspartic coatings top-coated bridge in Hamburg, Germany

Figure 5.24: Polyurethane marine coatings are typically used on steel above the water line[199]

Water-borne polyurethane systems for can coating applications consist of hydrophilically modified, thermally activated polyurethane crosslinkers and water-borne, hydroxy-functional polymers. These water-borne systems produce low emissions and yield coatings which are resistant to sterilization and equal to solvent-borne coatings in terms of their other properties. Due to VOC legislation, water-borne systems are expected to win market share from solvent-borne systems.

Outlook

Bisphenol A-containing, epoxy-based can coatings are currently under special focus of public discussions and regulatory decisions due to the potential endocrine disruptive properties of bisphenol A. In the European Union bisphenol A has been added to the candidate list of Substances of Very High Concern for Authorization in January 2018[15]. Therefore, there is an increased interest in using alternative systems for can coatings.

5.2.4 Protective and marine coatings

Corrosion of metal structures causes considerable economic damage with an estimated 3.4 % of the world's gross domestic product spent on corrosion prevention.[16] The field of anti-corrosion coatings in construction includes all applications in factory coating operations and those in the context of civil and hydraulic engineering as defined by ISO 12944.[17] Globally, protective coatings are applied to stationary objects of industrial and public infrastructure built with concrete or metal, while marine coatings covers maritime vessels and offshore structures (see Figures 5.22, 5.23 and 5.24). Protective and marine coatings can be divided into the following market segments:

- Protective coatings
- Infrastructure
  - bridge and highway;
  - potable water/wastewater facilities, storage and pipelines;
  - commercial architecture such as stadiums and transportation terminals;
  - power generation via fossil fuels, nuclear and wind.
- Oil and gas/petrochemical
  - petrochemical processing facilities;
  - oil and gas infrastructure including storage tanks and pipelines.
- Industrial processing/manufacturing
  - food and beverage manufacturing;
  - pulp and paper mills;
  - mining operations.
- Marine coatings
  - commercial, military, yachts and leisure ships;
  - offshore infrastructure.
- Pipline engeneering
  - external pipe coatings;
  - internal pipe coatings.

The 2017 world market for anti-corrosion coatings is estimated at 3,700 kilotons with a value of EUR 30 billion (see Figures 5.25, 5.26 and 5.27).[1] Coatings based on epoxy and polyurethane raw materials are the most widely applied in this market, oftentimes used in conjunction with each other in multicoat systems. While epoxy systems hold the lead in terms of volume, both systems generate approximately the same revenue.

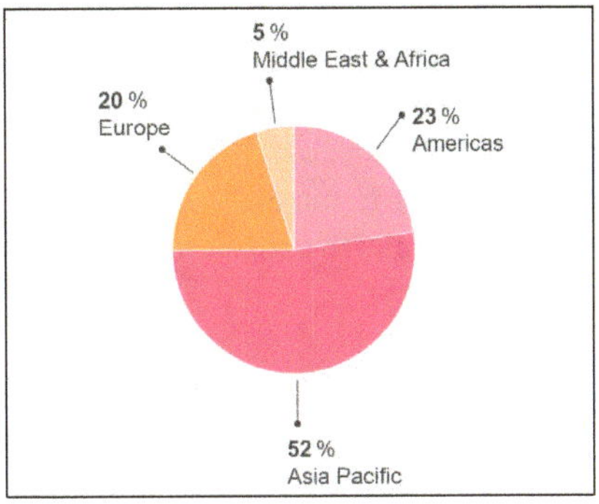

Figure 5.25: Protective and marine coating 2017: consumption by region, total 3,700 kt[1]

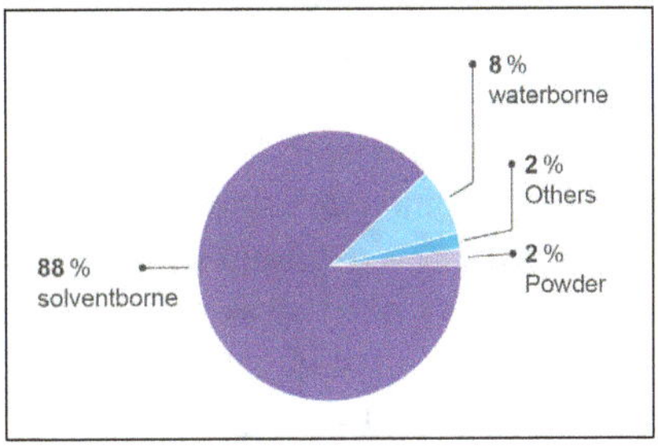

Figure 5.26: Protective and marine coating 2017: consumption by technology, total 3,700 kt[1]

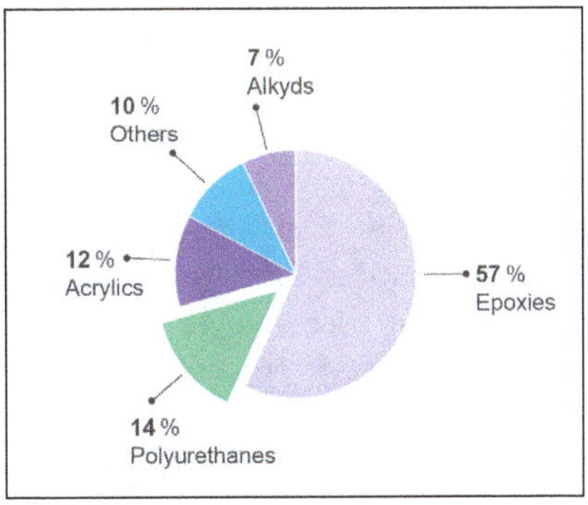

Figure 5.27: Protective and marine coating 2017: consumption by chemical systems, total: 3,700 kt[1]

Polyurethane coatings typically find application as primers and topcoats for atmospheric service conditions and coatings for large diameter pipes in buried service.[18] Coatings based on alkyd resins provide adequate corrosion resistance and weatherability for less severe service environments or on assets with shorter design lives. As global VOC restrictions continue to increase, solvent-based alkyds are largely being replaced by acrylic latex coatings either applied as direct-to-metal or over zinc-rich primers for more severe environments.

Technical aspects of polyurethane anti-corrosion coatings

Aliphatic polyurethane topcoats protect the substrate by providing chemical/abrasion resistance, barrier resistance to water, salt and oxygen, and resistance to UV degradation. Aromatic polyurethane provides enhanced corrosion, abrasion, and chemical resistance as compared to aliphatics at the expense of UV stability. Both aliphatic and aromatic systems are known to provide a balance of hardness and flexibility unattainable by most other coatings systems. These systems provide the aforementioned properties largely independent of climatic conditions, especially low temperatures.

Polyurethane topcoats are the most widely specified for atmospheric conditions due to their application robustness, affordability, and productivity. Polyaspartic acid ester coatings allow for even more productivity that is beneficial for new build and field maintenance painting when spray applied. While polyurethanes and polyaspartic coatings are limited in their maximum dry film thickness, epoxies do not typically have these limitations. However, other complications such as sweat-in time and amine blush can be problematic with epoxies.

Polyurethane coatings can be formulated as one- or two-component systems. In the case of two-component systems, the co-reactants for both aliphatic and aromatic polyisocyanate components are polyols, polyamines or polyaspartic acid esters, a special class of hindered polyamine with lower reactivity. The one-component NCO-functional prepolymers systems cure through a reaction with moisture.

Basic composition and application methods do not differ greatly from other commonly used anti-corrosion coatings. Typically, binders, pigments, extenders, co-resins, additives and solvents are present – unless they are solvent-free formulations. Conventional techniques such as brushing, rolling and airless single component or plural component spray units are used during application. Both one- and two-component polyurethane systems can be used as primers, intermediate and topcoats. Topcoats used in the protective and marine market are mainly lightfast aliphatic two-component polyurethane systems, whereas primer applications are one-component, aromatic zinc-rich systems, and both are applied using conventional single-component airless spray equipment. Due to their high reactivity, solvent-free aromatic two-component polyurethane systems, which are used as external coatings for pipelines, must be applied by plural component spray equipment.

Table 5.10: Typical formulation and technical data of a one-component, zinc-rich polyurethane primer

Raw material	**Generic name/trade name**	**Parts by weight**
Binder	e.g. "Desmodur" E 23, (Covestro)	11.0
Thixotropy agent	e.g. "Bentone" 34, 10% suspension, (Elementis Specialties)	3.9
Active pigment	e.g. Zinc dust VM 4 P 16, (Even Zinc SA/NV)	39.1
Barrier pigment	e.g. Micaceous iron oxide "Miox" AS, (Kärntner Montanindustrie)	39.1
Solvents	e.g. "Solvesso" 100, (various supplier)	5.4
Drying agent	e.g. "Additive TI", (OMG Borchers)	1.1
Additive to enhance storage stability	e.g. "Additive OF", (OMG Borchers)	0.5
		100.0
Technical data		
Pigment volume concentration [%]		59
Solid content by volume [%]		71
Solid content by weight [%]		91
Volatile organic compounds [g/l]		252
Theoretical cover rate at 80 µm [m²/kg]		3.1
Drying times (23 °C/50 % rel. humidity), DIN 53 150		
T_1 [min]		40
T_6 [h]		7

Corrosion protection via organic coatings is designed to provide protection via a multi-layer approach with each coating providing a specific function. The requirement profile is so varied that a single coat cannot yield all the properties required. As per ISO 12944, three-coat concepts with a total film thickness of 200 to 300 µm represent the standard for steel for severe atmospheric service conditions. Zinc or aluminum-rich primers provide sacrificial galvanic protection to the metal substrate, a thick barrier intermediate coating prevents the diffusion of water, salts, and oxygen to the substrate, and a lightfast topcoat yields protection against chemicals, abrasion, and UV light. Economic considerations are driving development toward two-coat concepts that provide the same level of corrosion protection, and ISO 12944 has adopted two-coat systems that provide equivalent protection as a three-coat system. The two-coat system consists of a galvanic zinc rich primer and

Table 5.11: Typical formulation and technical data of a two-component high-solid polyurethane topcoat

Raw material	Generic name/trade name	Parts by weight
Binder	e.g. "Setalux" D A 160 SN, (Nuplex Industries)	42.9
Thixotropy agent	e.g. "Aerosil" R 972, (Evonik)	0.5
Thixotropy agent	e.g. "Bentone" 38, 10 % suspension, (Elementis Specialties)	4.4
Dispersing additive	e.g. "Disperbyk" 161, (Byk-Chemie)	1.4
Deaeration agent	e.g. "Byk" 141, (Byk-Chemie)	0.4
Light stabilizer	e.g. "Tinuvin" 292, (Kremer Pigmente)	0.4
Pigment	e.g. "Bayferrox" 415, (Lanxess)	1.6
Pigment	e.g. Chrome Oxide Green GN-M, (Lanxess)	8.8
Pigment	e.g. Sachtleben R-KB-4, (Sachtleben Chemie)	13.3
Extender	e.g. Barytes EWO, (Sachtleben Chemie)	10.5
Solvents	e.g. "Solvesso" 100, (various suppliers)	4.9
Hardener	e.g. "Desmodur" N 75 MPA/X, (Covestro)	10.8
		100.00
Technical data		
Pigment volume concentration [%]		20.3
Solids content by volume [%]		54.4
Solid content by weight [%]		69.9
Volatile organic compounds [g/l]		403
Drying times (23 °C/50 % rel. humidity), DIN 53 150		
T_1 [min]		70

lightfast topcoat, with a total film thickness at least equal to that yielded by the three-coat concept, with the intermediate coat being replaced by a thick topcoat.

Primers

The primer's main functions are corrosion protection and to ensure the bond between the metal substrate and subsequent coats. Epoxy systems are often preferred because of their lower cost. However, polyurethane systems cure faster and more reliably, especially at low temperatures. Most polyurethane primers are moisture-curing, one-component systems based on aromatic TDI and MDI prepolymers. They are applied to blasted surfaces or on surfaces which have had minimal preparation, e.g. surfaces which have been derusted by hand or using power tools. Fast drying yields logistical benefits, especially when applied on new structures. In rehabilitation projects, a key advantage is their application at low temperatures and high humidity.

Depending on the corrosive conditions, primers can be formulated with or without galvanic pigments (zinc dust or anti-corrosive pigment) and/or barrier-forming lamellar pigments (aluminum flakes or micaceous iron oxide). The barrier pigments slow the migration of corrosion accelerators through the coating, delaying the corrosion process. In two-component polyurethane coatings, zinc phosphate provides better corrosion protection than zinc dust or zinc oxide. However, because of its high water content, zinc phosphate cannot be used in one-component polyurethane coatings and zinc dust is preferred.[19]
To ensure adequate storage stability of one-component polyurethane coatings, the moisture released from the pigment(s) must be removed during production by the addition of a chemical dryer such as p-toluenesulfonyl isocyanate. It is generally recommended that the pigments and extenders used have a water content of less than 0.2 %.

Intermediate coats

The intermediate coat has the function of protecting the primer, increasing the film thickness to enhance barrier properties and ensuring optimal adhesion between the primer and the topcoat. Formulations based on epoxy resins are primarily used in this application with the exception of three-coat moisture cure systems. These all polyurethane systems are valuable where the painting season needs to be extended due to low temperature or high humidity application conditions.

Topcoats

Two-component polyurethane-based topcoats are proven in this application over more than 50 years based on polyisocyanate (e.g. "Desmodur" N) in combination with acrylic and/or polyester polyols. Their particular advantages are application robustness, fast drying, long overcoat window, and resistance to weathering, water and chemicals. This type of system is the most commonly used topcoat for steel in both protective and marine markets.

Acrylic and polyester polyols bring specific attributes for polyurethane topcoats. Typically, polyol blends offer a compromise of properties acceptable to multiple markets. Acrylic polyols typically have lower viscosity and better weatherability while polyester polyols bring chemical and abrasion resistance at the expense of increased viscosity. With regulations moving to higher solids systems and coating lines becoming more global, low VOC acrylic polyols combined with low viscosity polyisocyanates are becoming more prevalent in the market.[19]

Fluorourethanes are a newer class of coatings that have found acceptance in water tank exteriors because of their unparalleled gloss/color retention and application robustness.

Fluorinated polyols (e.g. "Lumiflon") based on copolymers of vinyl ethylene and fluorinated monomers are formulated and applied similarly to acrylic or polyester polyols.[18]

Polyaspartic coatings

Polyaspartic coatings represent a relatively newer class of coatings, and are also referred to as polycarbamides, polyaspartic or polyaspartic urethane coatings. Polyaspartic acid ester resins are sterically hindered secondary-aliphatic amines (e.g., "Desmophen" NH) which exhibit moderate reactivity with aliphatic polyisocyanates (see Chapter 3.7).

Compared to solvent-borne two-component polyurethane systems, polyaspartic acid ester coatings are characterized by the following advantages:

- high-solid content with VOC contents lower than 250 g/l,
- fast drying (2 hour hard dry), and
- high blister-free dry film thickness in a single pass (up to 250 μm).

Polyaspartic coatings are applied using the same methods (spray, brush, or roll) as traditional epoxy or polyurethane coatings. Even with their fast cure they have a reasonable pot life, only slightly less than a polyurethane topcoat.

Polyaspartic coatings display several disadvantages whose effects can be minimized through proper formulation. At high film builds (> 300 to 400 μm) an internal cyclization reaction can impart brittleness if the formulation does not exhibit initial flexibility. Atmospheric moisture catalyzes both film cure and can reduce pot life; dry times are more uniform using a prepolymer polyisocyanate bringing both physical as well as chemical drying

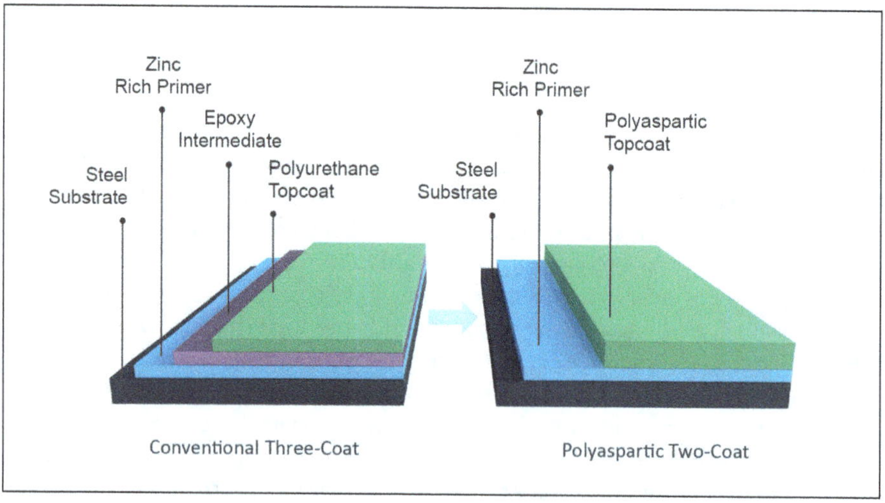

Figure 5.28: Highly efficient two coat layer for high corrosion resistance requirements

Table 5.12: Typical formulation and technical data of a polyaspartic coating for protective applications

Raw material	Generic name/trade name	Parts by weight
Binder	e.g. "Desmophen" NH 1420, (Covestro)	28.0
Thixotropy agent	e.g. "CAB-O-SIL" TS 720, (Cabot Corp.)	1.1
Yellow Pigment	e.g. "Bayferrox" 415, (Lanxess)	1.1
White Pigment	e.g. Sachtleben R-KB-4, (Sachtleben)	8.8
Extender Pigment	e.g. Barytes EWO, (Sachtleben)	16.5
Green Pigment	e.g. Chrome Oxide Green GNM, (Lanxess)	6.1
Barrier pigment	e.g. Microtalc IT extra, (Mondo Minerals)	4.7
Solvent	e.g. Methoxy propyl acetate, (Various)	8.4
Drying agent	e.g. UOP L-powder, (Kurt Obermeier)	2.4
Flow agent	e.g. "Byk" A 530, (Byk Additives)	1.0
Dispersant	e.g. "Disperbyk" 163, (Byk Additives)	1.0
Light stabilizer	e.g. "Tinuvin" 292, (BASF)	0.5
Hardener	e.g. "Desmodur" N 3600, (Covestro)	20.4
		100.00
Technical data		
Pigment volume concentration [%]		20
Solid content by volume [%]		84
Solid content by weight [%]		90
Volatile organic compounds [g/l]		149
Pot life [h]		1.5
Drying times (23 °C/50 % rel. humidity), DIN 53 150		
T_1 [min]		30
T_6 [h]		1

characteristics. Lastly, re-coatability issues are addressed using a lower functionality polyisocyanate prepolymer.

The ability to apply thicker film builds allows for the reduction in coating layers without sacrificing corrosion resistance (see Figure 5.28). For environments designated C4 and C5 per ISO 12944, traditionally a three-layer coating system consisting of a zinc-rich primer with an epoxy intermediate and polyurethane finish coat are required for adequate corrosion protection in severe environments. Polyaspartic coatings allow for the reduction from three layers to two layers in these corrosive environments. The epoxy and polyurethane layers are replaced with a 150 to 200 µm polyaspartic acid ester finish coat. This concept has been validated through both accelerated and real-world performance.[20, 21]

For less corrosive environments (C2–C3 High per ISO 12944), where a two-layer epoxy primer with polyurethane finish coat are typically deployed, a single-layer polyaspartic coating applied directly to blasted steel can be substituted without sacrificing long term corrosion protection (see Figure 5.29). This helps to improve painting efficiency and lower overall applied cost.

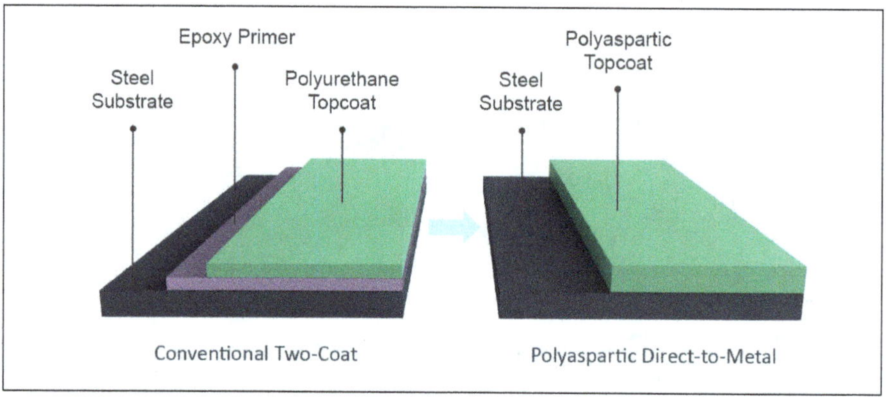

Figure 5.29: Reduction in coating layers in moderately corrosive environments using a direct-to-metal polyaspartic coating

Figure 5.30: Stadium in Arlington, Texas, steel was painted in the shop using two-coat polyaspartic coating system

Table 5.13: Minimum requirements for polyurethane coatings as per DIN EN 10290

Properties	**Minimum specification (summary)**
Film thickness	Class A: 800 µm; Class B: 1,500 µm
Shore D hardness	According to project specification
Pore test	12 kV for 1,500 µm film thickness; max. 20 kV
Impact resistance	≥ 5 J x mm dft x k (at 23 °C); ≥ 3 J x mm dft x k (at -5 °C)
Adhesion	Knife test, rate ≤ 3 (at 23 °C); rate ≤ 4 (at 80 °C)
Flexibility	Bending test at 23 °C and 0 °C
Elongation	≥ 10 %
Indentation resistance	≤ 0.2 mm (at 23 °C, 1,500 µm); ≤ 0.45 mm (80 °C, 1,500 µm)
Thermal aging	100 d value for 60 °C, 80 °C and 100 °C
Cathodic disbonding	Disbonding: ≤ 8 mm (23 °C, 28 d); ≤ 8 mm (60 °C, 2d)
Specific electrical insulation resistance	107 Ωm^2 (23 °C, 100 d); 104 Ωm^2 (at max. operating temperature, 30 d

Polyaspartic coatings are used in both new construction and field maintenance applications (see Figure 5.30). For shop-applied coating applications, such as the fabrication of steel bridges, stadiums or wind turbine towers, polyaspartic coatings add significant value in speeding up the painting process. The painting operation is a significant bottleneck in the fabrication process, and polyaspartic coatings have been documented to reduce painting costs by up to 14 % while increasing throughput by 60 % for steel fabrication shops.[22, 23]

For maintenance painting applications, polyaspartic coatings reduce the time an asset is out of service such as a bridge, storage tank, or railcar. The asset can be put back into service faster or in the case of bridge coatings, costly traffic restrictions are reduced. Two-layer polyaspartic coating systems have been documented to reduce maintenance painting costs of steel bridges by 20 %, while increasing painting efficiency by 30 %.[24]

Pipe coatings

A specialty in the segment of protective coatings is the exterior coating of pipelines installed in the ground or under water. The coating systems used for pipes can roughly be classified according to the application, depending on whether single pipe sections are coated in the factory or complete pipelines (or only weld joints) are coated on-site.

Factory coating

In the case of factory coatings (OEM), three-layer systems comprising epoxy powder primer, epoxy adhesive and polyolefin finish (3LPE, 3LPP) are often used in exterior pipe coating. In the US pure epoxy powder systems ("fusion bonded epoxy", FBE) are used.

Also, two-component polyurethane coating systems are being used increasingly because of some advantages they offer: The use of fast curing, solvent-free two-component polyurethane coatings eliminates the need for costly heating of the pipe sections to about 220 °C, which is necessary for baking the epoxy powder systems (FBE, 3LPE, 3LPP), followed by subsequent cooling the coated pipe down to about 80 °C, in order the pipes can be stacked quickly.

A flexible coating system makes it possible to cold bend the pre-coated pipe sections in the field without cracks forming.

On-site coating

For pipeline rehabilitation usually one-coat systems are used. This kind of coating is applied directly to the metal substrate without a primer and has the necessary resistance properties thanks to the high film thicknesses of 800 to 1,500 µm that are required. This ensures long-term corrosion protection of the pipelines after rehabilitation (see Figure 5.31 and 5.32).

Figure 5.31: Applying an OEM polyurethane coating on a pipeline

Figure 5.32: Rehabilitating a gas pipeline in the Ukraine

Mostly solvent-free two-component polyurethane or liquid epoxy coatings have proven to be ideal for on-site application. Two-component polyurethane systems offer the advantages of good mechanical properties coupled with rapid application due to their high reactivity (even at low ambient temperatures). This translates into significant cost benefits. Even on vertical surfaces, sag-free application in a single pass can be achieved with two-component polyurethane coating systems. The good sag resistance is yielded either by conventional rheological additives or by chemical thickening during application. Table 5.13 gives an overview of the most important requirements for underground pipelines as per EN 10290.

In the case of underground pipelines, the mechanical properties of the coating system are of greatest importance. For example, the high impact resistance of the coating prevents damage during transportation or subsequent installation. Good long-term flexibility, even at low temperatures, and good adhesion on the metal surface are essential for lasting protection of underground pipelines. These properties can be provided by appropriately formulated aromatic two-component polyurethane systems. As the protective current for pipelines serves as active corrosion protection, the pipe coatings have to pass the cathodic disbondment test as per ASTM G8 or ASTM G42.[25]

Given the many raw materials available for formulating solvent-free two-component polyurethane systems, mechanical properties, viscosity and reaction time can be optimized specifically to satisfy individual requirement profiles. Standard binder products for the development of reactive two-component polyurethane systems are polyols based on polyether, polymeric MDI isocyanates or polyurethane pre-polymers.[26]

Figure 5.33: New pipelines for sewage water in Al Dhakira, Qatar, interior coated with two-component aromatic polyurethane systems[27]

It should be mentioned that the adhesion properties of polyurethane pipe coating systems, especially if cathodic protection is applied, depends greatly on the pre-treatment of the substrate. In both OEM and repair coating work, the substrate has to be shot-blasted (SA 3.0 for OEM and SA 2.5 for on-site coating, as per ISO 8501-1). Particularly in the case of on-site coating, it is advisable to monitor ambient conditions.

Figure 5.34: Applying a polyurethane coating on the interior of a pipe

The film thicknesses of interior pipe coatings are very different to those of the exterior coatings and may vary greatly depending on specification and application. For the interior coating of gas pipelines for example, film thicknesses between 80 to 200 µm are required for OEM applications. Solvent-free two-component polyurethane systems (MDI/polyether-based) and solvent-based epoxy coatings are being used for this kind of application.

The depth of roughness Rz is in the region of 50 to 70 µm as per the given specifications (ISO 4287-1). For interior coatings of sewage water pipelines two-component aromatic polyurethanes are applied as Direct-to-Metal coating in a dry film thickness of approximately 1000 µm (see Figure 5.33).

In the case of drinking water pipelines, film thicknesses of the OEM interior coatings range between 300 to 500 µm, while repair coatings are in the region of 1,000 to 3,000 µm (see Figure 5.34). In this kind of application, solvent-free systems based on aliphatic but also aromatic polyisocyanates, polyols and amines have proven ideal due to the fast application and their outstanding mechanical properties.

Further information on materials in contact with drinking water can be found in Chapter 10.2.4.

Outlook

Macro trends that will shape future innovation and development for protective and marine coatings include increased labor cost, VOC reduction, and globalization of standards and product lines. As wages increase globally, painting efficiency will drive the use of coating technologies that require fewer layers such as polyaspartic coatings. Increasing VOC restrictions are already having a significant global impact, and this will drive the accelerated adoption of water-based and high-solid technologies. Water-based polyurethanes and high-solid polyaspartic coatings are well suited to meet the low VOC challenge. Globally recognized standards developed by organizations such as International Organization of Standardization (ISO), International Maritime Organization (IMO), The Society of Protective Coatings (SSPC), and National Association of Corro-

Figure 5.35: Armature winding of coated copper wire to produce an AC electric motor

sion Engineers (NACE) will continue to govern coating selection and global coatings companies will reformulate their product lines to meet these standards to produce one product for all regions.

Success in the pipe coating business often depends on delivering value that is a balance between cost, speed of application and performance. For example, the fast curing speed of solvent free 2K PU coatings at ambient temperature allows for shorter wait times to conduct inspections and back-filling of dirt/gravel after the application for field applied coating applications to pipes and joints. Additionally, in-shop OEM applications of polyurethanes to pipes can be conducted at ambient temperatures eliminating the cost of high temperature baking ovens.

5.2.5 Wire coating

The winding wires used for electric coils that produce magnetic fields, have in contrast to electric cables, only thin insulation layers in the range of 5 to 100 µm. The wires are wound tightly to yield effective current densities and coil performance in electrical components (see Figure 5.35).[28] The insulation of these winding wires is produced by coating up to 30 layers and layer thicknesses in the range of 1 to 2 µm. The coating is applied by wetting the wires in an enamel bath and then removing excess coating through a die or, in the case of fine wires, through a felt. The wire is dried in an oven after application of each coat. The oven temperature is at 500 °C or above. The contact times are in the region of seconds on account of the production speed and vary greatly with the diameter of the wire.

The wire is nearly always made from copper or, more rarely, aluminum. Diameters typically vary from 0.1 to 1 mm in circular wires. However, there are also very fine wires with diameters starting at approximately 0.018 mm and heavy rectangular wires with edge lengths in the centimeter range.

The coated wires have to meet many requirements:

- flexibility,
- coating adhesion,
- abrasion resistance,
- thermal resistance (softening range),
- temperature index,
- heat shock resistance,
- resistance to chemicals,
- resistance to oxidation,
- dielectric strength (breakdown voltage),
- threshold temperature of the operating range (tan delta), and
- direct solderability.

Coils which are used in generators, electric motors and transformers – all apparatus which operate at high energy densities or high ambient temperatures – are subjected to correspondingly high continuous temperature loads. Polymers with high thermal resistance are used as the insulation material because they can meet the requirement profile. Polyesters are most widely used, followed by polyester imides and amide imides. Imides and polyhydantoins, often blended to increase thermal endurance, were used for high temperature applications.[29]

For their application as coatings, however, it is essential that these polymers are completely soluble in organic solvents. N-methylpyrrolidone (NMP), cresols and phenols are often the only solvents suitable. During the coating process, the solvents which evaporate during drying are catalytically burned, thus contributing a considerable amount of energy for operating the drying oven.

If wires are to be joined by soldering or in some other way, the insulating coating must first be burned off or removed mechanically. This usually involves increased effort. It is generally perceived as a disadvantage that polyurethanes have lower heat resistance than polyester imides, but this has proven to be a significant advantage for certain wire coatings.

At the temperatures used in direct soldering (starting at around 320 °C), urethane groups are cleaved instantly, leading to evaporation of the entire coating film within a few seconds, depending on the wire diameter. Polyurethane coated wires can therefore be directly soldered without pretreatment, an undeniable advantage in the automated manufacture of electrical components. For this reason, these wires are used primarily in electronics applications in which very high component temperatures are not expected during operation. Standard polyurethane coated wires are used in applications with continuous temperature loads of 130 °C (heat class B) and 155 °C (heat class F).

The polyurethane coatings are processed as one-component systems and usually consist of a branched thermally activated polyurethane crosslinkers and a branched hydroxyl-bearing polyester. The hardener is normally an adduct of trimethylolpropane and toluene diisocyanate (TDI). Phenol or cresol is used as the blocking agent. Whereas the hardener was supplied as a solid which then had to be dissolved, it is now usually synthesized directly in solution by wire coating manufacturers. If higher thermal resistance is required, thermally activated TDI-trimers are also used.[30]

As well as TDI-based products, thermally activated MDI isocyanate crosslinkers are used to formulate wire coatings. These display better resistance to moisture, an important requirement in Asia especially, and tend to result in better thermal resistance. The downside, however, is reduced solderability. World binder consumption for the insulation of winding wires totals about 50 kilotons, with polyurethanes estimated to account for about 20 %.

Latest developments in this field encompass aromatic and aliphatic polyisocyanurate, e.g. trim trim coatings, which show very high thermal and chemical resistance in combination with easy preparation and fast crosslinking (see Chapter 4.6).

5.3 Automotive OEM finishing

The economic significance of raw materials for automotive coatings equates with the high economic importance of the automotive industry. In 2017, about 1,400 kilotons of coatings were used in automotive OEM finishing (excluding wax conservation, flange adhesives, seam sealing and underbody protection) (see Figures 5.36 and 5.37). Today approximately 55 % of paints used in automotive OEM finishing applications are formulated, at least partially, with polyurethane raw materials with growing tendency.[31]

Automotive OEM coating applications are one of the most technically sophisticated and most highly automated of all mass production coating applications. Through the use of paint robots with higher transfer efficiency and lean paint processes that use less material, the paint used per body has been reduced by 30 to 50 % over the last 10 years. This savings have been realized without any compromise on quality. Auto OEM manufacturers further reduced cost by optimizing air conditioning and waste air recycling which resulted in up to 50 % less energy usage per body than 10 years ago.[32]

Also, the transfer efficiency with modern application equipment is constantly being improved, helping to reduce overspray and thus the paint consumption and waste. Automotive coatings are baking systems that cure by chemical crosslinking within 20 to 30

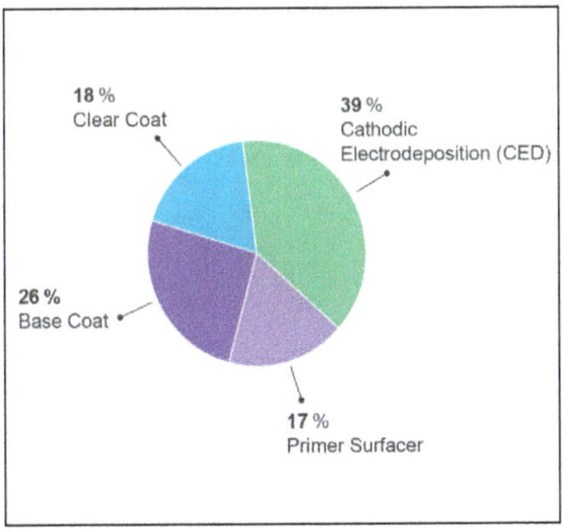

Figure 5.36: Global consumption of automotive OEM paints by coating layers (2017), total approximately 1400 kt[1]

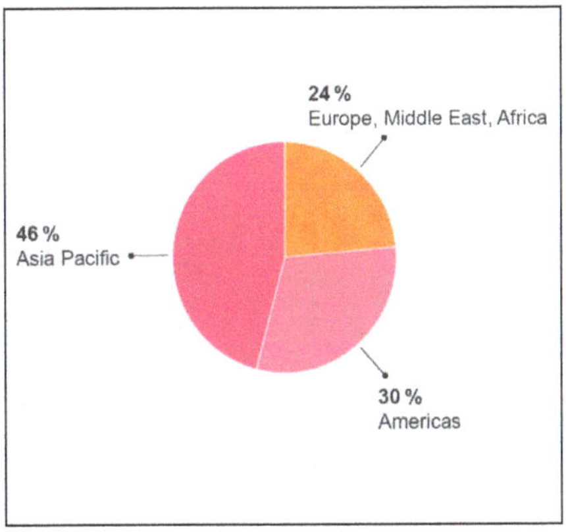

Figure 5.37: Global consumption of automotive OEM paints by regions (2017), total approximately 1400 kt[1]

minutes at temperatures ranging from 120 to 180 °C. With a few exceptions, all major coating systems available today are used in this segment in various forms: solvent-borne high-solid, water-borne and solvent-free liquid paints. Powder coating has not achieve a technology position in the market.

5.3.1 Automotive OEM finishing process

The modern automotive OEM finishing process is relatively uniform and can be divided into five phases (see Figure 5.38):

- cleaning and pretreatment of the uncoated car body;
- application of a cathodic electrodeposition coating (CED);
- application of a primer surfacer;
- application of a base coat;
- application of a clear coat;

The application of each layer (CED, primer surfacer, base coat and clear coat) is followed by a flash off or baking cycle. Cavity and exterior sealing is performed separately, usually after final assembly of the vehicle.

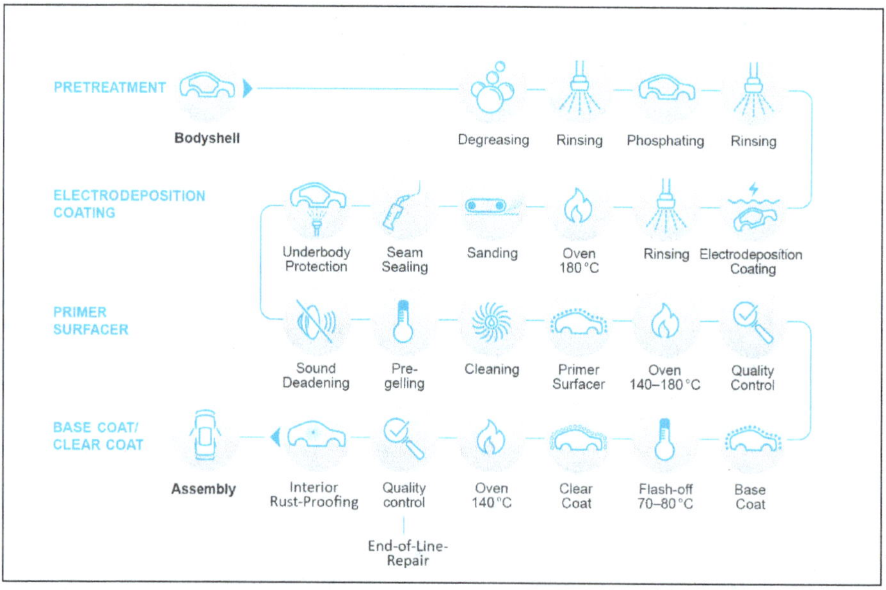

Figure 5.38: Standard coating process in automotive production

Traditionally, the coating process also includes the application of an underbody protection system and seam sealing compounds. Depending on the design of the coating line, underbody sealing and seam sealing systems are applied mostly after baking of the CED. The coats applied in each phase of the process fulfill specific functions in the overall finish (see Figure 5.39). The passivating pretreatment (phosphating) of the steel or aluminum and the CED are largely responsible for protecting the car body against corrosion. The primer surfacer compensates any remaining unevenness in the substrate and has the additional role of providing stone chip protection. The primer surfacer also protects the usually non-lightfast CED against UV degradation, which could cause delamination of the coatings over it if the primer surfacer did not have sufficient opacity.

The base coat brings color and often special effects – metallic effect, pearl effect – to the car body. As the final layer the clear coat is responsible for the overall appearance. It is also important that the clear coat be resistant to degradation from exposure to light and weathering, to various household, industrial and natural chemicals and to mechanical damages (scratching, stone chipping, car washes). All this means that only high-quality, lightfast and weather-stable binders may be considered for formulating clear coats.

5.3.2 Cathodic electrodeposition coating

The first true coating applied after cleaning and passivating (zinc phosphating or zirconia for multi metal car bodies)[47] of the car body is the water-borne dip primer applied electrophoretically. Today, this is almost always a cathodic electrodeposition (CED) system, which is applied by using the metal car body as the cathode. The anode is contained in electrodialytic cells that remove the generated acid via the anolyte loop. (see Figures 5.40 and 5.41).

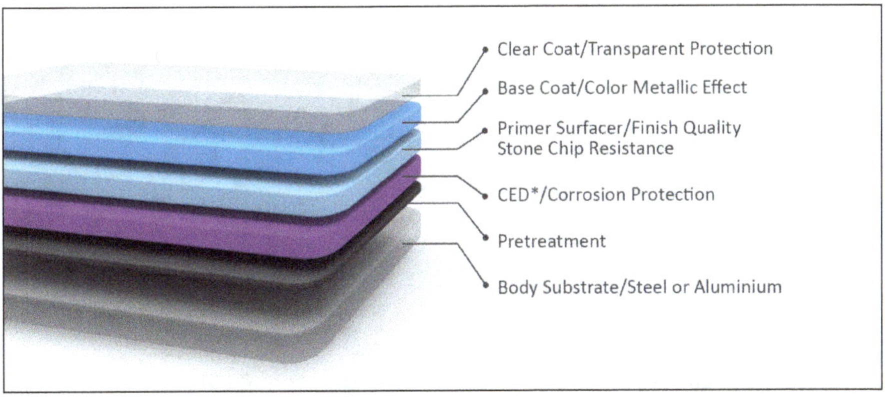

*Figure 5.39: Diagram of an automotive OEM finish, *CED: cathodic electrodeposition coating*

During electrolysis of water, the cathode will develop an alkaline layer of hydroxyl (OH^-) ions. Dispersed paint particles carrying a positive charge will be attracted by the negatively charged cathode. Once in contact with the alkaline layer, the positively charged binder polymers will coagulate as a deposit on the car body. With increasing layer thickness, electric resistance will increase until coagulation stops. Remaining water within the coagulated film is used up by further electrically driven osmosis.

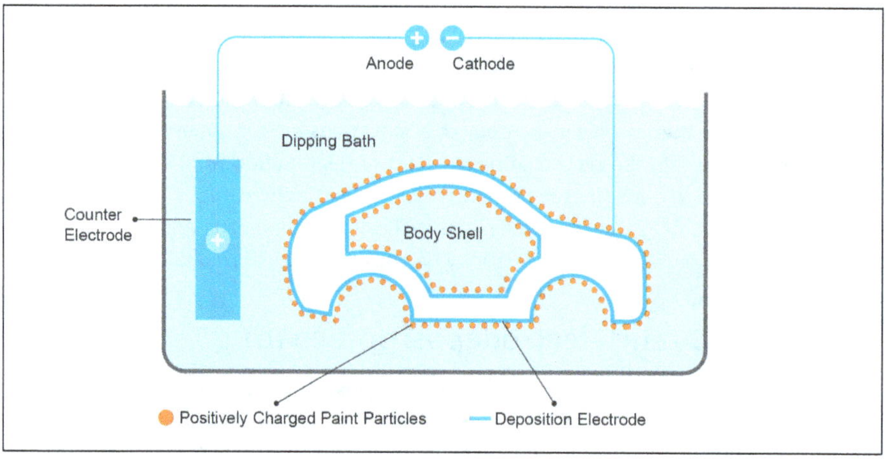

Figure 5.40: Schematic description of the cathodic electrodeposition (CED) process

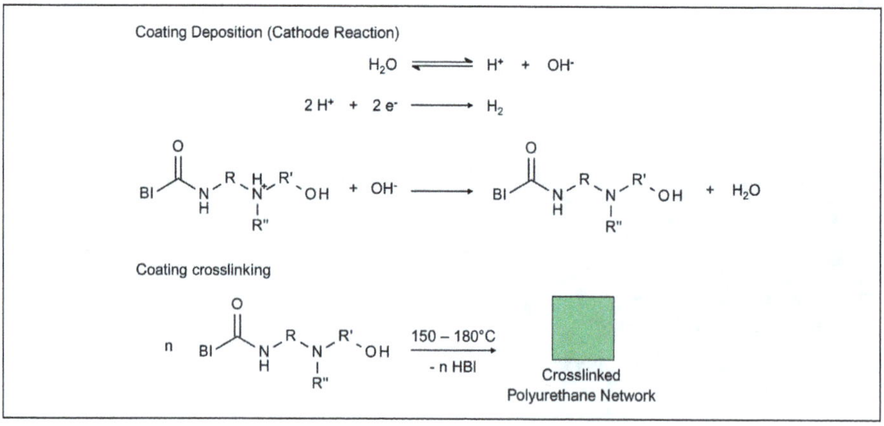

Figure 5.41: Polyurethane formation in a cathodic electrodeposition (CED) coating

Bl = blocking agent

Polyurethane/epoxy resin combinations dominate as binders in this application. As a rule, these are amine-extended, aromatic epoxy resins and thermally activated PU-crosslinkers. Quaternization of the tertiary amine function with organic acids yields the necessary positive charge on the particles.

The thermally activated PU-hardeners incorporated in the resin molecule allow subsequent crosslinking in the baking oven at temperatures of 150 to 180 °C. The di- or polyisocyanates used to synthesize the polyurethane/epoxy system are most commonly based on diphenylmethane diisocyanate (MDI), and sometimes toluene diisocyanate (TDI). Instead of monomeric MDI an oligomeric pre-polymerized form is often preferred. CED formulated with these materials is characterized by improved bath stability and increased stone chip resistance of the baked film.

CED systems formulated with aliphatic isocyanates based on HDI or IPDI (or their derivatives) are a specialty (mostly in Japan) which provide advantages of low baking temperatures (150 to 160 °C) and lightfastness. However, the corrosion protection properties of these products are inferior to MDI systems.

Among the blocking agents used for the thermally activated PU hardeners are methyl ethyl ketoxime, secondary alcohols such as 2-ethylhexanol or ether alcohols such as diethylene glycol monomethyl ether. When selecting the blocking agent, it is important that the resulting products have adequate stability, e.g. to hydrolytic degradation (bath stability), and relatively low deblocking temperatures . Recent developments have yielded coating systems that do not need lead or tin organic catalysts for the crosslinking reaction. Systems with low co-solvent content satisfy ecological requirements, while improved stone chip resistance and corrosion protection have been improved in response to technical demands. While systems with lower baking temperature are in use for general industrial applications they have not yet met Auto OEM performance requirements.[33]

5.3.3 Seam sealing, underbody protection and sound insulation

Seam and underbody sealing systems are usually applied successively after baking the CED. Owing to their very balanced application-specific property profile and their favorable cost/benefit relationship, PVC plastisols dominate both applications worldwide. Depending on the material-specific properties, each plastisol system is pre-gelled at about 100 to 120 °C. The primer surfacer is then applied and all three coats are baked together at temperatures between 140 to 160 °C.

Underbody sealing systems are applied in relatively thick layers on the underside of the vehicle. In seam and overlap zones, a thickness up to 2 mm may be achieved. This protects the CED from stone chipping. The underbody sealing system must have good adhesion on the CED and be highly resistant to abrasion. In order to improve the adhesion of PVC plastisols there is the possibility to add aliphatic or aromatic thermally activated polyurethane hardeners.

Seam sealers are used to seal the weld seams in a car body. They are applied on seams of doors, hoods and hatches and are predominantly made of PVC plastisols, but also acrylic plastisols or rubber derivatives. To provide adequate protection, a seam sealer must be sufficiently flexible and suitable for application in a thickness of several millimeters. Polyurethane systems meet both requirements and thus have specific advantages over the established PVC systems.

In contrast to PVC materials, solvent-free polyurethane systems can be applied in coats thinner than 100 µm which yield a compact film with outstanding adhesion on the CED. Despite the thinner film thickness, their abrasion resistance is as good as that of PVC systems (see Figure 5.42). For this reason, polyurethane underbody sealing systems can be applied in thin layers. This contributes to an overall reduction in the weight of the vehicle.

Depending on the polyol/polyisocyanate combination, one- and two-component polyurethane systems may be formulated to yield the desired flexibility. One particular advantage is that the flexibility at low temperatures is not reduced significantly, so the stone-chip resistance of the systems remains effective. Additionally, seam and underbody sealing systems are applied to provide sound insulation. These systems are made less flexible and with higher pigment and extender load to reduce the vibrations of the car body. In the application field of commercial vehicles like busses these are usually mats based on bituminous substances that are fitted by hand. The mats melt during pre-gelling of the underbody sealing system and adhere to the substrate. For passenger cars, sound insulation systems are also applied by high-viscous liquid coatings due to their automated and faster application.

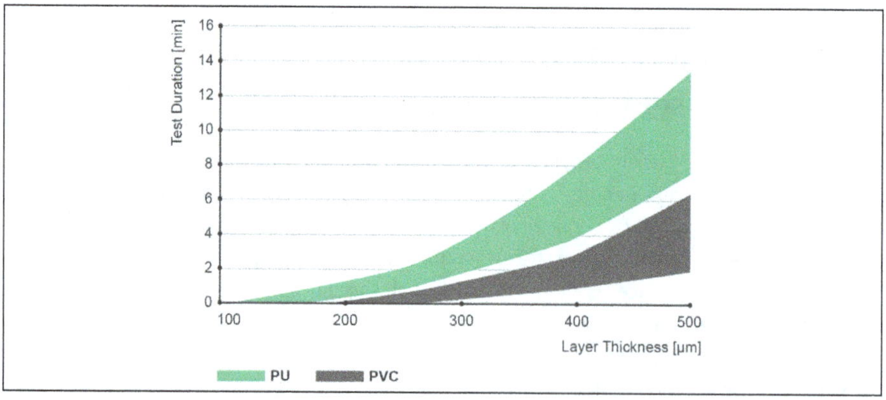

Figure 5.42: Abrasion resistance of PVC and polyurethane underbody sealing systems (Sablux test)

Suitable solvent-free polyurethane systems may also be formulated for this application. These coatings can be applied on the automated line and have excellent long-term adhesion to the substrate and satisfy the sound insulation requirements of the automotive industry. Also available are solvent-free, one-component polyurethane underbody and seam sealing systems containing microencapsulated hardeners, e.g. TDI-based. Depending on the type of microencapsulation, these systems can be baked at temperatures of 100 °C. The microcapsules are destroyed by the heat, releasing the hardener for the crosslinking reaction. However, because of its relatively high cost, this technology is used only in niche applications.[48]

One alternative technology to the use of coatings for underbody protection, as described above, is to fit plastic tubs on the undersides of cars. These provide stone chip protection and sound insulation. Due to their application efficiency underbody coating systems are focusing on wheel wells.

5.3.4 Primer surfacer

Originally conceived as an intermediate coat with filling properties and therefore sometimes called Filler, the primer surfacer has undergone considerable transformation in recent decades, with more change still to come. For example, it has increasingly assumed a color-imparting function in support of the base coat. In order to satisfy demands for solvent reduction, water-borne binders have been developed which have a property profile fully equivalent to solvent-borne primer surfacer resins.

Solvent-borne primer surfacer

The stone-chip protection properties required of a primer surfacer may be optimized using polyurethane resins, which has allowed polyurethanes to break into this field of application. Polyurethanes yield primer surfacers that combine toughness with flexibility yet retain good sanding properties. Moreover, the stone-chip resistance of primer surfacers formulated with alkyd and polyester resins crosslinked with melamine and/or urea resins can be improved considerably by adding thermally activated polyurethane hardeners. This improves the stone-chip behavior of the overall finish significantly.

The greatest effect is achieved using thermally activated crosslinkers based on hexamethylene diisocyanate (HDI) or diphenylmethane diisocyanate (MDI). Considering the baking temperatures of 130 to 160 °C that are typical for primer surfacers, established blocking agents such as methyl ethyl ketoxime, 3,5-dimethylpyrazole and malonic acid ester are suitable.[49]

A typical (solvent-borne) primer surfacer formulation is shown in Table 5.14 (left column). Solvent-borne primer surfacers contain thermally activated polyurethane hardeners either combined with melamine resin or used as the sole crosslinker. In this way, a very

Table 5.14: Comparison of two typical formulations for solvent-borne and water-borne primer surfacers

Component	Solvent-borne primer surfacer	Parts by weight	Water-borne primer surfacer	Parts by weight
Polyester	e.g. "Desmophen" T 1665 or "Desmophen" 670	21	e.g. Bayhydrol" PT 241	19
Polyisocyanate (blocked)	e.g. "Desmodur" BL 3175	7	e.g. "Bayhydur" BL 5140	7
Melamine resins		9		3
Pigments/fillers		30		21
Organic solvent		33		5
Water		0		45
		100		100

broad spectrum of requirements with respect to application reliability, leveling and stone-chip resistance can be satisfied. The solvent can be chosen to match the flash-off behavior of the primer surfacer to the conditions on individual coating lines.

Water-borne primer surfacer

The good performance of polyurethane binders is even more evident in the recent significant growth of the water-borne primer surfacer segment. Water-borne primer surfacers have experienced significant growth over the last three decades. These systems are based on pure polyester dispersions and hydroxyl alkyl melamine resins as hardeners showing a balanced price-performance ratio. Modifications with polyurethane will improve the chipping resistance especially at very low temperatures or under overbaking conditions, where pure polyester-melamine systems exhibit weak performance.

Polyurethane modification can be brought into the primer surfacer by two kinds of binders. On the one hand by use of a water dispersible, thermally activated polyurethane hardener, formulated either as sole crosslinker or, in combination with hydroxyl alkyl melamine resins. These PU-crosslinkers are hydrophilically-modified and are easy to incorporate into water.[50] On the other hand, a water-borne polyurethane dispersion can be used to modify the polyol component.

In the meantime, also so called self-crosslinking urethane resins are spreading in the market. These kind of binders combine OH-functional polyester-polyurethane dispersions and thermally activated PU hardeners within one molecule. Good conditions therefore exist to produce primer surfacers with properties typical of polyurethane systems, such as flexibility and stone-chip resistance.

To ensure reliable application of water-borne primer surfacers, the spray booths and flash-off zones must be air-conditioned to maintain the correct humidity and temperature.

The addition of ≤ 5 % of co-solvent to the formulation improves the flash-off behavior, making the resulting paint film less vulnerable to fluctuating conditions on the coating line. Despite the low solid content of around 50 %, compared to 70 % in the case of high-solid systems, solvent emissions from water-borne primer surfacers are halved. A basic water-borne primer surfacer formulation is shown in Table 5.14 (right column).

In 2017, the share of aqueous primer systems worldwide was about 28 %.[1] The largest penetration of water-borne primer surfacers is found in Western and Eastern Europe with a share of approximately 72 %. Although car producers in Japan and China are increasingly using water-based primer surfacers, the growth rate will flatten in worldwide automotive processes due to a further shift towards new OEM compact processes aiming increase process efficiency.

New OEM compact processes have been developed and introduced in the market in recent years with increasing market share. One example is a process in which, instead of applying and baking a primer surfacer in a separate operation, a functional coat of reduced film thickness is applied and baked together with the base coat and clear coat. For both economic and environmental reasons, these 3 coat/1 bake systems have attracted great interest globally, both in solvent-borne and water-borne systems[34]. These systems have to show a very good compatibility to the base coat layer due to their wet-on-wet application. They also need an improved physical drying, that is why the polyurethane binders are more comparable to a base coat than to a classical primer surfacer. In order to ensure the high-quality standard in terms of e.g. stone chip resistance at low film thicknesses these functional coatings require a higher share of polyurethane resin compared to conventional systems.

5.3.5 Base coat and clear coat

Approximately 90 % of the topcoats used in the automotive industry are two-coat systems consisting of a color-imparting base coat and a clear coat applied wet-on-wet (see Figure 5.43). The remaining systems are one-coat systems, i.e. pigmented topcoats, which – as the term implies – are applied and baked in a single layer.

Base coat

As car body shape/design and color are selling features for automobiles, various types of base coats (i.e. solid shade, metallic and pearlescent paints) have been developed to meet customer demand. Eye-catching base coats which use high brightness pigments are the trend in the automotive industry.

Today's base coats are created with the latest coatings technology, in combination with advanced application methods. Currently, there are various compact processes in place which employ specialized spray booth conditions, flash off techniques that use IR lamps to control water evaporation and spray methods having paint reciprocators and robots in combination.

Both solvent-borne and water-borne base coat systems are utilized in the market. In 2017, approximately half of the base coats used globally were water-borne systems,[1] and the adoption of water-borne base coat for new application facilities remains high due to increasing legislative pressures. Water-borne systems have different application requirements compared to solvent-borne paints (Figure 5.44). Water-borne systems have advantages with regards to special effects and stone-chip resistance.

Most water-borne base coat formulations contain combinations of polyacrylate and polyurethane dispersions, in combination with melamine resins. New dispersions, such as core shell emulsions based on acrylates/polyesters and acrylates/polyurethanes, have been recently introduced for water-borne base coats to improve the pigment wettability, flip-flop properties for special effects, and rheological stabilization in spray application as well as during the drying processes (sufficient atomization with no sagging). Polyurethane

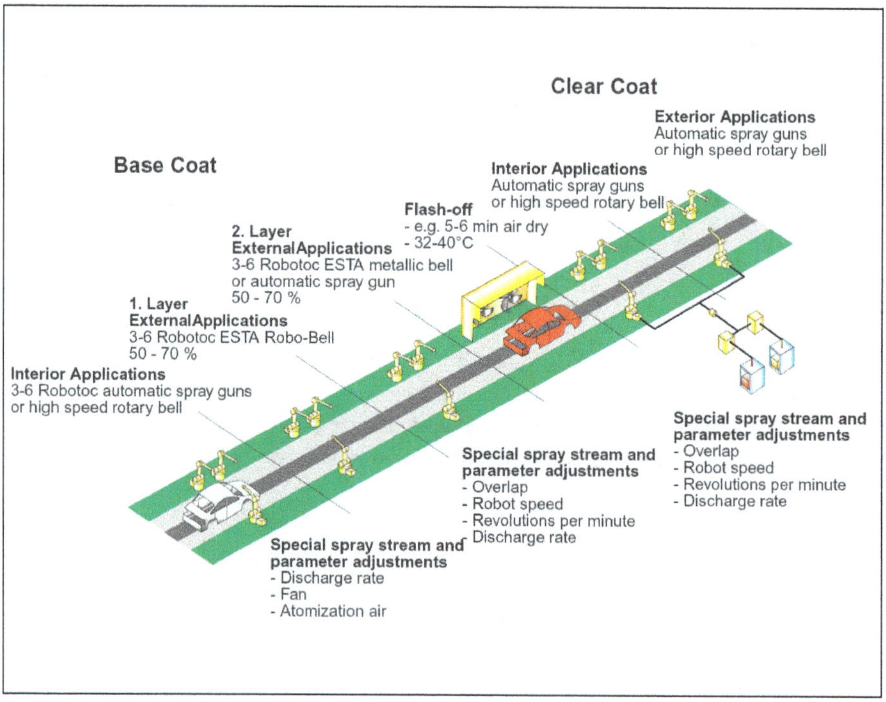

Figure 5.43: Coating line for water-borne base coats and two-component polyurethane clear coats in the automotive industry[51]

dispersions (i.e. "Bayhydrol" UH and UA, which are co-solvent-free, and high shear stable) greatly enhance important quality characteristics such as brilliance, adhesion, leveling, flexibility, and stone-chip resistance.

In order to save energy in the overall coating process, primerless concepts have become more important. In this case, the base coat takes over the function of the primer surfacer. Specifically, two layers of base coat are applied. The first can be a one- or two-component base coat and the other a one-component base coat. The two-component base coat is comprised of the same resins as the one-component base coat which is formulated with a hydrophilic polyisocyanate hardener to ensure interlayer adhesions and improved chipping resistance.

Clear coat

Clear coat is the final coating layer and provides aesthetics and appearance, scratch resistance, durability, lightfastness, mechanical properties and chemical resistance as the basic function for the protection of the car body.

Figure 5.44: Electrostatic application of a water-borne coating in the automotive industry[63]

Clear coat market

In 2017, 93 million vehicles were produced worldwide, and it is estimated that approximately 250 kilotons of clear coats were consumed globally[1]. Figure 5.45 illustrates the types of clear coats utilized from a technology perspective. Two-component

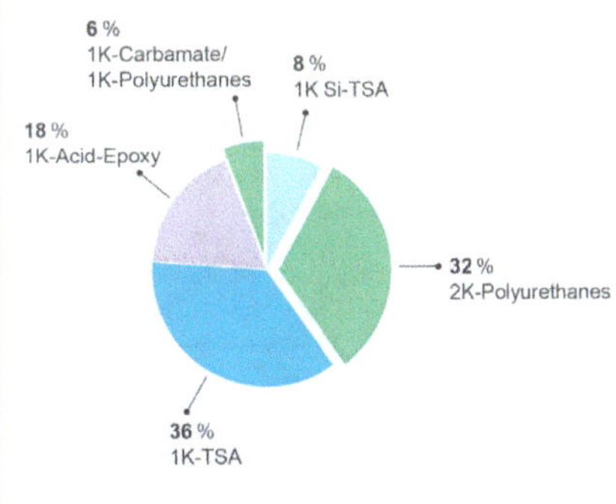

Figure 5.45: Global consumption of clear coats by chemical systems 2017, total approximately 250 kt[1]

polyurethane clear coats are used on about 30 % of all vehicles produced globally and are expected to grow by 9 % annually until 2021.

The majority of the binders used today in the formulation of automotive clear coats are based on acrylic resins crosslinked with various hardener systems. Typical clear coats

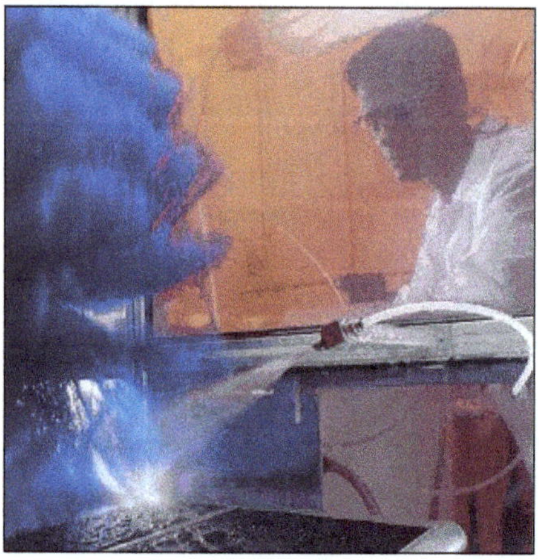

Figure 5.46: Scratch resistance testing in the laboratory car wash simulator[53]

Figure 5.47: Determining chemical resistance by the gradient oven method[53]

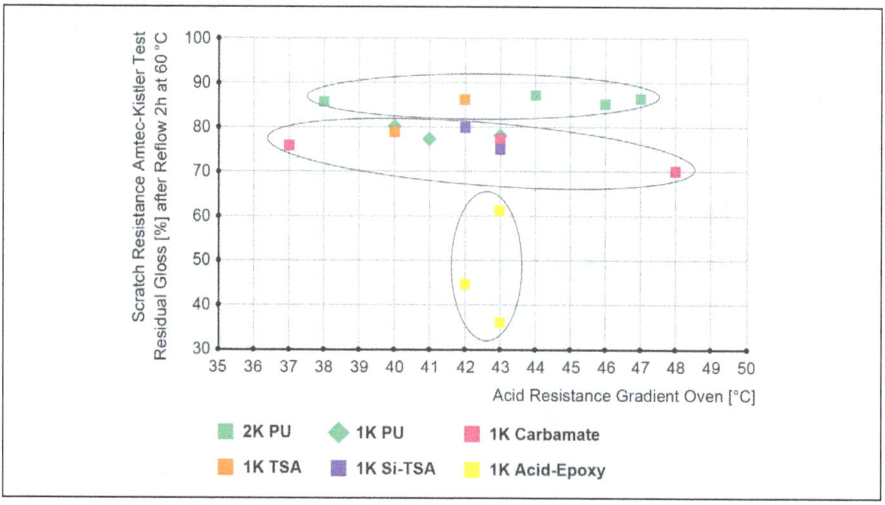

Figure 5.48: Clear coat benchmark study results showing the performance levels of various clear coat systems with regard to scratch and acid resistance

widely used for OEM automobiles include two-component high-solid polyurethane (2K PU), one-component PU modified thermosetting acrylics (1K PU), one-component carbamate-modified thermosetting acrylics (1K carbamate TSA), one-component acid epoxy (1K AcEp), one-component silane-modified thermosetting acrylics (1K Si-mod. TSA) and one-component thermosetting acrylics (1K TSA).

Clear coat technologies

Clear coats can be solvent-borne high-solid (solid content > 55 %), water-borne or powder. In the last two decades, water-borne and powder clear coats were adopted by some automotive manufacturers, however they have all disappeared from the market due to quality issues, application difficulties and economic reasons. Most of these clear coat technologies have been converted to two-component high-solid polyurethane systems. While two-component water-borne polyurethane clear coats have been developed in joint projects of the raw material and paint suppliers, with the automotive industry they have not been implemented so far[65]. Two-component high-solid polyurethanes as a clear coat already meet the current VOC-legislation and they exhibit similar performance and film build.

Intensive development work has been in progress for many years in an effort to improve resistance to scratching, etching and chemicals. For this purpose, a number of test methods have been developed which have a significant effect on the results. In particular, the crockmeter and nano-scratch tester as well as the Amtec-Kistler automated laboratory car wash simulator (see Figure 5.46) are widely used for testing scratch resistance,[35] while the gradient oven method is used to determine chemical resistance (see Figure 5.47).

Studies demonstrated the quality levels of the various clear coat systems. Figure 5.48 shows the results concerning acid and scratch resistance of various line clear coat materials. This chart demonstrates clearly that 2K high-solid polyurethane coatings show the best performances in terms of acid and scratch resistances in comparison to the one-component clear coats.

Two-component polyurethane clear coats

Since they were first used in automotive OEM finishing by Mercedes-Benz in the mid-1980s, polyurethane clear coats have achieved a significant and steadily increasing market share. The key factor for the market success is the superior performance. This applies to appearance properties such as gloss and leveling, as well as to mechanical and chemical resistance properties. Moreover, the high compatibility of two-component polyurethane clear coats with water-borne base coats has further boosted market growth in recent years. Today, almost all automakers in Europe, Americas and Asia-Pacific use polyurethane clear coats to some degree.

The long life weatherability of a two-component polyurethane clear coat is demonstrated in Figure 5.49. After 16 years and 240,000 km of everyday use, never been in a garage, this vehicle still has a 73 gloss reading (20°) on the vertical surfaces and 61 on the roof surface area. The original gloss readings were 83.

Two-component polyurethane clear coats are based on hydroxy-functional acrylates, and in some cases polyester resins, which are crosslinked with aliphatic polyisocyanates[54], such as HDI- and IPDI-based polyisocyanurates. Due to their high reactivity they are baked for approximately 20 to 30 minutes at 130 to 140 °C. Two-component polyurethane systems differ from conventional clear coat technologies in the following aspects:

- high-solid content;
- high reactivity;
- significant improved chemical and, in particular, environmental etch resistance;
- good mechanical properties (resistance to scratching and stone-chipping, as well as self-healing ability);
- outstanding appearance properties (levelling, brilliance, long-term gloss retention).

Two-component polyurethane clear coats based on linear or only slightly branched polyacrylic or polyester resins and an HDI-based hardener such as "Desmodur" N yield films with a relatively low glass transition temperature (T_g), high flexibility and excellent scratch resistance.

Figure 5.49: Vehicle coated with a two-component polyurethane clear coat after 240,000 km and 16 years of everyday use, never been in a garage

The addition of an IPDI-based hardener such as "Desmodur" Z increases T_g and hardness of the clear coat films. Their resistance to etching and chemicals is much higher[52, 53] but their scratch resistance is noticeably lower than that of the flexible formulations only based on HDI hardeners.

This inverse relationship between scratch and chemical resistance is derived from basic laws of polymer physics and is also observed in other crosslinking technologies. However, in the case of polyurethane crosslinking, both requirements can be satisfied to a high degree, thanks to the high chemical stability of the urethane bonds.

To optimize scratch resistance while retaining high chemical resistance, two approaches were being pursued in two-component polyurethane clear coats. One approach uses inorganic nanoparticles[36, 55–57] and has resulted in a marked improvement in resistance to dry scratching. The second approach involves clear coats with a high crosslinking density and flexible network. The high crosslinking density may be designed by using a high functional HDI-based hardener such as "Desmodur" N 3580, and the flexibility can be achieved by using high molecular weight polycarbonate diol blended with hydroxyl-functional acrylates.[37]

When there is a balanced relationship between flexibility and hardness, such two-component polyurethane clear coats are characterized by high reflow or self-healing behavior. This performance is made possible by the formation of hydrogen bonding between the urethane groups, which contribute to the stability of the network.[58, 59]

The formulation of coatings with high scratch resistance requires consideration of the levels of covalent bonds and hydrogen bonds. Hydrogen bonds may dissociate and re-form on heating, depending on their strength. 2K PU systems with high functionality polyisocyanurate curing agents show lower activation energy for recovery compared to standard

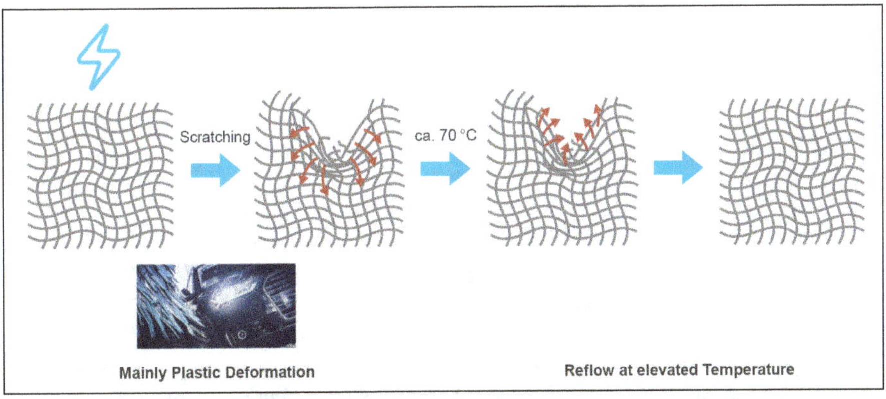

Figure 5.50: Self-healing two-component polyurethane clear coat with high reflow capacity

clear coats. The use of polyisocyanates differing in structure and functionality enables the adjustment of the clear coat performance.[38]

During the scratching of a coating layer, plastic deformation (plastical scratch) typically occurs first followed by brittle fracture of the coating layer if the object causing the scratch is forceful enough. If heat is applied to a plastically scratched polyurethane coating, the scratches are observed to reflow (see Figure 5.50). In practice, this self-healing effect can be tracked visually or by conducting gloss measurements.

The paint industry has been continuously developing clear coats with improved micro-scratch resistance in close co-operation with raw material suppliers and the automotive industry. These new clear coats launched in the market often are based on hybrid technologies comprised of inorganic and organic materials. The clear coats can be formulated with alkoxysilanes compounds or nanoparticles as additional components either in the polyol formulation, or in the hardener of the two-component polyurethane system.

Silanes-modified polyisocyanates combine the hard, glassy properties of silanes with flexible polyurethane structures. The fact that the silanes are evenly integrated into the network prevents the formation of brittle clusters. Especially after surface polishing, systems with silanes-modified polyisocyanates retain their scratch-resistance, unlike nanoparticles that are removed during polishing. This enables the formulation of long-term scratch-resist-

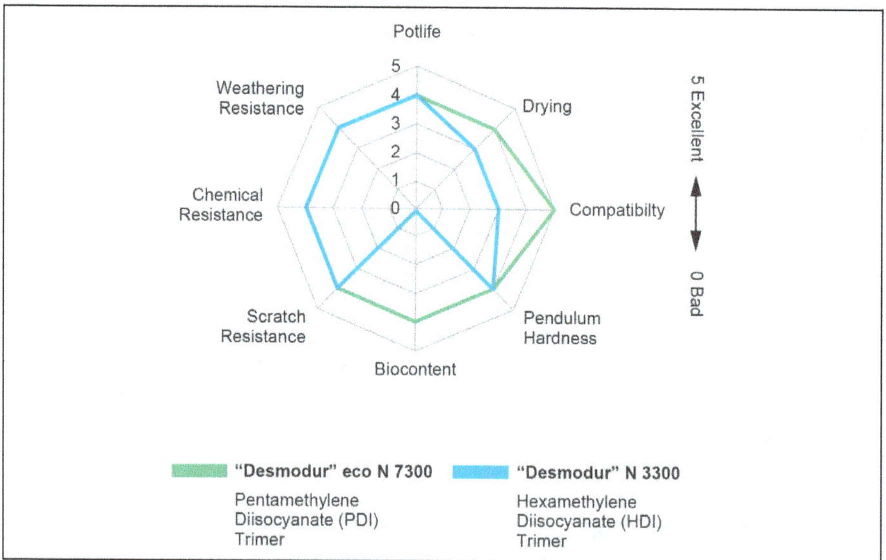

Figure 5.51: Property profile of a two-component polyurethane clear coat based on a pentamethylene diisocyanate (PDI) crosslinker in comparison to its HDI-counterpart[40]

ant coatings. Silanes-modified systems with their high network density are also the starting point to develop clear coats with less soiling behavior, the so called easy-to-clean coatings.[39]

Another approach is to use formulations based on polyacrylates with high hydroxyl numbers. In stoichiometric crosslinking with a flexibilizing HDI polyisocyanate hardener they lead to a high crosslinking density with improved chemical resistance. However, the limit of such formulations is the compatibility of both components. Interesting is that the isocyanurate crosslinker based on the bio-based pentamethylene diisocyanate (PDI) shows advantages in compatibility over its HDI counterpart (see Figure 5.51).

The demand for more sustainability and lower carbon foot print binders has intensified in the past few years. First trials with a two-component PU clear coat using a crosslinker based on the bio-based PDI have been successfully conducted in the automotive industry. This hardener contains 71 % of renewable carbon, and is a contribution reducing the consumption of fossil resources.[41]

One-component polyurethane clear coats

The two-component application equipment is more complex and expensive to maintain than one-component equipment. By modifying thermo-setting acrylic (TSA) coatings with thermally activated polyurethane crosslinkers it is possible to formulate one-component polyurethane clear coats that represent a compromise between two-component polyurethane and TSA coatings in terms of their performance, while still being applied using one-component equipment. The addition of thermally activated IPDI-based crosslinkers may greatly increase the acid resistance of solvent-borne melamine-crosslinked clear coat systems. Similarly, blocked HDI-based polyisocyanates may increase the scratch resistance. Due to their tendency to yellow on exposure to heat, thermally activated PU crosslinkers were long considered unsuitable for use in automotive OEM topcoats. Today, hardeners blocked with 3,5-dimethylpyrazole are used almost exclusively since they are virtually non-yellowing.[60] Generally, the performance properties of one-component polyurethane clear coats are not of the same high level as those yielded by two-component polyurethane coatings in terms of scratch and chemical resistance and weatherability.

Compact processes

In compact OEM coating processes currently in use (i.e. 3 wet- or primerless processes), the clear coat is applied on the top of the one or two layers of base coat in a wet-on-wet manner and then these two, or respectively three layers, are simultaneously baked. Two-component polyurethane clear coats have advantages over other types of clear coats due to the fact, that the polyisocyanate in the clear coat has the ability to migrate into the base coat layer.[42] This migration strengthens the adhesion between the base coat and clear coat layers, as well as building cohesive strength within the base coat itself, both of which result in improved water resistance of the coatings system.

To maximize these properties, a hardener combination should have specific characteristics which promotes, on one hand, the ease of penetration into the base coat layer, and, on the other hand, the strength of the clear coat. Low viscosity HDI-based hardeners like the uretdione hardener "Desmodur" N 3400 or the asymmetric trimer "Desmodur" N 3900 in combination with high functional hardeners like "Desmodur" N 3580 provide the best solution.[43] This technology will become even more interesting, when low temperature baking compact processes are utilized. These low bake OEM painting processes are expected to be implemented increasingly do to the higher expected usage of multiple materials with higher temperature sensitivity in the automobile construction.

Latest developments for advanced polyurethane clear coats

The automotive industry has already begun installing compact paint processes for reducing environmental impact, as well as for improving production efficiency. The industry is now seeking to increase the use of lightweight materials such as thinner gauge metals, plastics or composites for body manufacturing. Some OEMs are actively developing low temperature baking in-line processes (i.e. 80 °C) for painting body and bumper/fascia together. This concept could eliminate the paint shop for plastic parts and would make it easier to match colors between plastic parts and body by applying the paint in the same line. Furthermore, these lower temperature processes would give benefits for multi-material constructions, which should be baked at lower temperature due to heat sensitivity and thermal expansion differences of these materials.

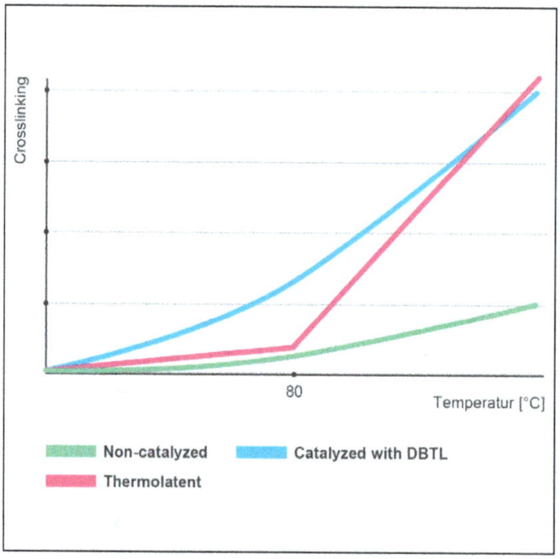

Figure 5.52: Principle curing behavior of a two-component polyurethane clear coat using thermolatency in comparison to a standard two-component polyurethane clear coat, catalyzed and none-catalyzed

To meet these requirements, thermo-latent hardeners have been developed. The reactivity of such two-component polyurethane clear coats is accelerated when heated above 80 °C. Below 80 °C the reactivity is comparable to conventional non-catalyzed two-component polyurethane clear coats. This enables an improvement of the flow properties of the clear coat. As soon as the baking temperature reaches 80 °C the clear coat will be activated leading to the crosslinked polyurethane network. This allows a shortened baking time and/or reduced baking temperatures without sacrificing the crosslinking reaction compared to standard baking conditions (i.e. 140 °C for 30 minutes) (see Figure 5.52).[44, 45] In addition, this technology should pave the way for further improvements in appearance of the coating by minimizing the orange peel effect.

5.3.6 Outlook

The global demand for automotive OEM coatings is estimated to be 1,400 kilotons in 2017 and is expected to grow at 3 % CAGR over the period from 2016 to 2021.[1] This growth is boosted by increasing population and increasing economic status of the middle-class family.[46] On the other hand, autonomous vehicles and car sharing are appearing in the market, which are expected to affect the automotive paint industry due to less car production expected in the future.[46]

The automotive industry has largely achieved VOC emission targets directed by regional/country legislations on environmental protection. Increasing cost pressure worldwide, as well as environmental requirements and sustainability goals of automotive manufacturers are expected to result in further increases in efficiency in the paint application. Also, the proportion of primerless 3 coat/1 bake systems will continue to increase. Both lead to an overall lower paint consumption per body.

The ongoing trend of light-weighting vehicles with multi material use in car bodies increases the move to low-temperature curing for all coating layers. Reactive two-component polyurethane systems offer a great opportunity for the technological breakthrough. Already today there are solutions possible for a low bake process after CED. Other options could include polyaspartic-based coatings which cure even at room temperature, or UV-curing technologies, but there are still big challenges in coating performance and application processes to be overcome.

OEMs are now focusing on sustainability aspects, especially those with a reduced carbon footprint in relation to materials, processes and applications. This will create new opportunities for innovation. The use of paints based on renewable raw materials in the automotive OEM coating is becoming more likely. A first bio-based crosslinker for two-component polyurethane clear coats has been introduced to the market and has been successfully applied under close-to-production conditions.

5.4 Automotive refinish and transportation coating

Automotive refinish coatings differ from automotive and transportation OEM coatings in that the former are typically cured at ambient or slightly elevated temperatures (a maximum temperature of 60 °C), while OEM coatings commonly cured at much higher temperatures (typically 130 to 180 °C). In automotive refinishing, coatings are applied to complete vehicles, including temperature sensitive plastic components, filled gas tanks, and car tires, which is why curing at high temperatures is impossible. The situation is similar in other transportation sectors (i.e. buses, trucks, airplanes, and rail vehicles). Due to the size of these vehicles, it is not economical to cure the coatings in drying booths at temperatures significantly above room temperature. As the property profiles and application conditions are similar, slightly modified or original automotive refinish coatings are often used for transportation coating applications. Table 5.15 shows a comparison of requirements automotive and transportation OEM coating and refinishing.

5.4.1 Automotive refinish

The history of automotive refinish coatings dates back to the early 20th century as the popularity of the automobile grew. The coach enamels used at that time, based on natural resins, were succeeded by nitrocellulose lacquers around 1925. A few years later, medium-oil alkyd coatings were developed. Starting in 1940, spray coating application, the technical innovation of the day, replaced conventional brush application. Polyurethane chemistry came into refinishing in form of the reactive crosslinkers for medium-oil alkyd resins. These crosslinkers were diluted solutions of IPDI adducts, later IPDI-trimers, which were added to the coatings before their application. They greatly improved the over-coatability, through-drying and solvent resistance of the alkyd coatings.

Figure 5.53: Application of two-component polyurethane coatings in automotive refinish

Solvent-borne two-component polyurethane coating systems which came into the market around 1970 became the dominant resin systems for automotive refinish coating. In order to meet regional/national VOC regulations, water-borne

Table 5.15: Comparison of requirements in automotive and transportation OEM coating and refinishing

	Automotive OEM finishing	**Transportation OEM coating**	**Automotive and transportation refinishing**
Size of paintshop	Industrial scale	Industrial scale	Bodyshops
Objectives	Identical sized bodies	Identical sized bodies	Very different sized bodies
Paint consumption per unit	High volume	High volume	Very low volume
Spray application method	Automated, robot application	Partly automated, partly manual application	Manual application
Quality control	Precise control of film thickness	Partial control of film thickness	Variable film thickness
Reqired formulations	Special, OEM manufacturer-specific formulations	Special, OEM manufacturer-specific formulations	Universal formulations needed
Curing condition	130–180 °C	RT–80 °C	RT–60 °C

base coats and high-solid clear coats for automotive refinish coatings were developed in the early 21st century and afterwards followed by water-borne primer surfacers in some regions. The change from solvent-borne to water-borne two-component polyurethane clear coat is expected as the next step.

In 2017 the global consumption of paints in the refinish sector was 816 kilotons.[1] The market is expected to grow by approximately 1 % annually within the next five years.

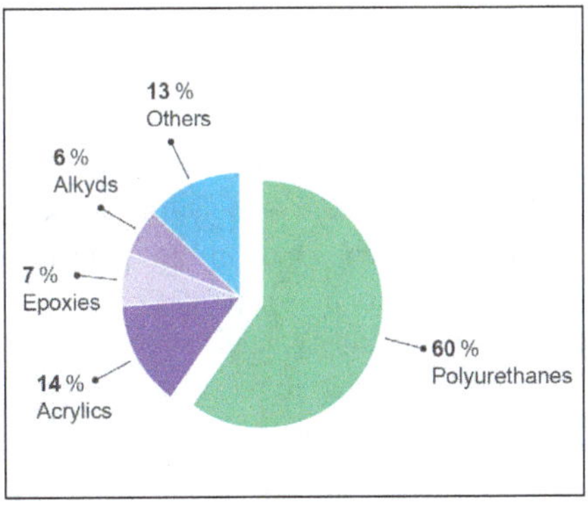

Figure 5.54: Global refinish paint market by technology 2017, total: ~ 816 kt[1]

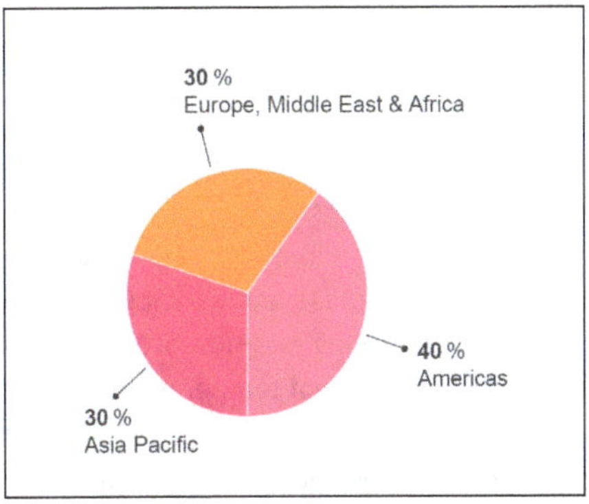

Figure 5.55: Global refinish paint market by regions, 2017, total: ~ 816 kt[1]

Table 5.16: Typical formulation and technical data of a two-component polyurethane clear coat for automotive refinish application

Raw material	Generic name/trade name	Parts by weight
Component A		
Binder	e.g. "Setalux" D A 870BA, (Allnex)	43.3
Levelling agent	e.g. "Tego" Flow 370, (Evonik)	0.3
Defoamer	e.g. "Tego" Glide 415, (Evonik)	0.2
HALS, 50 % in BA	e.g. "Tinuvin" 292, (BASF)	0.9
UV absorber, 50 % in BA	e.g. "Tinuvin" 1130, (BASF)	1.8
Catalyst, 1 % in BA	e.g. "DABCO" T-12, (Evonik)	1.8
Solvents	e.g. BA/Xyl/MPA(1/1/1), (various suppliers)	26.7
Component B		
Hardener	e.g. "Desmodur" N 3390 BA, (Covestro)	16.3
Solvents	e.g. BA/Xyl(1/1), (various suppliers)	8.7
		100.0
Characteristic data		
Solid content on formulation [%]		45
Mixing ratio components A : B by weight		3 : 1
Initial viscosity, DIN 53211 #4 mm cup, 23 °C [s]		15
Mechanical properties		
Drying times (20 °C/58 % relative humidity), GB 1728		
T_1 [min]		15
T_2 [h]		6
Konig pendulum hardness, GB 1730 [s]		195
Gloss 20°/60° [%]		91/95
Cupping test [mm]		> 8.7

This slow rate of growth in developed markets is due to a number of factors, including advancements in driver assist technologies.

Automotive refinish is dominated by polyurethane technology, in 2017 approximately 60 % of the coatings (clear coat, base coat, primer) used were based on polyurethane, followed by acrylic/cellulose acetobutyrate, epoxy, alkyds and others. These include polyvinyl butyral, nitrocellulose and acrylic/melamine systems (see Figure 5.54).

The dominating position of PU technology is defined by the excellent ability to cure sufficiently at ambient temperature or only slightly elevated temperatures. Despite the different conditions in which automotive refinish and OEM systems are applied, the coated surfaces are expected to differ little in terms of general quality and, above all, appearance[64]. Figure 5.56 shows the accelerated weathering behavior of two-component polyurethane in comparison to NC and alkyd-systems.

Table 5.17: Characteristic data of a solvent-borne two-component polyurethane clear coat for automotive refinishing application

Coating/film parameter	**Characteristic data**
Solids content [%]	45
Mixing ratio of polyol base to hardener [parts by volume]	3 : 1
Pot life [h]	4
Drying at 23 °C	
T_1 (sand-dry) [h]	< 1
T_3 (touch-dry) [h]	1–2
Drying after 30 min at 60 °C	
T_1 (sand-dry) [h]	direct
T_3 (touch-dry) [h]	direct
Blister-free dry film thickness [µm]	>100
Pendulum hardness after 30 min at 60 °C	
1 day [s]	100
7 days [s]	180
Condensation resistance (240 h/40 °C)	passes
Chemical resistance	passes
Outdoor weathering, 24 months in Florida Gloss 20°, before/after	88/84
Accelerated weathering, 1,000 h QUV (313 nm) Gloss 20°, before/after	88/84

The two-component polyurethane coatings currently in use for automotive refinishing fully satisfy these requirements as well as customer demands for excellent durability comparable to the original finish. Table 5.16 shows a basic formulation of a two-component polyurethane clear coat for refinish application.

Table 5.17 summarizes the typical characteristic data of a two-component polyurethane refinish coating dried at 60 °C. The key features are outstanding optical properties (gloss,

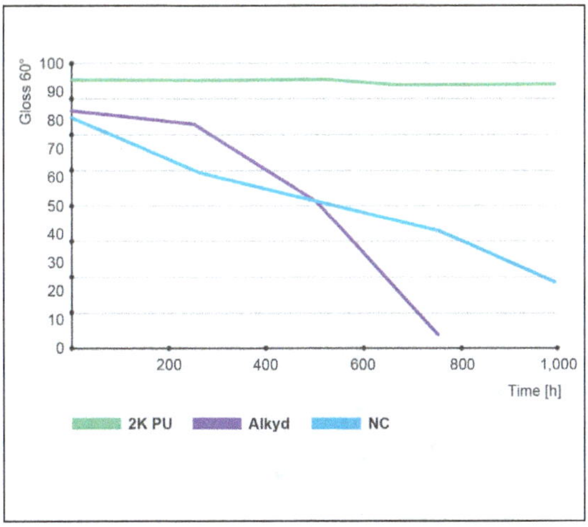

Figure 5.56: Comparison of weather stability of various clear coats for automotive refinishing (UV-A weathering)

leveling) and retention of these properties over long periods, mechanical resistance (resistance to scratching and stone chipping), gasoline resistance and resistance to chemicals and etching (acid rain, caustic cleaning agents, bird droppings etc.). The coatings are formulated to yield blister-free films at the high film thicknesses (more than 100 μm) which may occur in refinishing.

Steps in automotive refinishing

Figure 5.57 shows the typical sequence of layers used in automotive vehicle refinishing. The damaged areas of the car can be returned to original appearance by either hammering

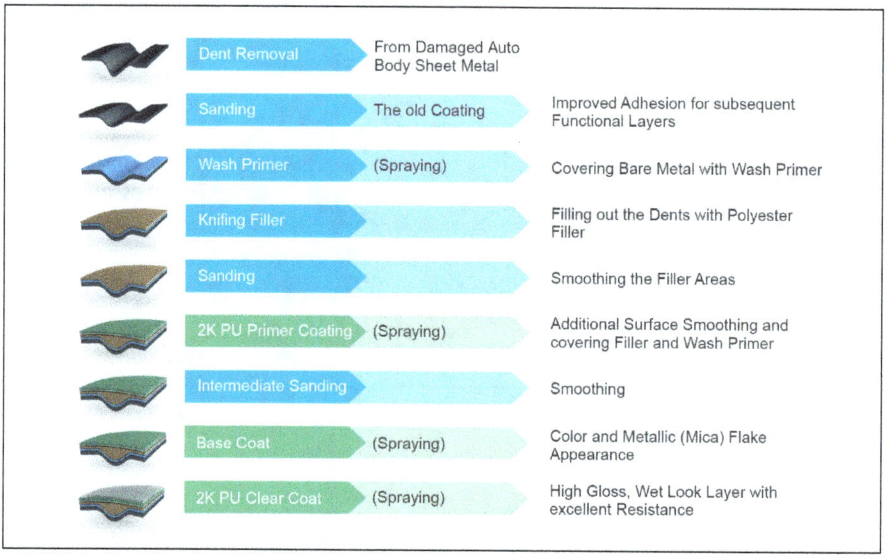

Figure 5.57: Process steps in automotive spot repair

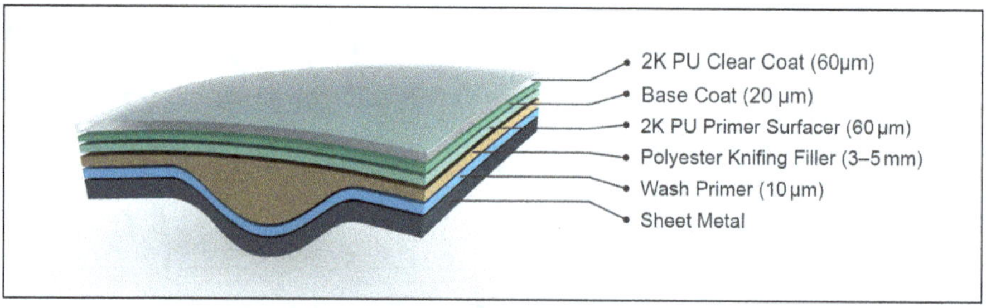

Figure 5.58: Cross-section of an automotive refinish coating

out minor dents, or through the use of aftermarket replacement parts. Typically, the repair area is sanded to improve the adhesion of subsequent functional layers. A one-component anti-rust primer, known as a wash primer, is often applied with a thickness of about 10 μm to prevent the corrosion of any areas of bare metal that may have been revealed by sanding. This primer is commonly based on polyvinyl butyral and contains phosphoric acid. Dents deeper than 1 mm can then be filled with an unsaturated polyester knifing filler (i.e. body putty), which is then sanded to a smooth surface.

A two-component solvent-borne or water-borne polyurethane primer surfacer is then applied at a thickness of 60 μm to further smooth the surface and protect the wash primer and filler. After sanding the primer surfacer, a combination of a one-component solvent or water-borne base coat and a two-component polyurethane clear coat is applied. The resulting refinish system (see Figure 5.58) has an average total film thickness of approximately 150 μm (excluding the knifing filler) – roughly twice the film thickness of an OEM automobile coating (Refer to Chapter 5.3.5).

Water-borne primers and base coats contain combinations of water-borne polyacrylic, polyester, and/or polyurethane dispersions. Polyurethane dispersions will be used due to their excellent wet color matching performance. For water-borne two-component primer sytems, water-dispersible PU hardeners such as "Bayhydur types" are used to cure hydroxyl functional binder dispersions in water. Currently, there are efforts underway by leading automotive refinish coating suppliers to develop water-borne two-component polyurethane clear coats to serve the future needs of refinish body shops with complete water-borne systems.

5.4.2 Transportation coating

The demand for high quality also applies to transportation coatings. In 1965, the first use of lightfast solvent-borne two-component polyurethane coatings was commercialized for the exterior of airplanes. Later, this technology was also utilized for rail applications. In 2017 the global paint consumption for transportation coatings was approximately 373 kilotons.[1] Polyurethane coatings now dominate the transportation sector (including trucks and buses, in addition to aerospace and rail applications) worldwide (see Figure 5.59 - 5.61).[64]

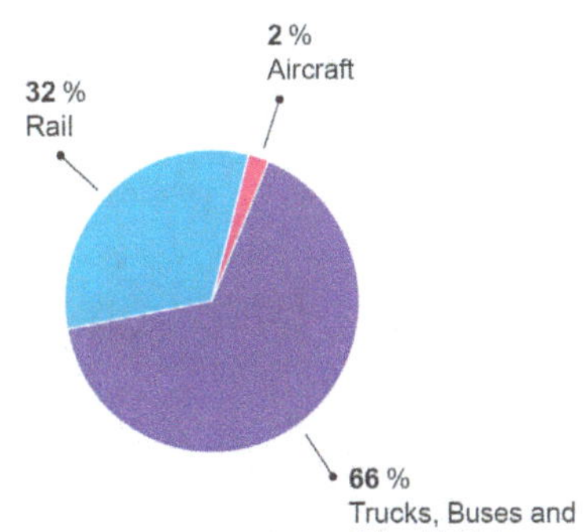

Figure 5.59: Global paint market for transportation by market fields, 2017, total: ~ 373 kt[1]

Table 5.18: Performance of water-borne two-component polyurethane coatings versus specifications for transportation coatings

	Specifications for transportation application	**Water-borne two-component polyurethane coatings (white)***
Drying time Dust-dry [h] Tack-free [h]	≤ 2 ≤ 24	< 2 < 8
Pot life [h]	≥ 4	≥ 6
Gloss 60° [%]	≥ 85	≥ 90
Pencil hardness (7 days at RT)	≥ HB	≥ H
Erichsen Indentation [mm]	≥ 4.0	≥ 8.0
Impact [cm]	≥ 50	≥ 50
Adhesion (cross cut)	≤ 1	≤ 0
Flexibility [mm]	≤ 2	≤ 1
Water resistance (40 °C, system) [days]	≥ 2	≥ 10
Acid resistance (0.1 N H_2SO_4, system) [h]	≥ 24	> 48
Alkali resistance (0.1 N NaOH, system) [h]	≥ 24	> 48
Humidity resistance (47 °C, 96 % RH) [h]	≥ 240	≥ 480
Gasoline resistance (system) [h]	≥ 6	> 24
Weathering resistance		
QUV B (313), 1000 h	Gloss retention ≥ 70 %60° ΔE ≤ 6	Gloss retention ≥ 90 % 60° ΔE ≤ 1.5
QUV A (340), 2000 h	Gloss retention ≥ 70 % 60° ΔE ≤ 6	Gloss retention ≥ 94 % 60° ΔE ≤ 2.0
Xenon 1500 h	Gloss retention ≥ 70 % 60° ΔE ≤ 6	Gloss retention ≥ 94 % 60° ΔE ≤ 1.0
Blister free film thickness (air drying) [µm]	≥ 80	≥ 110
VOC (excluding water) [g/l]	≤ 420	≤ 250

* based on "Bayhydrol" A2470/A2645, "Bayhydur" XP 2655, "Bayhydur 401-70, "Desmodur" N3900 without any UV absorbers

The reasons for these successful adoptions of PU systems were due to high performance requirements of weatherability, high chemical and mechanical resistance properties.

Water-borne two-component polyurethane topcoats have been widely adopted in different transportation applications (such as buses, trucks, agricultural/construction equipment and rail applications) in order to meet high environmental protective regulations, (i.e. lower VOC, especially in Europe and China). These systems fulfill technical specifications of the transportation industry (Table 5.18).

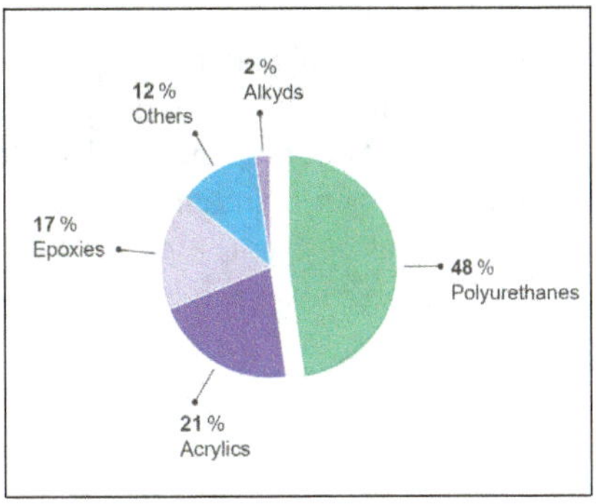

Figure 5.60: Global paint market for transportation by systems, 2017, total: ~ 373 kt[1]

Application steps in transportation coating

In OEM transportation coating (i.e. for rail, truck, bus, aircraft, agricultural/construction equipment, etc.), the first coating layer applied is typically a primer (often epoxy-based). This is either applied directly to metal, body filler (if it is used), or to metal pretreatment (i.e., zirconium or iron phosphate types). Next, a two-component polyurethane primer surfacer is often applied at a thickness of about 100 µm to further

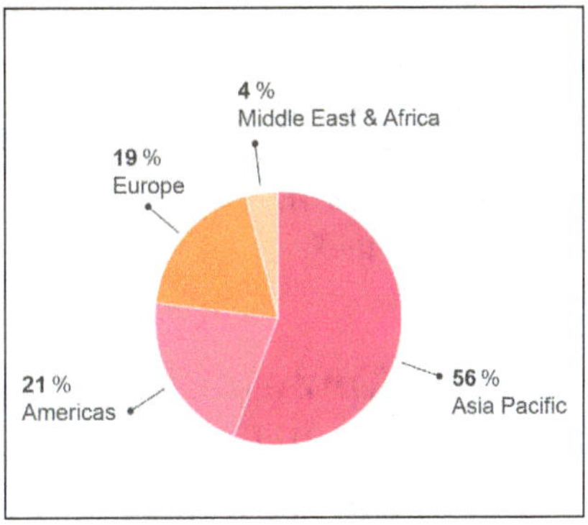

Figure 5.61: Global paint market for transportation by regions, 2017, total: ~ 373 kt[1]

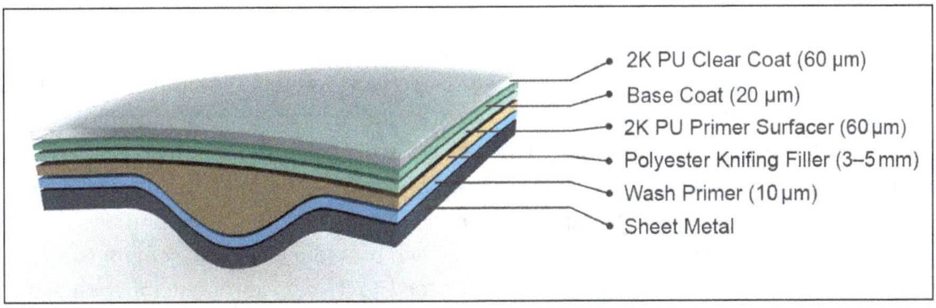

Figure 5.62: Cross-section of an OEM transportation coating

smooth the surface and protect the filler and primer. After intermediate sanding of the primer surfacer, a high-quality pigmented two-component polyurethane topcoat is applied. Increasingly, the transportation sector is using a two-coat finish of a base coat (20 µm) and a two-component polyurethane clear coat (60 µm) on some vehicles such as buses, passenger trains and truck cabins. The overall coating system in transportation applications has an average total film thickness about 300 µm in the case of OEM transportation coating, making it much thicker than an automotive refinish coating. The main reason for this is that the system does not include a cathodic electrodeposition (CED) primer so corrosion protection has to be ensured mainly by application of a higher film thickness (see Figure 5.62).

Coatings for transportation applications – refinishing

In the case of transportation vehicle repair or maintenance operations, the old coating is removed either by sandblasting (rail vehicles, airplanes) or by aggressive solvent mixtures (airplanes). Optionally, any damages deeper than 1 millimeter can be filled with polyester filler, which is ultimately sanded smooth. Bare metal is coated with a direct-to-metal primer in a thickness of about 60 to 100 µm, which protects against corrosion and ensures surface smoothness. This primer is then topcoated in a similar fashion to transportation OEM finishing with either a pigmented topcoat, or a base coat/clear coat system. The resulting coating system has an average total film thickness of approximately 200 to 300 µm (excluding body filler).

5.4.3 Application and characteristic data of 2K PU coatings for automotive refinish and transportation coatings

The two-component polyurethane coatings used in automotive refinishing are mixed shortly before application. The size of the repair (i.e., spot, panel or full body repair) determines the amount of material to be mixed. The polyol and hardener (polyisocyanate) components are measured in parts by volume (either using a graduated metering stick or mixing cups), and then mixed thoroughly. The coatings are adjusted to spray viscosity before application by thinning with suitable solvents (or perhaps water in the case of two-component water-borne polyurethane coatings) based on the temperature and humidity conditions for spray and drying application. The typical viscosity of an automotive refinish coating is 20 seconds flow time at 23 °C in a DIN 4 cup (Zahn#2 cup of 18 to 20 seconds flow time). The coatings are normally applied using HVLP (high volume, low pressure) spray guns, which greatly reduce spray mist formation and overspray. The two-component polyurethane primer surfacer is sanded after drying for 1 to 3 hours at room temperature and then overcoated with a base coat/clear coat or a two-component polyurethane topcoat. In the case of automotive refinish coatings, the base coat/clear coat is dried at room temperature, or in a drying booth for 20 to 30 minutes at 60 °C. When refinishing smaller areas, an IR lamp can be used to accelerate drying.

Due to the larger coating volumes that are processed in transportation coatings, stirring units (rail vehicles, buses) or even two-component mixing equipment (OEM truck coating) are often used. The ideal mixing ratio for topcoats is 2 : 1 (polyol base : hardener), but for primer surfacers it is usually 4 : 1 due to the lower hardener requirement. The spray application is usually performed using airless or airmix method.

In the transportation sector, drying is often done at room temperature. When coating rail vehicles, the maximum oven temperature is 60 °C. Truck cabins can be finished with baking coatings that are dried at temperatures similar to those used in automotive OEM finishing (130 to 160 °C), however, trends here continue towards lower drying temperatures.

5.4.4 Raw material selection for conventional solids automotive refinish and transportation coatings

There are many similarities between the formulations of 2K PU coatings for automotive refinish and transportation applications. As such, the raw materials (polyols and polyisocyanates) used to formulate these coatings will be examined collectively in this section. Various polyol components are available on the market for the formulation of two-component polyurethane primer surfacers and topcoats/clear coats. As a rule, these are hydroxy-functional polyacrylates or polyesters with various characteristics (hardness, flexibility, hydroxyl content), which are combined with suitable polyisocyanate hardeners.

The polyol resins used have a hydroxyl content between 2 to 8 % (calculated on the solid resin) and an average molecular weight between 5,000 to 20,000 g/mol. The relationship between the molecular weight and specific coating properties is of particular significance when drying at room temperature. Although resins with a high molecular weight produce fast drying properties desired, a correspondingly large amount of solvent is needed to adjust the coating to spray viscosity. This is often not in line with the overall demand for environmentally friendly coatings which are common around the globe. In principle, the solvent content of a ready-to-spray coating can be more easily reduced when using polyester resins than when using polyacrylates. However, polyacrylates generally exhibit much faster drying and better weather stability than polyesters, which has made polyacrylates the dominant binder type in automotive refinish clear coats. The conflict between achieving fast drying and a high-solid content remains the focus of continued research activity, which has led to the development of higher solid systems based on newer polyacrylic technologies and reactive thinner technologies such as polyaspartic acid esters. These resin materials will be examined in more detail in the following section.

Formulations for the various coating layers can be customized by choosing the right polyol components. Using polyacrylates with a high glass transition temperature and high molecular weight yields faster drying, increased hardness, good optical properties, but relatively low elasticity. These formulations are therefore ideal for use as primer surfacers with early sandability or as fast-drying clear coats and topcoats. The addition of polyesters with a low molecular weight flexibilizes two-component polyurethane systems. This type of more flexible combination is preferred for the refinishing of plastic components. Polyesters are also used to formulate topcoats for transportation coating, because their outstanding pigment wetting properties result in high-gloss films. Their slower drying results in better overcoat absorption, which means that a second coat can be easily absorbed by the first coat in overlapping zones, without causing any appearance problems in the paint film. Such systems are used, e.g. as topcoats for bus and rail vehicles.

The properties obtained in a 2K PU coating formulation are largely influenced by the polyol choice, but the polyisocyanate crosslinker component also plays a role as well. As automotive refinish and transportation systems are required to have good weather stability, only aliphatic polyisocyanates can be used as the hardener component. The most important products by far are biurets and isocyanurates (trimers) of hexamethylene diisocyanate. In automotive refinish coating, trimers are increasingly favored over biurets because they allow the formulation of higher solid coatings and improved long-term weathering resistance properties of the paint films. IPDI-based polyisocyanates or high functionality HDI-based polyisocyanates are added to improve physical drying speed. Catalysts and special high and low boiling solvents allow the formulations to be customized to meet application requirements:

- mixing with IPDI-based or high functionality ($F > 4$) HDI polyisocyanates – yields faster drying (spot repair, winter applications, and air drying scenarios);
- addition of catalysts (see Chapter 4) – yields improved chemical through-curing (spot repair, winter applications, and air drying);
- use of solvents with differentiated evaporation rates depending upon weather conditions – addition of high boiling solvents (full refinishing, summer applications) or addition of low boiling solvents (spot repair, winter application).

Although aromatic (i.e. non-lightfast) polyisocyanates were once used for primers, today lightfast hardeners are used almost exclusively. The main reasons for this are the simplification of inventories, the elimination of mix-ups when various hardeners are available, and the improved stone-chip resistance.

5.4.5 Low emission polyurethane coatings

Whereas in the past the emphasis was primarily on improving the quality of automotive refinish coatings, the focus today is also on environmental protection and the preservation

*Table 5.19: VOC values of low emission automotive refinish coatings; * VOC = volatile organic compounds*

Automotive refinish system	VOC* value [g/l] according to DPD**	Technology
Knifing filler	250	UPE
Primer	540	Two-component polyurethane MS/HS
Primer surfacer		Water-borne two-component polyurethane
Base coat	420	Water-borne one-component
Clear coat	420	Two-component polyurethane HS
		Water-borne two-component polyurethane
Solid-color topcoat	420	Two-component polyurethane HS
		Water-borne two-component polyurethane

* VOC = volatile organic compounds, ** Decopaint Directive

of resources. Legislative initiatives in Europe and the United States (particularly California) aim to reduce the use of organic solvents and thus cut emissions in automotive refinishing. In some Asian countries, e.g. in Japan, Korea, and especially China, there is some evidence of similar approaches.

In 1999 and 2004, the European Union passed directives (**S**olvent **E**missions **D**irective 1999/13/EC and **D**ecopaint **D**irective 2004/42/EC) which define limits for the maximum content of volatile organic components (VOC) in coatings. These directives entered into force in 2007, and maximum permissible emissions are set depending on the application and the type of coating. From January 1, 2007, the following thresholds must be observed in automotive refinishing (see Table 5.19).

Likewise in the United States, the Environmental Protection Agency (EPA) passed the National Rule for Automotive Refinish Coatings in 1998. This rule established maximum VOC limits for various coatings across the country. As of yet, this National Rule has not been updated, but certain parts of the United States have established more restrictive VOC limits. For example, in 2005, the California Air Resources Board (CARB) established VOC limits for the automotive coatings industry. In 2010, they added several additional categories of coating types. Table 5.20 summarizes VOC limits established by CARB.

It should be mentioned that the United Sates has adopted the usage of a limited number of VOC exempt solvents (such as methyl acetate, acetone and p-chlorobenzotrifluoride). These are solvents which have been tested and found to have negligible reactivity with ozone. The usage of these solvents has in many cases allowed formulation of VOC compliant coatings with relatively small changes to the actual formulations (aside from replacement of conventional solvents with ones that are VOC exempt). At the present time, there is some consideration to reverse the VOC exempt status of certain solvents, although

Table 5.20: VOC limits established in 2005 and supplemented in 2010 by the California Air Resources Board (CARB)[62]

Coating category	VOC regulatory limit, as applied [g/l]	
	Effective January 1, 2009	**Effective January 1, 2010**
Adhesion promoter		540
Clear coating	250	
Color coating	420	
Multi-color coating	680	
Pretreatment coating	660	
Primer	250	
Primer sealer		250
Single-stage coating		340
Temporary protective coating	60	
Truck bed liner coating	310	
Underbody coating	430	
Uniform finish coating	540	
Any other coating type	250	

no final decisions have been made as of yet. This has prompted coating manufacturers in the United States to begin development efforts for higher solid or water-borne coatings.

Higher solid solvent-borne coatings and water-borne coatings are somewhat in competition with each other. In many regions, solvent-borne coatings (especially clear coats and primers) with a high-solid content are preferred for automotive refinish coating because they are easier to handle and because VOC limit values are higher. In many ways, high-solid solvent-borne coating formulations perform similarly to their conventional counterparts (i.e. in terms of application properties for the applicator). In the case of base coats, it is generally more difficult to formulate at a higher solids level, while still maintaining proper metallic flake orientation and good appearance properties. As a result, water-borne base coats with a solid content of approximately 20 % are growing in usage in virtually all the major regions of the world. While the usage differs significantly by region, water-borne base coats already have a market share (in 2017) of > 50 % in Europe, 40 % in NAFTA and 10 % in China.[31] While there are some water-borne primer surfacers and clear coats already on the market, the usage has yet to see any significant market success.

The situation for transportation coatings is totally different depending on the region. Some legislation has set much lower VOC thresholds than those for automotive refinishing, depending on the surfaces being coated. Moreover, transportation coating is often a (semi-)industrial process, which allows better control during the preparation and application of the coatings. In the past, certain companies (e.g. Deutsche Bahn, Germany) switched to the use of water-borne coating systems for better protection of their employees from solvent emissions and to promote a green image. These companies significantly furthered the market penetration of water-borne formulations in transportation coatings.

Germany, the largest Western European market, often uses water-borne two-component polyurethane primers and almost always water-borne two-component polyurethane topcoats in refurbishing passenger rail vehicles and locomotives. Water-borne two-component polyurethane topcoats have been used with great success in the OEM coating of truck chassis for some years. Nowadays (2017) in China, some bus OEM manufacturers (e.g. Beijing Foton) switched to the use of water-borne coatings like two-component polyurethane primer surfacers and topcoats due to strict VOC legislation from local governments.

Figure 5.63 compares a conventional solvent-borne automotive refinish coating build-up with a modern concept system with 60 % lower solvent emissions.[66] The implementation and practical use of such a technology in refinish is entirely feasible. In the transportation sector, the situation is somewhat more complex due to legislative requirements. In addition to high-solid topcoats, which are used for buses and airplanes, water-borne topcoats and, more recently, water-borne base coat/clear coat systems have been used increasingly, especially for rail vehicles and trucks. Water-borne primers (direct-to-metal) and primer surfacers are already available and in use.

Two-component polyurethane high-solid coatings

Two-component high-solid polyurethane coatings are based on similar chemistry as the established solvent-borne two-component polyurethane coatings. The co-reactants used,

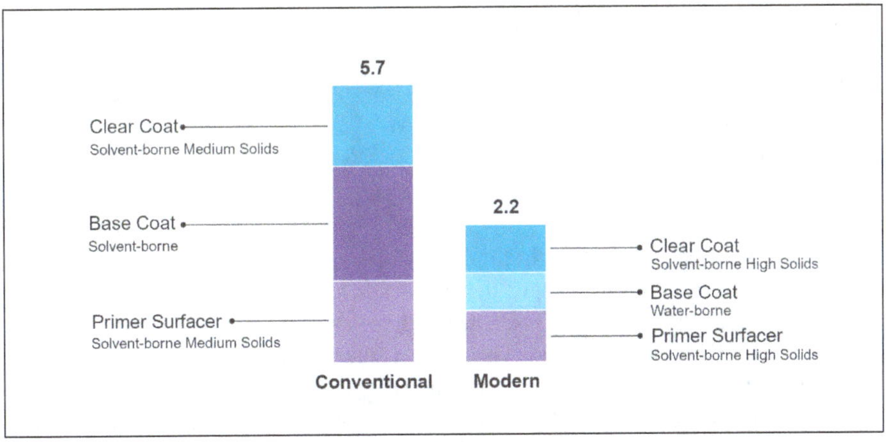

Figure 5.63: Comparison of solvent emissions in kg per car body from different coating concepts used in automotive refinishing

Table 5.21: Basic formulation of a solvent-borne two-component high-solid clear coat; VOC = 400 g/l

Component A		**Parts by weight**
Binder	e.g. "Setalux" 1907 BA-75, (Allnex)	35.00
Binder	e.g. "Eterac" 7397 MS-80, (Eternal)	30.00
Levelling agent	e.g. "Byk" 358, (Byk)	0.50
Wetting agent	e.g. "Byk" 300, (Byk)	0.10
Levelling agent, 10 % in BA	e.g. "EFKA" FL 3600, (BASF)	0.20
HALS	e.g. "Tinuvin" 292, (BASF)	0.50
UV absorber	e.g. "Tinuvin" 1130, (BASF)	0.50
Catalyst, 10 % in BA	e.g. "DABCO" T-12, (Evonik)	1.50
Pot life retarder	e.g. Benzoic acid, (various suppliers)	2.00
Solvents	e.g. Butyl glycol acetate/MAK/BA (17/17/66), (various suppliers)	29.70
		100.00
Component B		
Hardener	e.g. "Desmodur" N 3600, (Covestro)	18.40
Solvents	e.g. Butylacetate, (various suppliers)	4.60
		123.00
Characteristic data		
Solid content on formulation [%]		57
Mixing ratio components A : B		100:23
Initial viscosity, DIN 53211 #4 mm cup, 23 °C [s]		18
Mechanical properties		
Drying times (20 °C/58 % rel. humidity), GB 1728		
T_1 [min]		30
T_3 [h]		> 6
Konig pendulum hardness, GB 1730 [s]		115
Gloss 20°/60° [%]		95/96
Cupping test [mm]		> 8.7

however, are lower molecular weight products with a narrow molecular-weight distribution that requires less solvent to adjust the coatings to application viscosity. A large range of polyisocyanates and polyols are available to achieve that goal. The end properties of the paint films are of a high quality which is required. Table 5.21 shows an example of a 2K polyurethane high-solid clear coat formulation with a VOC level of 400 g/l.

In the case of high-solid acrylates the average molecular weight is only about half of conventional acrylates. Various polyols are available depending on the desired property profiles of the film. As automotive refinish coatings are also expected to yield scratch-resistant paint films, suitable polyols have been developed. One approach is to formulate self-healing clear coats via an increased crosslinking density and a low glass transition temperature (T_g)[58, 59] (see Chapter 5.3).

The molecular weight of low viscosity HDI isocyanurate is about 20 % lower than a standard HDI isocyanurate. The viscosity is thereby reduced from 3,000 to 1,200 mPa s. Additionally, using a special catalyst technology, the trimerization of hexamethylene diisocyanate can be optimized to produce a trimer product with a significantly higher level of asymmetric trimer compared to symmetrical trimer. This molecular structure greatly reduces the viscosity of the HDI isocyanurate to 700 mPa s, while the functionality of approximately 3 is still retained.[67]

Two-component very high-solid polyurethane coatings

Low emission two-component polyurethane coatings with a VOC value of < 250 g/l require the use of additive resins, such as reactive thinners, which have a strong viscosity-reducing effect. These substances are either mixed with the polyols or can be used alone. Extremely low viscosity polyisocyanates can be used as hardeners. Suitable reactive thinners include polyaspartic acid esters as previously introduced in Chapter 3. These amine reactive thinners crosslink via reaction between their secondary amino groups and polyisocyanates. They are sterically hindered to reduce their reactivity, making them easy to handle.

The low viscosity symmetric and asymmetric HDI-trimers described above are used as the polyisocyanates in very high-solid (VHS) coatings. Other HDI allophanates and uretdiones having extremely low viscosities can also be used.

By combining these low viscosity polyisocyanates with high-solid acrylates and reactive thinners, it is possible to achieve VOC values of 250 g/l or less. These products are compatible with each other and, despite the presence of amino functionality, the pot life is generally adequate for normal application conditions (approximately 0.75 to 1 h pot life). The films through-dry quickly at room temperature and when force-dried for 30 minutes at 60 °C. The

Figure 5.64: Freight cars with a very high-solid coating based on "Desmophen" NH: half the coating time, low emissions and high coating quality

reduction in viscosity also results in a reduction in functionality of the polyisocyanates discussed here – to the detriment of the crosslinking density. For this reason, attention must be paid to the solvent and chemical resistance of these formulations to meet the required specifications.[68, 69]

In Europe and the United States, these products are being used to formulate two-component polyurethane primer surfacers and clear coats for automotive refinish and transportation coatings. While these polyaspartic esters can be used to formulate very high-solid coatings for these applications, care should be taken to watch out for certain negative effects which can be present when using this technology for automotive and transportation applications. For instance, coatings formulated with polyaspartic esters may have a tendency to show in-can yellowness; additionally, they can form hydantoin structures in the resulting film which can lead to property changes. Figure 5.64 shows a freight car finished with a one-coat system based on polyaspartatic ester/polyisocyanate. The coating time was cut in half compared to the use of a conventional two-coat system (epoxy primer and polyurethane topcoat), without incurring any reduction in coating quality.

Water-borne polyurethane coatings

Two-component water-borne polyurethane coatings

Water-borne coating systems perform very differently compared to solvent-borne systems in terms of their application and drying properties. This is due largely to the different evaporation behavior and usually higher susceptibility to blistering of water-borne systems compared to solvent-borne coatings. Ambient conditions such as temperature and humidity play a larger role in the application of water-borne coatings compared to solvent-borne counterparts. Furthermore, water-borne two-component polyurethane coatings[70, 71] are a two-phase system, consisting of an aliphatic polyisocyanate crosslinker dissolved in an organic solvent and a water-borne polyol dispersion, which are mixed shortly before application. In automotive refinishing, and to some extent in the transportation sector, it is typical to mix the polyisocyanate crosslinker and the polyol component by hand using a measuring stick. Only hydrophilic polyisocyanates with good emulsification properties in water yield the desired fine droplet size of less than 200 nm. This is essential to achieve transparent films after spray application and drying.

Water-borne two-component polyurethane coatings are also expected to have a pot life of at least 3 h. It must be possible to apply the premixed two-component coating within this time without any significant reduction in coating properties. This is only possible if the polyol dispersion stabilizes the fine polyisocyanate droplets and prevents their agglomeration. Often a slight increase in viscosity is observed after mixing the polyol dispersion and polyisocyanate hardener, but it is less critical because the viscosity generally remains within an acceptable range for application.

When using water-borne two-component polyurethane formulations in automotive refinish or transportation coatings, particular attention must be given to the mixing quality (mixing fluctuations are particularly hard to avoid in bodyshop applications), as well as to the impact of temperature and humidity on the paint film. The maximum blister-free dry film thickness is particularly important as it is far more dependent on application conditions when applying water-borne coatings than when working with solvent-borne coatings. Leading water-borne two-component polyurethane coatings have drying speeds and resistance properties which are equal to those of solvent-borne formulations.

The situation in some specific areas of transportation coating is different. In the case of coating rail vehicles, where comparatively few vehicles can be coated at the same time (due to having to be dried overnight), the effectiveness of the drying process does not play the central role. Mechanical stirring units are often used to mix the hardener and polyol components. After all, the requirements of the coating's appearance properties are lower in rail applications compared to automotive refinish coating (see Figure 5.65).

Truck coating is an industrial process somewhat similar to automotive OEM finishing, which often allows acceptable control over temperature and humidity during coating and drying. Drying temperatures above 60 °C and fully automated two- or three-component (polyol plus hardener plus water) mixing equipment are frequently used. Water-borne

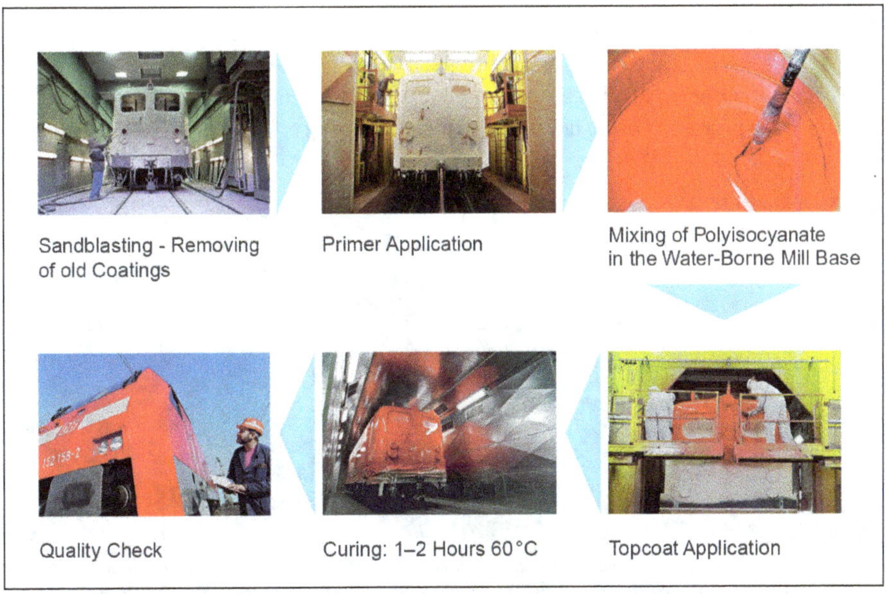

Figure 5.65: Water-borne two-component polyurethane coatings in rail vehicle refinishing

systems have successfully prevailed in recent years in truck coating applications. Figure 5.66 shows various transportation application areas where water-borne two-component polyurethane coatings are being used in Europe.

Table 5.22 contains a typical formulation for a two-component, water-borne polyurethane clear coat. Developments over the last decade have resulted in two-component water-borne systems which perform at similar level to solvent-borne counterparts. Table 5.23 (see page 224) shows a comparison of the two systems. Of particular note is the considerable reduction in solvent emissions (≤ 200 g/l compared to > 500 g/l) from the ready-to-spray coating. Other key parameters such as the pot life, drying, appearance, chemical resistance and weather stability have been improved to a level that has greatly increased market acceptance, especially in transportation coating segments. Water-borne two-component polyurethane clear coats and base coats based on binders and polyisocyanate hardeners from the latest product generation are already superior to the solvent-borne systems in some respects. For example, they are more resistant to the penetration of dyes (graffiti resistance, see Figure 5.67) and in some cases display better reactivity and faster drying. Although, water-borne two-component polyurethane clear coats for automotive refinishing are now available commercially, market penetration remains low. Often the reason for this low acceptance so far is due to the variability in properties which are obtained under the wide range of temperature and humidity conditions.

Two-component water-borne polyurethane primer surfacers

Water-borne primers and primer surfacers based on acrylates, epoxy resins or polyurethanes are either being utilized already in the automotive refinish market (i.e. Europe), or they are in the process of being developed in other regions (i.e. United States). They are also

Figure 5.66: Water-borne two-component polyurethane coatings have successfully been used to coat a variety of transportation related vehicles, including heavy truck[200] and high speed rail vehicles[201]

Table 5.22: Typical formulation and characteristic data of a two-component polyurethane clear coat for transportation coatings

Component A		**Parts by weight**
Binder	e.g. "Bayhydrol" A 2470, (Covestro)	53.34
Binder	e.g. "Bayhydrol" A 2646, (Covestro)	35.56
Surface control agent	e.g. "EnviroGem" AD-01, (Evonik)	0.89
Wetting agent	e.g. "Tego" 510, (Evonik)	0.89
Levelling agent	e.g. "Tego" Glide 100, (Evonik)	0.44
Co-solvent	e.g. PGDA, (various suppliers)	8.88
		100.00
Component B		
Hydrophilic hardener (HDI-based)	e.g. "Bayhydur" XP 2655, (Covestro)	10.63
Hydrophilic hardener (IPDI-based)	e.g. "Bayhydur" 401-70, (Covestro)	13.02
Hydrophobic hardener (low viscous HDI-based)	e.g. "Desmodur" N 3900, (Covestro)	10.63
Co-solvent	e.g. PGDA, (various suppliers)	9.11
		143.39
Characteristic data		
Solid content on formulation [%]		50
Mixing ratio components A : B		100 : 43
NCO/OH index		1.5
Initial viscosity, DIN 53211 #4 mm cup, 23 °C [s]		25
Pot life at 23 °C [h]		< 5

already used in transportation coating applications. Preference is given to crosslinkable (two-component) systems with greatly reduced thermoplasticity, which are less likely to clog the abrasive sandpapers.

Due to their good drying properties and pronounced hardness (resulting in good sandability), emulsion polymers are often used in the formulation of these systems. Mixtures of hydrophobic and hydrophilic polyisocyanates or purely hydrophobic, low viscosity polyisocyanates are used as the hardeners.

Figure 5.67: Water-borne two-component polyurethane topcoats improve the graffiti resistance of coating systems

Table 5.23: Comparison of solvent-borne and water-borne two-component polyurethane clear coats

Parameter	**Two-component polyurethane, solvent-borne**	**Two-component polyurethane, water-borne**
Solid content [%]	45	45
Co-solvent content [%]	55	≤ 10
VOC [g/l]	> 500	≤ 200
Pot life [h]	6	3
Gloss 20°	90	89
Drying T_1 (sand-dry) [h] T_3 (touch-dry) [h]	 ≤ 1 ≤ 7	 1 < 7
Force-drying* (30 min/60 °C)	0	0
Blister-free film thickness [µm]	≤ 100	80
Solvent resistance* 7 d at RT 7 d at RT + 16 h at 50 °C	 1 0	 1 0
Water resistance* 7 d at RT 7 d at RT + 16 h at 50 °C	 0 0	 0–1 0
Pendulum hardness 7 d at RT [s] 7 d at RT + 16 h at 50 °C [s]	 100 180	 160 160
Outdoor weathering, 24 months in Florida Gloss 20° before/after	88/86	87/85
1,000 h QUV (313 nm) Gloss 20° before/after	88/86	87/85

* rating from 0–5; 0 = excellent; 5 = unsatisfactory

To further improve the adhesion to unsaturated polyester filler, often various kinds of silanes additives are used in the formulation of hardeners. The resulting systems satisfy requirements with respect to other key properties, such as a pot life > 3 h, blister-free films > 100 µm, sandability after just 1 to 3 h at 23 °C, and over-coatability with either solvent-borne or water-borne two-component topcoats or base coat/clear coats.

Water-borne two-component polyurethane primers are replacing solvent-borne epoxy systems traditionally used to coat rail vehicles. This is due in particular to the better sandability and homogenous film provided by the polyurethane primer, and due to the ability of applying a two-component polyurethane topcoat in a wet-on-wet manner. This has enabled entirely water-borne systems to be used in rail vehicle coating.

Table 5.24 shows a basic formulation of a water-borne two-component polyurethane primer surfacer for transportation coatings.

Table 5.24: Basic formulation of a water-borne two-component polyurethane primer surfacer for transportation application

Raw material	Generic name/trade name	Parts by weight
Component A		
Water		11.70
Neutralizing agent, 50 % DMEA in water	e.g. DMEA in water, (various suppliers)	0.20
Wetting agent	e.g. "Surfynol" 104 BC, (Evonik)	1.00
Defoamer	e.g. "Tego Airex" 901W, (Evonik)	0.20
Dispersant	e.g. "Byk" 190, (Byk)	1.50
Dispersant	e.g. "Orotan" 731A, (Dow)	1.00
TiO_2	e.g. "Ti-Pure" R-960, (Chemours)	10.00
Talc	e.g. Talc 1250 mesh, (various suppliers)	15.00
Extender	e.g. $CaCO_3$, (various suppliers)	15.00
Rheology controlling agent	e.g. "Acrysol" RM-8W, (Dow)	0.20
Disperse at 2500 rpm for 15 min. then grind to achieve fine particles < 30 µm		
Resin	e.g. "Bayhydrol" A 2470, (Covestro)	20.50
Resin	e.g. "Bayhydrol" A 2846 XP, (Covestro)	20.50
Defoamer	e.g. "Byk" 022, (Byk)	0.30
Pigment	e.g. "Colanyl" Black N-131CN, (Clariant)	0.20
Co-solvents	e.g. "BC"/"DpnB"(1/1), (Dow)	2.40
Rheology controlling agent	e.g. "Acrysol" RM-8W, (Dow)	0.30
		100.00
Component B		
Hydrophilic hardener (HDI-based)	e.g. "Bayhydur" XP 2655, (Covestro)	10.63
Hydrophilic hardener (IPDI-based)	e.g. "Bayhydur" 401-70, (Covestro)	13.02
Hydrophobic hardener (low viscous HDI-based)	e.g. "Desmodur" N 3900, (Covestro)	10.63
Co-solvent	e.g. PGDA, (various suppliers)	9.11
		143.39
Characteristic data		
Solid content on formulation [%]		50
Mixing ratio Comp. A : B		100 : 43
NCO/OH Index		1.5
Initial viscosity, DIN 53211 #4 mm cup, 23 °C [s]		25
Pot life at 23 °C [h]		< 5

Water-borne base coats

Water-borne base coats have an increasing market share in the automotive refinishing segment globally and this continues to grow. Like solvent-borne base coats, one-component water-borne technology has become established in this segment, although crosslinking with (often hydrophilically modified) polyisocyanates can be used as well. The share of base coat/clear coat systems in transportation coatings is increasing compared to pigmented topcoats, thus water-borne base coats are also gaining importance in this segment.

The resins used for water-borne base coats are generally acrylic dispersions, urethanized polyesters, polyurethane dispersions and combinations of these. Despite their higher cost, polyurethane dispersions are typically used as co-resins to improve coating quality, and have the following advantages over acrylic dispersions:

- excellent wet color matching properties,
- greater hardness and toughness (T_g between 10 to 30 °C) with a low minimum film-forming temperature (< 0 °C),
- improved pigment wetting,
- improved substrate adhesion, and
- improved elasticity, and thus increased stone-chip resistance.

The particular advantage of water-borne base coats lies in the reduced organic solvent emissions. If the application process is included in the emission calculation, a reduction of up to 60 % can be achieved compared to solvent-borne base coat. In addition, the better hiding power of these materials means that two spray passes, instead of the usual three, can be sufficient to ensure good coverage. Experience has shown that the quality (storage stability, effects, shade matching) and cost-effectiveness (spray passes, coating times, drying cycles) of water-borne base coats are equivalent to those of solvent-borne base coats. In the case of special colors, or when working with solvent-sensitive substrates, their performance is in fact superior to solvent-borne systems.

Resin selection for water-borne two-component polyurethane systems

Automotive refinish water-borne coatings with a performance profile similar to that of OEM finishes, but with drying temperature requirements at a maximum of 60 °C, present significant challenges. The same can be said of transportation coatings due to the high surface area to be coated. On the polyol side acrylate-based emulsion polymers (primary dispersions) and secondary dispersions are available for the formulation of water-borne two-component polyurethane systems. They yield paint films with the necessary hardness, appearance and chemical resistance properties. Secondary dispersions are characterized by medium molecular weights, and a co-solvent content as high as 10 %. After curing with suitable polyisocyanate hardeners, they yield paint films with excellent appearance and resistance properties. The spectrum of secondary dispersions ranges from grades that produce good industrial coating quality (a lower hydroxyl group content) to high-quality products for refinishing that satisfies automotive OEM standards. Faster drying grades satisfy the requirement for high coating process efficiency coupled with retention of good film property profiles. Polyester polyacrylate hybrid dispersions can also be used.

Primary dispersions have much higher molecular weights than secondary dispersions and can be produced without co-solvents. They can be used to formulate primer surfacers or sanding sealers, e.g. which require very fast drying and sandability. They are generally not suitable, however, for high-grade topcoat applications because they typically cannot achieve required high gloss levels.

Especially suitable on the polyisocyanate side are low viscosity hardeners, which may be matched to the viscosity of the water-borne polyol component by dilution with organic solvents which makes them easier to mix. These hardeners are based on low viscosity isocyanurates and iminooxadiazinediones of hexamethylene diisocyanate (HDI), HDI and IPDI allophanates or HDI uretdiones. Iminooxadiazinediones (asymmetric HDI-trimers) in particular are excellent hardeners for water-borne two-component polyurethane coatings due to their high isocyanate functionality (> 3) and their good dispersibility in water-borne systems (as a result of their very low viscosity). On the other hand, polyisocyanates modified with some type of emulsifier (hydrophilically-modified polyisocyanates) are also used in the formulation of water-borne two-component polyurethane coatings for automotive refinish and transportation coating due to their easy incorporation into water-borne systems (see Chapter 3.6).

The products yielded by reacting mono-hydroxy-functional polyethers with an excess of HDI- or IPDI-trimer are characterized by good miscibility with polyol dispersions on application of low shear (manual processing). In the polyether allophanate-modified polyisocyanates, the hydrophilic polyether segments are connected to the base polyisocyanate via allophanate structures, resulting in higher isocyanate functionality and reduced hydrophillic content of the hardener (Chapter 3.6).

Such hardeners are particularly suitable for transportation coating where good hardness and high resistance properties are required. The polyisocyanate hardeners of the third generation, which are hydrophilically-modified with ionic amino sulfonate units, are characterized by a significantly reduced content of hydrophilic groups – with a comparably good dispersibility. The chemical resistance, drying, and hardness development of films cured with this new type of hardener offer further benefits over previous generations and significant property improvements. Due to the fast drying and excellent appearance properties, these hardeners are particularly suitable for formulating clear coats and topcoats for automotive refinish and transportation coating, due to the very high demands in terms of film resistance properties.

When selecting a polyisocyanate hardener, it is important to find the best compromise between easy and reliable mixing into the polyol dispersion on one hand (possibly even considering manual mixing) and achieving the necessary coating properties (especially hardness, gloss/haze, water resistance). Good compromises can be achieved by mixing low viscosity with hydrophilically-modified hardeners, typically in a ratio of 1 : 1. Also, as is common in the formulation of solvent-borne coatings for automotive refinish and transportation coating, mixtures of an HDI-trimer with an IPDI allophanate or trimer can be used to accelerate drying and prolong pot life.

The current trends in water-borne two-component polyurethane systems for automotive refinish and transportation coating can be summarized as follows:

- improvement of cost-effectiveness (faster drying, fewer coats),
- improvement of robustness (blister-free film thickness),
- further reduction of co-solvent content,
- further improvement of technical properties such as adhesion on metal, intercoat adhesion and chemical resistance,
- development of "wet on wet" process (such as only 5 minutes forced drying at 40 to 60 °C), and
- expansion of the use of water-borne systems into additional layers (such as water-borne primers/surfacers being developed to replace solvent-borne counterparts).

The first two targets will only be achieved through intensive cooperation between the manufacturers of raw materials, equipment and coatings. New mixing, drying and application concepts are keys to boosting market penetration of water-borne two-component polyurethane systems in the automotive refinish and transportation coating segments. It is anticipated that greater advances will be made in these directions with the continued global efforts to reduce VOC of systems used.

5.4.6 Radiation curing coatings

Since high drying speed and body shop efficiency are two of the most important criteria for automotive refinish systems, there has been interest in the past to develop radiation-curing automotive refinish coatings. Ultraviolet (UV) light curable systems based on polyurethane acrylates were among the preferred binders for developments due to a number of factors:

- good weather stability and good overall film properties,
- extremely fast drying and through-curing (the reaction completed within minutes),
- one-component systems – no metering or mixing of the components is necessary, thus the systems have no pot life to worry about, and
- curable with low-energy UV-A hand lamps (see Figure 5.68) which were favorable over higher intensity UV light sources from an occupational safety perspective.

Despite these excellent characteristics, these UV-curable coatings often had a number of limitations. It is particularly difficult to prevent inhibition of the free radical polymerization reaction by atmospheric oxygen at the coating surface (surface inhibition). This inhibition resulted in the coating not curing sufficiently in many cases. Moreover, the shrinkage which occurs during the polymerization can cause adhesion problems which cannot always be overcome. Due to these limitations, the few UV-curable products which were introduced did not find widespread acceptance by the market so far.

5.4.7 Outlook

It is anticipated that continued global movements towards lower VOC levels for both automotive refinish and transportation coatings will continue the established trend from conventional solids systems to alternatives based on either higher solids or water-borne materials. Time will tell which of these two VOC compliant technologies will emerge in various global regions as the dominant future technologies. For instance, developments in Europe and China appear to be favoring the usage of water-borne systems for these applications, while NAFTA tends to favor more high-solid systems.

Given the global interest in more sustainable technologies, future developments in these markets may also include new resin raw materials based on bio-based feedstocks. To this end, new polyisocyanate crosslinkers (both conventional and hydrophillically modified types) have been developed based on 1,5-Pentamethylendiisocyanate (PDI). Raw materials to produce PDI are sourced from sustainable bio-based feedstocks.

There is also considerable speculative discussion ongoing regarding the future market trend for automotive refinish systems.[61] On one hand, vehicle miles travelled have continued to increase (especially in regions where vehicle ownership is growing at high rates) which generally favors increased vehicle collisions, and thus repairs. Distracted driving due to smartphone usage has also contributed to higher collision rates in some markets. This is contrasted by vehicle collision avoidance technologies, such as advanced driver assistance systems (ADAS), which are designed to reduce or eliminate driver error, which is estimated

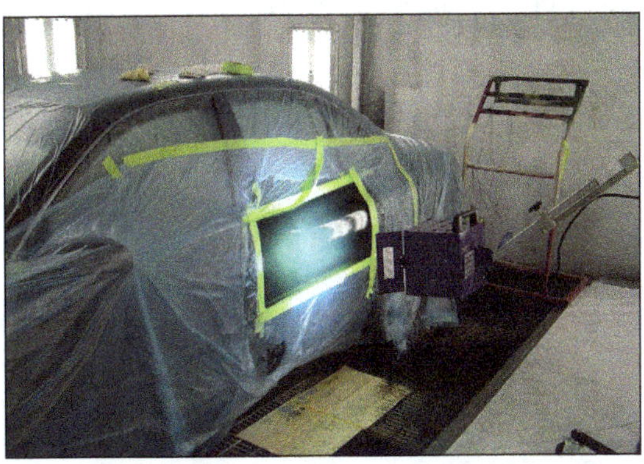

Figure 5.68: UV-curing polyurethane coatings for fast refinish operations

to cause 94 % of all collisions. While various studies generally predict that automotive refinishing market size will contract in future years, the timing and magnitude of these contractions differ greatly from study to study.

5.5 Plastics coating

Plastics have become indispensable, whether in industrial applications in the automotive,[72] electronics and household appliance sectors or in the leisure segment. In short, countless items we use every day are made from plastics. Over the last 20 years, it has also become common practice to apply coatings on plastics to enhance their appearance and technical properties:[73–75]

- to cover flow lines in unfinished moldings so as to produce certain shades and color effects,
- to protect plastics from weathering, chemicals and mechanical impact[76], or
- to give plastics surfaces certain haptics.

A large number of systems are suitable for coating plastics. Acrylic coatings, for example, have proved particularly suitable for finishing rigid plastic components such as housings for home stereo equipment. The choice of suitable coating systems becomes more restricted with increasing flexibility of the substrate and stringency of the quality requirements, e.g. automotive bumper.

Coatings based on polyurethane raw materials satisfy the technical requirements to a high degree.[77] In the automotive industry, which has always set very high standards in terms of light fastness, weather stability, flexibility, chemical resistance and scratch resistance, polyurethanes have become the systems of choice for coating plastics. Polyurethane technology is being used increasingly in industrial plastics coating.

5.5.1 Market evaluation

Plastics substrates

The world market in 2017 for plastics currently stands at about 350 million tons. About 3.8 % (13.4 million tons)[78] is used by the automotive industry, around 70 % of which are coated. Thermoplastics (mostly with low heat resistance) are used in the manufacture of both exterior and interior automotive components such as fenders, mirror housings and instrument panels. These plastics include polypropylene (PP), polycarbonate (PC), acrylonitrile butadiene styrene (ABS), polybutylene terephthalate (PBT) and their blends. Polystyrene (PS), polyamide (PA), polyvinyl chloride (PVC) and polyethylene terephthalate (PET) are also used.

Plastics coatings

In 2017, some 740,000 t of formulated coatings were used for coating plastics (see Figures 5.69 - 5.71). Around one third of this amount was used in the automotive industry. Coating systems with a high baking temperature (> 120 °C) are not suitable for use on plastics with low heat resistance (with low T_g). Two-component polyurethane coatings with bake temperatures around 80 °C are mainly used because of constantly increasing quality demands like chemicals resistance and scratch resistance.

In 2017, polyurethane coatings (solvent-borne, water-borne and UV-curing) accounted for pprox. 68 % of the formulations used in plastics coating worldwide, with annual growth of about 3 %. Polyurethane coating is the perfect technology for plastic coatings because it can be cured at low temperature and it has good chemicals resistance, good appearance and scratch resistance. Thermoplastic acrylic coatings are mainly used for pigmented base coat in exterior parts like bumper and mirror houses. UV-curing coatings are mainly used for electronics plastics coatings which require higher hardness and scratch resistance, but UV curing can only cure flat substrates. Ultimately, the pressure to reduce solvent emissions, will lead to a further increase in the use of water-borne coatings.

5.5.2 Coating process

When coating plastics, the formulation and application technology must be matched to the substrate. Pretreatment, formulation design and the choice of raw materials are of great importance.

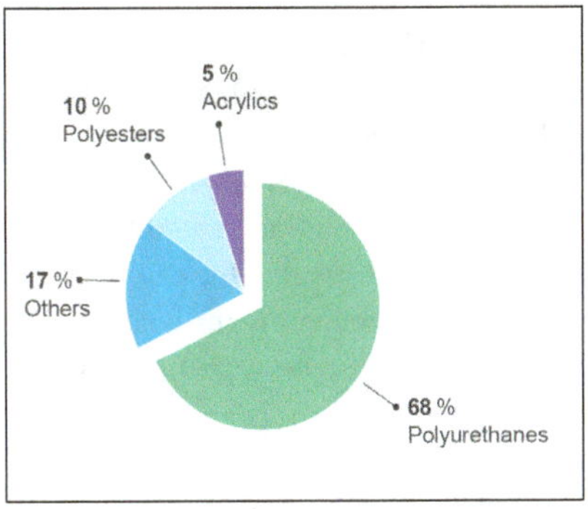

Figure 5.69: Global consumption of plastic coatings by chemical systems (2017), total: ~ 740 kt[1]

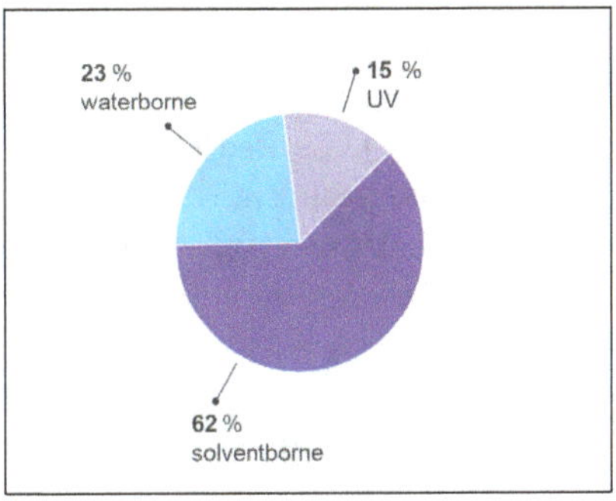

Figure 5.70: Global consumption of plastic coatings by technologies (2017), total: ~ 740 kt[1]

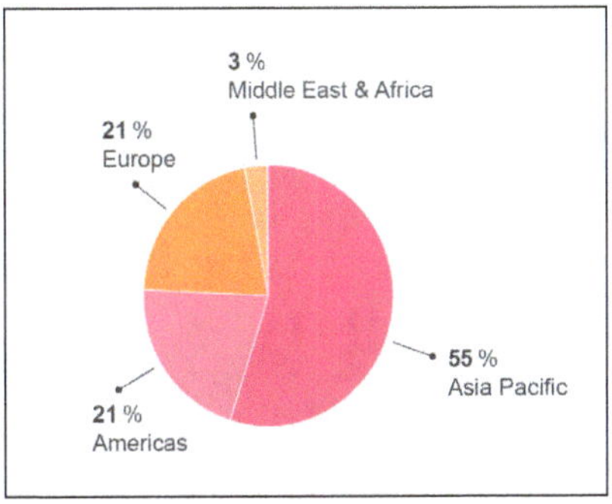

Figure 5.71: Global consumption of plastic coatings by regions (2017), total ~ 740 kt[1]

Pretreatment

As plastic surfaces are often contaminated with release agents, oil, grease or dirt, they must be cleaned before coating, e.g. using an acid or alkaline cleaner in a power washing station (see Figure 5.72).

A cleaning solution heated to 60 °C is pressure-sprayed onto the plastic components. After rinsing with water, the components are dried and blown down with ionized air to prevent dust contaminating the coating process. In the case of plastics with a low surface tension (e.g. polypropylene (PP)), additional surface activation may be necessary.[79, 81] Without this treatment, the coating may not adhere properly. Surface activation can be achieved by flame[79], corona[82] or plasma[79] pretreatment.

Flame treatment involves passing a flame briefly over the component. Polar groups form on the surface as the result of oxidation. In the corona method, a high voltage electrode is passed over the earthed plastic component. The corona formed results in a bombardment with ions or electrons that activates the surface in a similar way to flame treatment. In the plasma process, usually performed in dilute gases, the surface is activated by the effect of ions released from the gas, i.e. not only air ions as in case with the corona process. The plasma process is used less often due to the relatively high capital investment required.

Corona pretreatment is used above all to activate planar surfaces (e.g. films), while flame treatment is the standard process for three-dimensional components such as fenders.

In addition to surface activation methods described above, it is also possible to use adhesion primers or adhesion promoters based on chlorinated polyolefins.[79, 83] These also ensure good coating adhesion on non-polar polypropylene substrates.[84] However, the use of organochlorine compounds is being viewed increasingly critically in some countries.

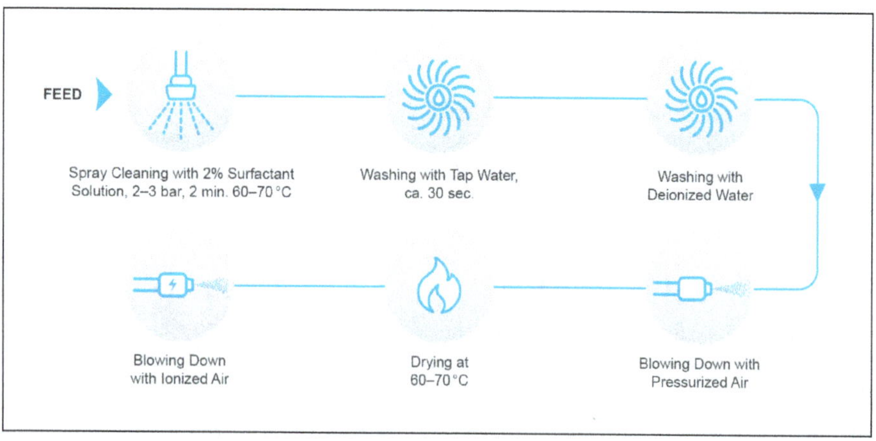

Figure 5.72: Flow chart of a three-stage power washing process

Table 5.25: Coating concepts for plastics

	Film thickness [μm]	Drying
Three-coat system		
Primer	20–30	10 min at RT, 30 min at 80 °C
Base coat	10–15	5–10 min at RT, 5 min at 80 °C
Clear coat	25–50	10 min at RT, 30 min at 80 °C
Two-coat system I		
Primer	20–30	10 min at RT, 30 min at 80 °C
Topcoat	30–50	10 min at RT, 30–45 min at 80 °C
Two-coat system II		
Base coat	10–15	5 min at RT, 5 min at 80 °C
Clear coat	25–50	5 min at RT, 30 min at 80 °C
One-coat system		
Topcoat	30–50	10 min at RT, 30 min at 80 °C

Application conditions:
solvent-borne: 30–70 % rel. humidity at 15–30 °C; water-borne: 45–60 % rel. humidity at 18–25 °C

The application of primers is often combined with other methods of pretreatment to enhance the general reliability of the plastics coating process.

Coating concepts

Three different coating concepts are normally used for plastics:

- three-coat systems (primer/base coat/clear coat),
- two-coat systems (primer/topcoat or base coat/clear coat), or
- one-coat systems (topcoat).

Growing environmental awareness and increasingly stringent VOC legislation have resulted in an increasing switch from solvent-borne to water-borne coating systems, particularly in Europe. It can be assumed that water-borne coating systems will be used increasingly in North America and the Asian-Pacific area as well. Typical film thicknesses and application conditions are summarized in Table 5.25.

In a three-coat system (see Figure 5.73, see page 234), the primer is applied directly on the cleaned and possibly pretreated plastic component. After a flash-off time of 5 to 10 minutes, the coating is dried for about 30 minutes at 80 °C. Dry film thicknesses of 20 to 30 μm can be achieved. The base coat is applied on the primer to give a dry film thickness of between 10 to 15 μm. After a flash-off time of about 5 minutes, the clear coat is applied and then force-dried together with the base coat for 30 to 45 minutes at 80 to 110 °C. The target dry film thickness for the clear coat is between 25 to 50 μm.

The three-coat system is mainly used in coating automotive components based on PP blends, e.g. fenders. The two-coat system (normally system II shown in Table 5.25, see page 233) is used in the coating of thermoplastic resins such as ABS and PC and their blends. This process is widely used for the coating of automotive and motorcycle components, especially in Asia. One-coat topcoats are used primarily on plastics in non-automotive applications, e.g. electronic housings, household appliances and toys.

5.5.3 Raw material selection

The selection of raw materials is as important as the application conditions. The film properties must be matched to the substrate, especially in terms of flexibility. Hard coatings on a flexible plastics substrate have been proven to result in splintering and the risk of injury e.g. in the event of a crash. This is not true of coatings whose flexibility has been matched to the substrate.[85]

The impact penetration test according to DIN 53443 is suitable to investigate this relationship. In America and the Asia Pacific region there are no specific test methods in use. High-speed photographs of an impact penetration test of coated PBT/PC show how a brittle topcoat results in splintering of the component (see Figure 5.74a). If a flexible coating such

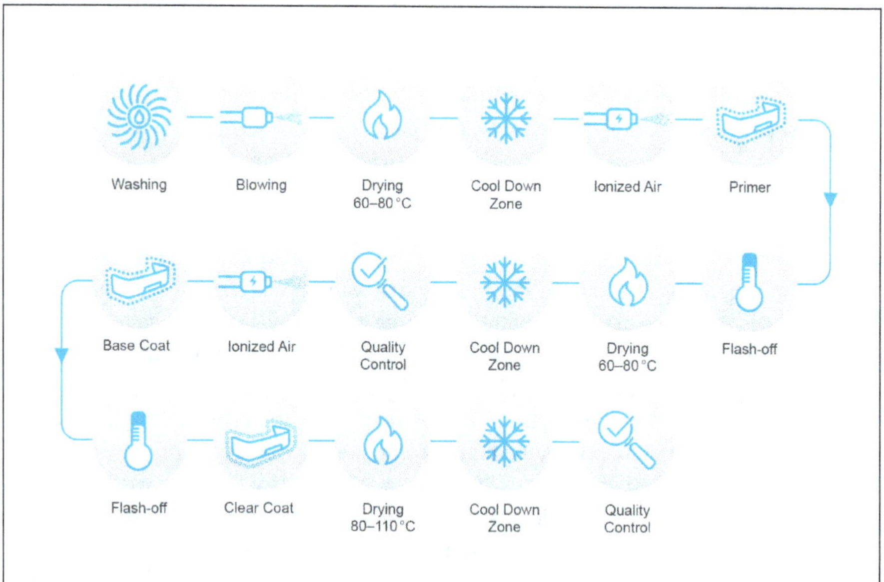

Figure 5.73: Coating line for a three-coat plastics coating system

as a two-component polyurethane formulation is used, the mechanical properties of the plastic remain unaffected (see Figure 5.74a and b).

On account of the major influence of temperature on the mechanical properties of both coated and uncoated plastics, the impact penetration test is normally performed in a temperature range of -40 to +30 °C. At temperatures down to -40 °C, uncoated PBT/PC components show no significant change in their energy absorption, as shown by Figure

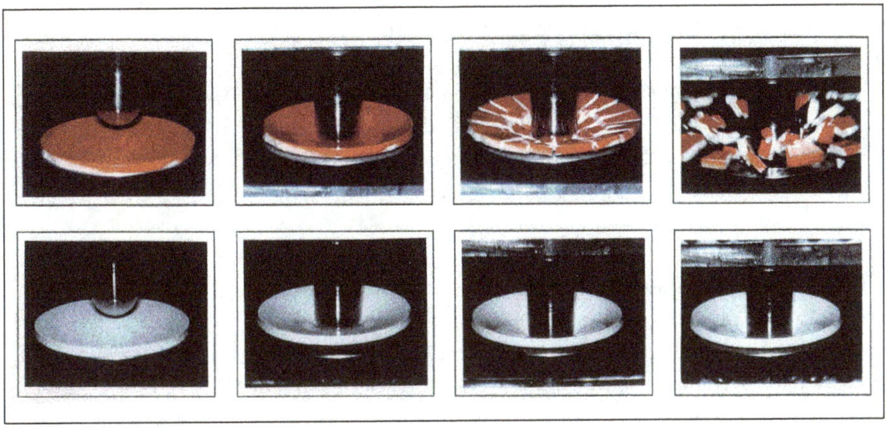

Figure 5.74: High-speed photographs of an impact penetration test of coated "Pocan" (PBT/PC); a) (above) a hard coating causes splintering; b) (at the bottom) a flexible two-component polyurethane coating prevents splintering

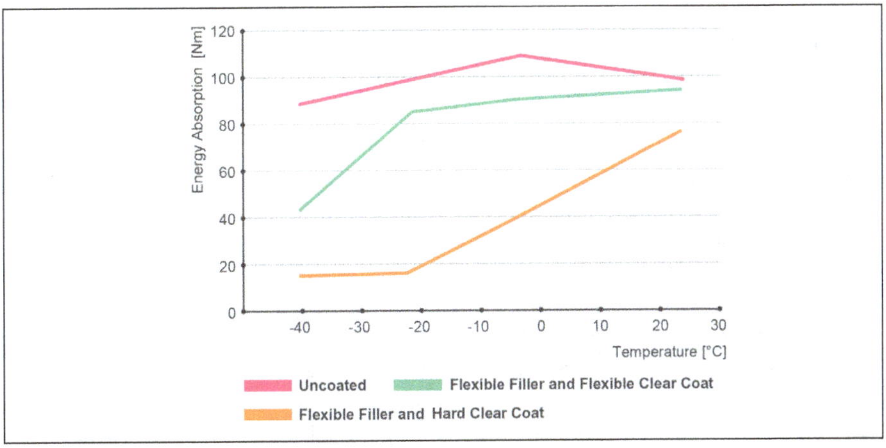

Figure 5.75: Impact penetration test on coated PC/PB

5.75. The ability of the coated component to absorb energy without fracturing is an indication of its splintering tendency. For example, a hard topcoat on a flexible primer reduces the energy absorption to 40 Nm at 0 °C, which would indicate a risk of brittle fracture in the plastic. In contrast, a flexible two-component polyurethane topcoat retains its energy-absorbing properties at -20°C.

Even better results are achieved using a water-borne polyurethane primer.[86] The notching effect of a brittle topcoat is buffered by the polyurethane primer. It should be emphasized that water-borne primers allow the use of relatively hard topcoats, e.g. solvent-borne OEM clear coats – an aspect which is important in the context of online and in-line coating of plastics.

Polyurethane clear coat systems for plastics coating are often based on aliphatic polyisocyanates such as HDI-trimer or biuret e.g. from the "Desmodur" N family in combination with flexible polyester polyol allows the flexibility of the paint film to be matched to the plastic substrate.

Coating concept – primer

The use of primers is state of the art in plastics coating.[87] Primers cover defects on the surface of the molded component and improve adhesion to the substrate. Good sandability, stone-chip resistance and over-coatability are required here and are achieved using polyurethane systems. HDI biuret e.g. "Desmodur" N 75 combined with polyester or polyether polyols is used for a solvent-borne formulation. Corresponding water-borne primers are formulated using polyurethane, polyester or polyacrylic dispersions or blends ideally combined with a hydrophilically-modified HDI-trimer, e.g. "Bayhydur" 3100 or "Bayhydur" 2655.

Most European coatings lines for exterior plastics, however, use the standardized hardener from the clear coat also to crosslink the hydroprimer, yet with lower crosslink density. Most water-borne primers in Europe also contained small amounts of N-methyl pyrrolidone (NMP) as co-solvent improving wetting and adhesion. Recent changes[88] in labelling rules for mixtures containing such co-solvent led to a phase out of NMP and reformulation of respective hydroprimers.

Conductive primers are required for the electrostatic coating of plastic components. They are formulated using conductive pigments.[79, 89] The most important of these is carbon black, not least because of its low price.

Coating concept – base coat

The base coats used on plastics are expected to meet similar requirements to those used on metal bodies.[79, 87] The special aspects of matching color and effects have already been discussed. Apart from greatly reducing the solvent content, the use of water-borne base coats has a further advantage. The parallel alignment of the metallic flakes is more pronounced due to the greater shrinkage of water-borne coatings and is thus reproducible.

Coating concept – clear coat

The clear coats used on plastics must be of high quality and match the flexibility of the substrate. These requirements are satisfied by two-component polyurethane systems, even at the low bake temperatures normally used in coating plastics. In contrast to the mostly hard melamine-crosslinked systems (thermosetting acrylics (TSA)), the two-component polyurethane coatings are characterized by the fact that their flexibility can be can be varied infinitely to match the flexibility of the substrate through the choice of raw materials. Even a flexible two-component polyurethane system, force-dried at just 80 °C, meets all standards for leveling, gloss (wet look) and resistance to weathering, etching and chemicals. This is not true of one-component TSA and one-component TPA (thermoplastic acrylic) coatings, even if they are baked at higher temperatures.

Two-component polyurethane clear coats are characterized by a more or less pronounced reflow effect. This means that scratching caused mechanically heal due to a recovery effect[80, 90]. As a result, there is relatively little scratching on component finished with this kind of clear coat, even after many years of use. In order to satisfy the demand for higher flexibility coupled with good resistance properties, many two-component polyurethane formulations are based on combinations of flexible polyester polyols such as "Desmophen" 670 with hard chemically resistant polyacrylates. Suitable polyisocyanates are HDI-trimers and -biurets. The use of flexible polyisocyanates e.g. "Desmodur" N 3800 also yields highly flexible coatings suitable for plastics.

The transfer of solvent-borne two-component polyurethane technology to water-borne systems has been achieved with equivalent polyols such as polyester or polyurethane dispersions, hydrophilically-modified HDI-polyisocyanates. For exterior components, solvent-borne technology is still predominantly being used.

Work is under way to develop special thermally activated PU-hardeners which deblock at low temperatures between 80 to 100 °C, thus facilitating one-component formulations with high resistance properties for plastics coating .[91] The application conditions for water-borne systems are more strictly defined than those for solvent-borne coatings, in particular due to the differences in drying behavior.[87]

5.5.4 Coating concepts for automotive add-on components

When coating plastic components for automotive applications, various processes can be used (see Figure 5.76, see page 238). Unless contrasting colors are used, it is particularly important that the coatings applied on the different components, whether interior or exterior, and on the car body all match in terms of color and effect.

- Offline: The add-on component is coated by the supplier and then fitted to the coated vehicle by the car manufacturer. This is the current state of the art in the automotive industry.
- Inline: The uncoated add-on component is fitted to the car body after application of the CED primer and passes through the remaining stations of the online coating process. In other words, the primer surfacer, base coat and clear coat are applied in the usual way. Dust may cause problems as it is virtually impossible to avoid dust contamination after application of the CED primer. Many of the plastics used are at the limits of their heat resistance and fracture strength at the baking temperatures of 130 to 160 °C. In addition, the differing coefficients of thermal expansion of plastics and metal must be considered when fitting of the add-on components. These are challenging process conditions for the plastic substrates. There are several concept studies on

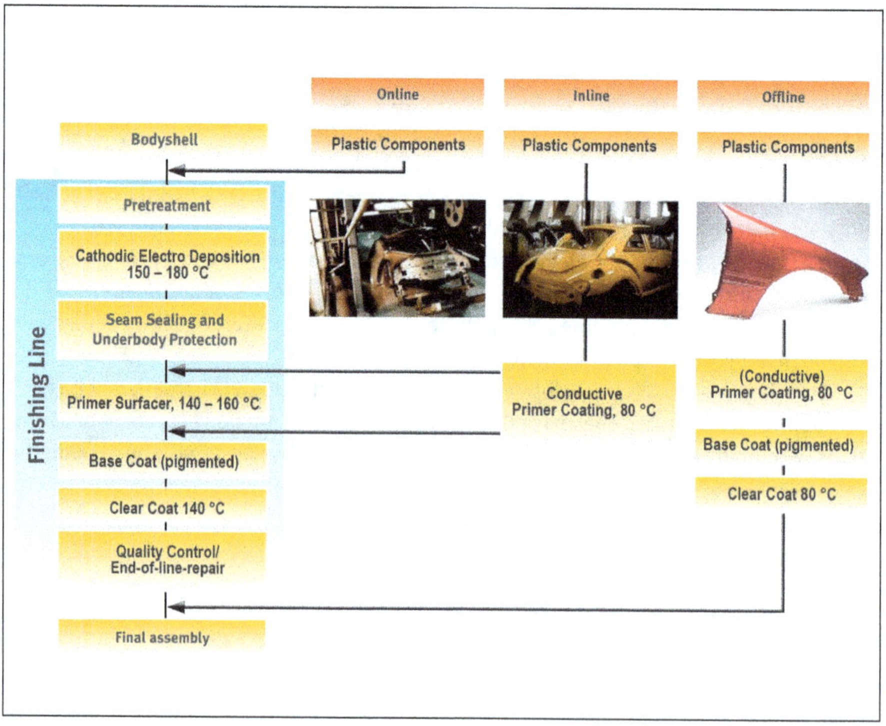

Figure 5.76: Coating concepts for the automotive industry: online, inline, offline

low temperature paint processes that use versions of inline processes to at least apply base and clear coats on metal, plastic and composite parts which have been preprimed elsewhere.[92]
- In an online process, the plastic components are fitted to the uncoated car body and pass through the entire coating process. The plastic component has to withstand the relatively high baking temperature used for the CED coating without deformation (see Chapter 5.3.1). Until now, this level of thermal resistance was found in only a few plastics such as sheet mold compounds (SMC), polystyrene/polyphenylene oxide (PS/PPO) or reinforced reaction injection molded polyurethane (PU-RRIM).

Given the difference in conditions between offline and online coating, a difference in shade is virtually unavoidable, even if it is only slightly visible in the best case.

In the case of online and inline coating, the differences in conductivity between plastic and metal play a major role, even if a conductive primer has been applied on the plastic component. Just slight variations in the electric field density may cause variations in shade in an electrostatic coating.

Two-component polyurethane systems are widely used in the offline coating of automotive add-on components. Inline and online coating are the object of ongoing development work. Two-component polyurethane technology is also ideal for these processes. Recent advances in thermolatent catalyzed two-component polyurethane coatings will facilitate the development of low bake paint processes. Low bake two-component polyurethane clear coats applied by the current bell systems can only be lightly catalyzed. Any more catalyst would also get active during film formation and solvent flash-off, thus impacting appearance. A new generation of catalysts that develop their full activity only at oven temperature (80 °C) can be added to hardeners in higher concentrations leading to more efficient cure.[93]

Due to tightening of VOC legislation in Western Europe, switches to water-borne technology took mostly place in the 1990s and the first decade of this century. In central Europe most of the primer surfacers and base coats used on car exteriors are now water-borne systems. In 2006, a water-borne clear coat was used for the first time in the OEM finishing of automotive add-on components. Usage of water-borne coatings in automotive applications is still expected to grow mainly driven by global concerns of interior emissions from VOC and other volatiles.

In NAFTA there is still no strong trend towards water-borne visible. In APAC, usage of water-borne coatings in automotive interior applications is driven by concerns of interior emissions from VOC and legislations. Usage of water-borne in automotive exterior applications like bumper is driven by concerns of VOC. The first switches to water-borne technologies took place around 2015 in APAC.

5.5.5 Soft-feel coatings

Haptics in car interiors

For some years, efforts to reduce the weight of cars have gone hand in hand with an increased use of plastics, also in the vehicle interior.[94, 95] The surfaces of molded plastic components, even if they have a grain pattern, appear inferior. The required surface finish can be produced using a soft-feel coating,[96–98] which gives plastic components a pleasant leather-like feel (see Figure 5.77). The coating not only provides protection against scratching and graying caused by UV exposure, but also prevents irritating noises such as the squeaking of component assemblies.

Figure 5.77: A water-borne two-component polyurethane soft-feel coating

Polyurethane soft-feel coatings also reduce fogging inside vehicles, which is caused by the emission of low molecular weight constituents from the plastics used. In contrast to back-foamed plasticized PVC laminating films, soft-feel coatings contain few low molecular weight constituents that can contribute to fogging. In addition, the coating acts as a barrier layer, thereby slowing the migration of low molecular weight substances.

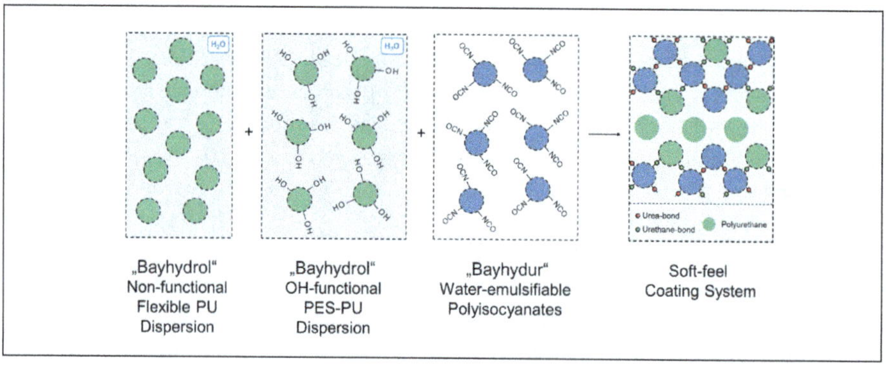

Figure 5.78: The reactions in a water-borne two-component polyurethane soft-feel coating

Water-borne formulations are usually based on highly flexible, non-functional polyurethane dispersions. Chemical resistance can be improved by adding hydroxy-functional polyester/polyurethane polyols, crosslinked with polyisocyanates, e.g. the hydrophilically-modified HDI-trimer (see Figure 5.78). Recent advances in water-borne soft touch-coatings can be found in industry trade journals.[99]

In practice, the gloss of soft-feel coatings is set very low by the addition of matting agents (see Figure 5.79). This reduces reflection on instrument panels, thus increasing driving safety. Furthermore, the matte surface underscores the luxurious appearance. Thanks to significantly reduced solvent emission during application and almost comparable film properties, water-borne soft-feel coatings are preferred to solvent-borne formulations. They were first used in the automotive industry in 1992 and are in use worldwide.[98, 100]

There are three main components that drive the balance between soft-feel and coating performance in 2-component water-borne soft-feel formulations:

- hydroxyl-functional polyester or polyurethane or polyacrylic dispersions,
- non-functional polyurethane dispersions, and
- water dispersible polyurethane crosslinkers.

The reaction of hydroxyl-functional dispersions with water dispersible crosslinkers works to build up the crosslinking of the film – this is important for chemical and scratch resistance properties. The non-functional polyurethane dispersions work to impart soft-feel characteristics.

High-quality surfaces in general industrial applications

Soft, leather-like to velvet dry coatings are also very suitable for finishing plastics in general industrial applications. Soft-feel coatings are in demand for control elements, and

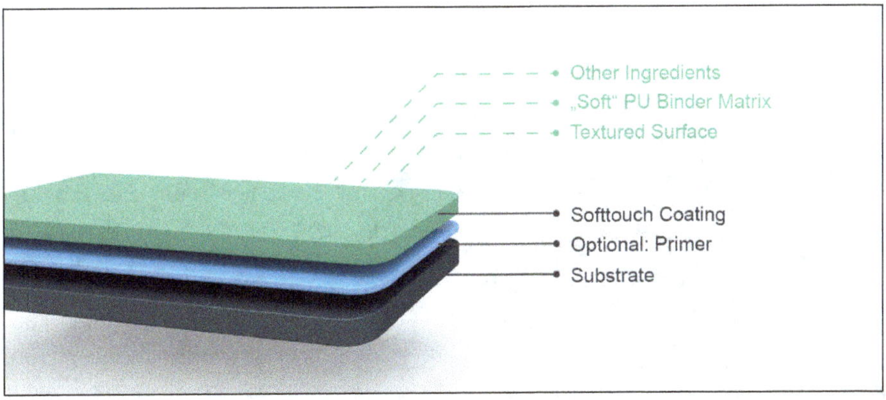

Figure 5.79: Film built of a polyurethane soft feel coating

Table 5.26: Typical formulation and characteristic data of a solvent-borne soft-feel coating;

		Parts by weight
Component A		
Resin	e.g. "Desmophen" 670 BA, 80 % supply form, (Covestro)	7.7
Resin	e.g. "Desmophen" 1652, 100 % supply form, (Covestro)	24.7
Solvent	e.g. *MPA/DAA=1:1 (various supplier)	49.1
Catalyst	e.g. "DABCO" T-12, 10 % in butyl acetate, (Evonik)	1.0
Leveling	e.g. "Byk" 333, supply form, (Byk)	0.5
Leveling	e.g. "Glide" 410, supply form, (Evonik)	0.5
Matting	e.g. "Acematt" TS 100, (Evonik)	4.5
		88.0
Component B		
Hardener	e.g. "Desmodur" N 75, 75 % supply form, (Covestro)	12.0
		100.0
Characteristic data		
Solid content [%]		~ 45
NCO/OH-Ratio		1.2 : 1
Initial viscosity (23 °C/DIN 4) [s]		~ 20

* MPA: methoxy propyl acetate, DAA: diacetone alcohol

Figure 5.80: Industrial soft-feel coating with water-borne two-component polyurethanes

this demand is growing with the continuing trend for mobility in many applications. Consumer electronic devices are coated with soft-feel coatings, as are hair dryers, food mixers and coffee machines. The applications are as varied as the consumer goods industry. Office furniture, chairs and cabinet handles, computer accessories (external hard disk drives, computer mice, monitors, notebooks), televisions and remote controls are all available with this warm and pleasant finish.

Table 5.27: Typical formulation and characteristic data of a water-borne soft-feel coating

		Parts by weight
Component A		
Resin	"Bayhydrol" U 2698, 55 % supply form (Covestro)	32.6
Resin	"Bayhydrol" UH 340/1, 40 % supply form (Covestro)	29.8
Resin	"Impranil" DLC-F,40 % supply form (Covestro)	15.0
Water		15.2
Defomer	"Byk" 093, supply form (Byk)	0.5
Wetting agent	"Byk" 348, supply form (Byk)	0.4
Feel agent	"DC"-51 (Dow Corning)	1.0
Add these materials in the sequence, and disperse at 1000 rpm (~ 2.1 m/s) for 10 min		
Matting agent	"Acematt" OK 607 (Evonik)	1.5
Matting agent	"Acematt" TS 100 (Evonik)	4.0
Added into vessel little by little and then disperse at 2000 rpm (~ 4.2 m/s) for 20 min		
		100.0
Component B		
Hardener	"Bayhydur" 304, 100 % supply form (Covestro)	5.6
Thinner	MPA	1.4
		7.0
Characteristic data		
Mixing Ratio of Component A : B by weight:		14 : 1
Spray viscosity [s]		~ 30
Solid content [%]		43.4
Pot life at 23 °C [h]:		~ 4
NCO/OH		1.5-2.0

Solvent-borne soft-feel coatings yield a surface with a rather sticky or rubbery feel, because solvent-borne soft-feel coatings are using low T_g polyester polyols and polycarbonate polyols with low T_g HDI polyisocyanates. Water-borne systems can be formulated to give dry and velvety soft, but also more leather-like, haptics. The surface grain also plays an important part. A deep grain reduces the contact surface, so a softer coating formulation is needed to produce a corresponding effect.

Because of their soft formulations, both solvent-borne and water-borne soft-feel coatings are usually less durable than other hard, highly crosslinked industrial coatings. To ensure that they still have good resistance properties, all soft-feel coatings are formulated with polyurethanes. The additional hydrogen bridge bonds of this chemistry make it the only crosslinking technology capable of uniting elasticity and robustness.

5.5.6 Industrial plastics coating

The coating of plastic components in the entertainment electronics and consumer goods industries has become indispensable. Successful marketing today includes product differentiation, which depends not only on technical performance data, but increasingly on design characteristics. High-quality finishes are in demand and are an important component of medium- and high-value products. Pearl effects, iridescent colors and, more recently, multicoat systems for high-gloss wet-look finishes are all used alongside traditional metallic finishes.[89] The further miniaturization of devices facilitates a mobile lifestyle and is accompanied by much more stringent specifications in terms of surface resistance properties. Portable devices must satisfy high requirements with regard to scratch resistance so as to preserve their appearance in continuous use. Steam jet and hot water immersion tests as well as resistance to solvents, skincare products, UV radiation, and climatic influences are further challenges for the coatings formulator.

Many substrates are used, with PP, PS, ABS, ABS/PC and PC/PBT and PA and their blends dominating in the production of molded components. PVC and PET are important materials for the production of decorative films (e.g. used in the furniture industry). PVC is also used widely in the manufacture of floor coverings. Polyurethane coatings adhere well on these later substrates, and also on others pretreated by standard processes. Polyurethane coatings are therefore also used as primers for polyurethane foams, glass fiber reinforced polyesters and, e.g. PPO blends.

Objects which are subject to intensive mechanical wear, such as portable computers, remote controls, medical instruments, computer mice, wheel rim covers, digital cameras, plastic handles or plastic furniture must be far more durable. Two-component polyurethane systems dominate in these applications. They are also used for aesthetic reasons, as only these coatings yield high-gloss films. Typical examples here are controls, buttons, optical displays or LCD and plasma flat screens. The plastic containers used in the cosmetics industry are also expected to be highly durable, making two-component polyurethane coatings popular here as well.

Polyurethane systems can also be formulated as textured coatings. The microstructure can be varied, even to produce suede-like surfaces. Both solvent-borne and water-borne polyurethane formulations are used in industrial plastics coating. The latter are of growing interest if there is a need to comply with statutory emission limits or if specifically required by the end customer.

5.5.7 UV technology in plastics coating

Important uses for UV-curing polyurethane systems on plastics are roll-to-roll application for flooring (PVC, natural rubber, polyolefin, linoleum), automotive components such as polycarbonate diffuser disks for head lights, cell phone casings made of polycarbonate or polycarbonate blends, plastic household goods, while floors are usually roller-coated with solvent-free polyurethane or water-borne coatings (Figure 5.81), diffuser disks and cell phones are spray-coated with solvent-borne formulations.

Due to the high productivity of UV-curing, the number of applications using solvent-free, solvent-borne, or water-borne UV-curing polyurethane systems[101] is growing strongly worldwide. Polyurethane is superior for applications requiring tough, flexible or particularly weather-resistant UV-curing coatings.

5.5.8 In-mold coating

The aim of in-mold coating (IMC) technology is to perform production and surface modification of molded components in a single step. The following methods are included under the term IMC:

- inserting molded components (mostly plastics, but also wood, for example) into a mold and subsequently injecting coating;
- back-molding of films (inserting film and injecting plastic or foam);
- back-foaming of films (inserting film and injecting foam);
- initially applying coating in the mold, then closing the tool and injecting plastic;
- injecting plastic, opening the tool a crack, injecting coating, closing, and pressing the coating onto the substrate;
- injecting plastic, cooling, and then injecting coating into the gap created by shrinkage;
- injecting plastic, cooling, displacing into second cavity, injecting coating.

Figure 5.81: Non-slip coating based on UV-curing urethane acrylates

IMC in the narrower sense usually refers to the methods by which a liquid coating system is applied to a plastic component still in the mold, produced immediately prior to this in an injection molding method. This is direct coating in the case of a coating application or direct skinning if polyurethane skins or foams are applied (see also Chapter 4.8). Possible applications for these technologies include:

- uses in the interior of automobiles (e.g. switches, glove compartment lids),
- uses on the exterior of automobiles (e.g. exterior mirrors, plastic lenses),
- electronics (e.g. cell phones, entertainment electronics),
- household appliances, and
- furniture, sports, leisure equipment.

Compared to conventional manufacture in two separate steps, IMC methods exhibit major advantages in terms of efficiency, surface design and ecology. In particular, due to the omission of a spatially separate coating application, there are cost advantages in terms of investment, operation and maintenance. Further savings potential results from the significantly reduced transport and logistics expenditure. In order to perform the IMC procedure economically, it is important to synchronize the timing of the injection molding and coating processes.

As IMC represents an imaging coating method, the surface structure of the coating can be adjusted reproducibly via the graining of the mold. Mold surfaces which are polished to a high gloss yield high-gloss paint films. In contrast, a rough mold surface produces matte films. In addition to the degree of gloss, the haptics of the coating can also be varied specifically over a wide range. Grained surfaces, for example, in combination with soft polyurethane coatings produce soft-feel haptics. The film thickness of the coating is determined via the mold gap. It can be varied from a few millimeters in the case of polyurethane skins and foams to a few hundred micrometers in the case of polyurethane coatings. In IMC methods, where the liquid coating is cured in a closed mold, only solvent-free coating systems are used.

5.5.9 Polyurethane gelcoats

Introduction

In recent years, the composites market has grown to a volume of 2,000 kilotons. Gelcoats play an essential role in that market by representing a unique methodology to decorate composites in ways that would not otherwise be possible with conventional coating systems. Gelcoat is term often used to describe a high-quality finish applied to a fiber reinforced composite. Gelcoats are modified resins typically based on thermosetting epoxy, polyurethane or unsaturated polyester/styrene resins which are applied to molds in the liquid state. They are cured to form crosslinked polymers and are subsequently backed with thermoset polymer matrix composites reinforced with fiberglass or other reinforcing fibers.

Gelcoat processing, performance and application techniques are all very specialized. To better understand the partnership shared by gelcoats and composites, a fundamental understanding of composite technology is useful.

Composites overview

Many products have component structural parts that are required to exhibit a combination of light weight and high strength. The wing of an aircraft is a classic example of just such an application. Conventional building materials typically provide strength at the price of weight and mass. Lightweight materials, on the other hand, are often incapable of meeting rigorous performance demands. Fortunately, this strength/weight dilemma can be successfully addressed through the use of composites. By definition, composites are materials made up of at least two-components whose combined physical strength is greater than that of either component individually. They are also typically lighter and stronger than alternative materials that could be selected for a particular structural application. Composites are often referred to as reinforced plastics because they consist of a fibrous reinforcing network that is embedded in a cured resin matrix. The resin is typically polyurethane, polyester or epoxy, and the reinforcing fibers can be chopped glass strands (fiberglass) or carbon fiber.

Gelcoat history

Due to the manner, in which they are processed, glass-reinforced composites cannot always be coated in a manner typically associated with conventional coatings. Production processes used to manufacture composites, as well other physical and performance requirements demand that alternative methods need to be employed. Therefore, if there is a need for a pleasing aesthetic appearance to a structural composite part, gelcoats are the technology of choice. In addition to appearance, gelcoats are expected to:

- provide a low-maintenance protective surface finish,
- enhance durability appropriate to the part end use,
- reduce fiber pattern on the surface of the component, and
- provide a finished surface of sufficient quality out of the mold so that the need for coating is eliminated.

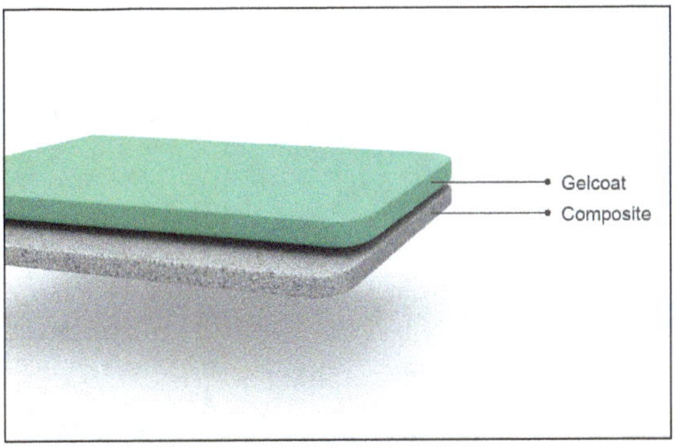

Figure 5.82: Gelcoat and composite backing

When the composites industry was in its early stages, laminating resin was mixed in the shop with thixotropy to provide a resin-rich surface on the part. However, these proved to be inadequate when demands grew for reduced air entrapment and colored finishes. This led to the

development of more sophisticated gelcoats based on unsaturated orthophthalic polyester reacted with styrene, and catalyzed with peroxides.[102] The advantages they offered include:
- reduction of the number of components mixed at the manufacturing site (only peroxide needs to be added to a pre-formulated batch),
- correct dispersion of additives such as surfactants and pigments,
- excellent air release characteristics, and
- exacting color match.[103]

Although there were cost penalties associated with these newer style gelcoats, they were offset by savings realized in labor and waste elimination of improperly mixed materials. The marine industry was among the first markets to adopt this technology on a large scale for structural component construction (see Figure 5.83). Therefore, early gelcoat development efforts focused on systems that could function well in environments where resistance to water is essential. This eventually led to the use of isophthalic-based polyesters. The formulation change provided greater water resistance, lower rates of water absorption, and improved blister resistance. Toughness was also enhanced as the tensile elongation at break of isophthalic-based polyesters is greatly improved.

Figure 5.83: Marine gelcoats offer outstanding water resistance

Figure 5.84: Gelcoats provide durable and attractive finishes

As the use of composites grew for other applications, gelcoat development also evolved to meet the performance and application demands for those markets (Figure 5.84). Improved resins and UV-resistant additives were introduced that provided gelcoats with the ability to better maintain color and gloss retention, even when subjected to severe weathering conditions. Today's gelcoat systems, which can be applied by brush or spray, can even be formulated to provide fire resistance.

Gelcoat challenges

Although gelcoat technology has made significant technological strides and has grown to an 80 kilotons market in North America, it faces serious new challenges. Among the most critical are those related to the use of styrene in gelcoat formulations. Health and environmental concerns have led to legislation that demands reduced styrene emission levels, forcing gelcoat manufacturers to develop systems that contain no more than 30 % styrene. The emission issue poses significant challenges to both gelcoat formulators and gelcoat end users alike. No substitute for styrene has yet been found to act as an acceptable co-reactant for unsaturated polyester. This severely limits the options formulators have at their disposal in developing a gelcoat that will meet government regulations and performance and application requirements. Some end users even face the prospect of a mandatory shutdown of their production lines if they exceed defined styrene emission limits.

Polyurethane gelcoat technology

The problems noted are forcing the gelcoat industry to look in new directions. Poised to meet these challenges and ready to take the industry on its next evolutionary step are polyurethane gelcoats. Unlike conventional gelcoats, polyurethane gelcoats are based on a near-zero VOC formulation. Thus, not only are concerns about styrene emissions eliminated, but the technology offers an overall environmentally friendly product.

Health and environmental issues are not the only reasons to consider polyurethane gelcoats. High performance is inherent to the nature of polyurethanes, and systems based on aliphatic isocyanates offer outstanding toughness, weathering characteristics and long-term durability. This is critically important for applications such as wind energy blades and marine/transportation vehicles that use gelcoat for outdoor applications. Those gelcoats must be resistant to embrittlement, chalking, down-glossing, fading, and yellowing. The latter is especially important to the marine market where white is the predominant color used.

Another benefit to the high performance of polyurethane gelcoats is the potential for reduced warranty liability. Whether the gelcoat application is marine pleasure craft or tub and shower surrounds, the use of higher quality materials results in the likelihood of fewer field problems or failures. This translates directly into fewer warranty claims and increased customer satisfaction. Not to be forgotten is the potential for reduced in-plant warranty claims. For example, damage can occur when a boat hull gelcoated with a conventional polyester/styrene formulation is demolded in the factory. The toughness and resiliency exhibited by polyurethanes can help to resist this kind of damage.

Application of conventional gelcoats has traditionally been a time-consuming and labor-intensive process. This translates to higher manufacturing costs, and has the industry developing better methodologies. Polyurethane gelcoats are typically rapid-curing systems but are also extremely versatile. Formulation adjustments and the use of catalysts allow the cure rate to be modified to meet the specific needs of the end use application. Controlled application time along with automated application equipment offers an opportunity to increase productivity and remove cost.

Polyurethane gelcoats attain their optimum properties when used in conjunction with polyurethane composite materials. By combining these two polyurethane systems into a single part, exceptional performance and a high degree of toughness can be expected that is not achievable with any other currently available technology. Polyurethane gelcoats bring a toughness that cannot be matched by the polyester/styrene chemistry. This increased toughness is critical to wind energy blade applications where longer blades are under consideration by design engineers.

Chemistry of polyurethane gelcoats

There are two chemistries that can be utilized when producing a polyurethane gelcoat. Both approaches have advantages and disadvantages.

One approach is to modify an existing unsaturated polyester gelcoat by the addition of a urethane acrylate. The system may be cured by conventional peroxide curing methods. This benefits the user by increasing the flexibility of the resin system, thereby improving the toughness and reducing the potential for warranty claims due to cracking. However, there is no reduction or elimination of styrene from the formulation, which is the major drawback of conventional unsaturated polyester gelcoats.

A second approach is to use two-component polyurethane chemistry. This approach can utilize either urethane or urea chemistries. The main benefits of the two-component system are the elimination of styrene and improved flexibility and weather stability. A drawback of this approach is the likelihood of higher raw material cost. However, cost becomes less of an issue when process and performance improvements are considered. For example, cost savings can be realized by fewer warranty claims, faster curing times, potential to eliminate post-mold coating, and higher factory output.

A typical two-component urethane could be comprised of a branched short-chain polyester polyol and an aliphatic polyisocyanate resin based on hexamethylene diisocyanate (HDI). The polyester polyol binder will contribute to a high-quality, hard, and weather-stable film. The aliphatic polyisocyanate will contribute weathering and chemical resistance while maintaining excellent gloss retention.

A standard two-component polyurea coating is the reaction product of an amino-functional reactant (e.g. "Desmophen" NH) and an aliphatic polyisocyanate based on HDI. The amino-functional system offers the benefit of decreased cure times and the ability to customize pot life. This system will also yield a hard, weather-stable and chemically resistant film.[104]

For both the urethane and polyurea systems, formulations could also contain pigments, fillers, extenders, diluents, plasticizers and surfactants. The pigments contribute to visual appearance, color and masking of the backing material. The remaining additives can be used to modify viscosity, volumes of reactive compositions, flow properties of the formulated coating, bulk volumes, and costs.

Conclusion for gelcoats

The use of gelcoats continues to grow along with the rapidly expanding number of applications for composite materials. At the same time, industry challenges such as legislated styrene emissions reductions and demands for improved performance and reduced warranty costs will force gelcoat technology into new directions. Polyurethane gelcoats stand ready to meet these challenges head-on and offer solutions to an evolving market.

5.5.10 Outlook

Continued efforts toward painting efficiencies, especially in automotive OEM coating processes will drive innovation in the plastic coatings market segment. The need to increase efficiency in both the speed of the coating process and in energy consumption will require coating chemistries that can cure faster and at lower temperatures, all while still providing outstanding appearance and performance properties. Polyurethane chemistry is particularly suited to these requirements and it is at the heart of new developments in wet-on-wet coating applications, in-mold coating applications, inline painting of plastic parts, and other state of the art coating processes for plastics.

The recent progress made in the development of autonomous driving vehicles has pointed out the need for coating materials that ensure a sensor detectable surface. Easy-to-clean or self-cleaning coatings are just a few of the innovation needs for this segment.

Light weighting efforts in the automotive and other transportation segments will continue to drive manufacturers to use more composites and plastic parts in the design of their products, thus leading to increased consumption of coatings for plastics.

Polyurethanes have played an important role in recent years leading to improvements in coatings for automotive interior plastic parts, as well as, consumer electronics and other non-automotive plastic applications. Additional innovation in these application areas utilizing bio-based polyurethanes appears to be a natural course of action.

5.6 Application on glass

Glass plays an important role in our daily life. It is available in wide range of colors, shapes and designs, and is used in a broad variety of fields from containers for beverages, household and consumer articles, to flat glass for glazing in buildings and cars. The following two examples show the importance of glass coatings which can contribute to significant improvements in glass quality. In general, glass can be coated with a variety of different coating systems; polyurethane coatings are especially suited, as they can be tuned to match the set of demands coming along with this substrate.

5.6.1 Coatings for glass containers

Colored glass is produced either by the addition of pigments directly into the glass itself, or by coating the glass surface with pigmented organic materials.

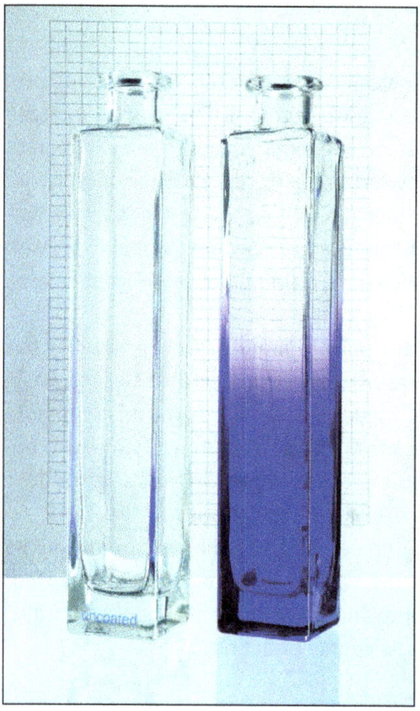

Figure 5.85: Decorative glass coating with water-borne one-component polyurethane systems

Pigmenting glass

The production and shaping of glass is generally performed at temperatures in the range of 800 to 1,500 °C. Colored glass is produced by adding metal oxides to the glass melt. The color results from the incorporation of the metal ions into the glass matrix.[106] One of the most important applications for colored glass is in the manufacture of containers for beverages. In addition to the standard browns and greens used for glass bottles, the market is increasingly requiring additional colors. These shades, ranging from yellow, orange, blue or red, can only be achieved by using the salts of heavy metals such as cadmium and cobalt, or with expensive additives such as gold.[105]

It is estimated that, during the processing and especially shaping of glass, there is a breakage rate of about 40 %. Such material must be sorted by color so that it can be remelted separately and reused – a costly and laborious process. The same problem is encountered with the different-colored

Table 5.28: Typical formulation of a water-borne polyurethane for transparent glass coating

Component A		**Parts by weight**
Aqueous hydroxyfunctional polyacrylic dispersion	"Bayhydrol" A 2770 (Covestro)	55.33
Aqueous thermally activated PU crosslinker dispersion	"Bayhydur" BL 2867 (Covestro)	42.05
Mix the components by stirring		
Component B (silane solution)		
Solvent: dipropylene glycol		2.10
Mercapto silane: γ-mercaptopropyltrimethoxysilane	"Silquest" A 189, (Momentive Performance Materials)	0.26
Amine silane: 3-aminopropyltriethoxysilane	"Dynasylan" AMEO (Evonik)	0.26
		100.00

disposable containers for beverages. They have to be sorted into clear, green and brown glass for recycling.[107]

Coating glass

Coating glass with pigmented organic materials has considerable advantages compared to pigmenting the glass itself. For example, clear glass objects coated with different colors can be recycled without sorting by melting and burning off the organic layer at temperatures between 400 to 600 °C. Such temperatures are exceeded in the melting procedure anyway, so no further energy input is required, and the process yields only clear glass.

Moreover, the coatings provide new properties which enhance the usefulness of the glass. They provide protection against shattering and give designers new options for using color and special effects. Design effects like color-gradients, mix of transparent and frosted finishes, soft-touch surfaces or metallic effects are easy to achieve. In addition to the unique shape of the containers, these effects help to convince the end user of the high value and benefits of the product.

Glass can be coated using UV-curing acrylates[108] or solvent-borne polyurethane coating systems[109]. Each of these technologies requires specific coating formulations and equipment. Examples here are special spray coating equipment, UV lamps, drying ovens and off-gas incinerators. Apart from differing in the application processes, the coatings also vary particularly in terms of their performance, e.g. chemical resistance, weather stability and appearance improvements by filling surface structures of the glass.

In addition to the systems described here, water-borne polyurethane systems are becoming more important in the market. These consist of hydrophilically-modified thermally activated polyurethane crosslinkers and hydroxyl-bearing polyurethane dispersions (see Table 5.28 component A)[110]. The formulations have a very low co-solvent content and are comparable to the above mentioned conventional systems in terms of chemical or humidity resistance, UV stability and hardness. Conventional application processes such as dipping, or spraying are used. Because of an insufficient coating stability, the silane solution (see Table 5.28 component B page 245) should be added shortly before processing. The coatings are typically baked for 30 minutes at about 180 °C.

These coatings have proved to be highly resistant to the heavy wear associated with sterilization, repeated rinsing (alkali resistance – important for use in the dishwasher) and weathering and are thereby very suitable for decorative glass coating. With their combination of hardness and durability, the coatings ensure excellent protection of the glass over a long period of time (see Figure 5.85 page 252). As the coatings can be pigmented easily, different colors and effects can be achieved economically and ecologically. This water-borne technology has opened up new perspectives for glass coating, because it is environmentally friendly, and at the same time satisfies the performance requirements specified by the industry.

5.6.2 Glass fiber sizing

Glass fiber reinforced materials play an important role for lightweight structures in automotive, electronics and construction applications. In this context, the glass fibers are typically used to reinforce the mechanical properties of a matrix material (e.g. plastic or concrete). As these composites have superior mechanical properties compared to the single matrix material, it is generally possible to lower the total weight of a specific component part, and therefore serving megatrends of e.g. mobility (i.e. lighter cars with lower environmental impact due to reduced fuel consumption or electric cars, see Figure 5.86) and digitalization (rapid growth in number of electronic devices, such as lightweight cell phones).

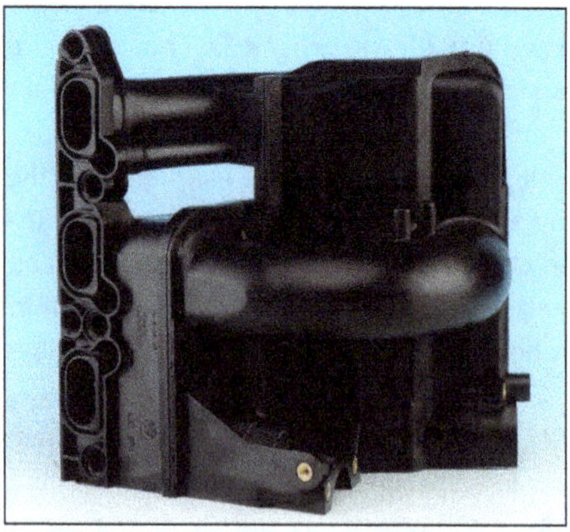

Figure 5.86: High-performance, glass fiber reinforced plastic in engine construction; component of an intake manifold made using a "Baybond" polyurethane dispersion sizing

Sizing application

Glass fibers are produced in the melt, whereas individual glass mono filaments are drawn out, coated and then combined into a fiber bundle. The diameter of the glass fiber is usually between 5 to 25 mm, and the bundles are further processed into chopped fibers, endless fibers, wovens or non-wovens. The coating process of the mono-filaments is referred to as sizing and this sizing has two important major functions: First, it protects the fibers against abrasion and breakage and gives them product-specific tunable properties. These include stiffness, cutting properties, weaving properties and in the case of chopped strands, flow properties. Secondly, the sizing is an important differentiator for these products as it determines the compatibility to the polymer or other substrate matrix and therefore the reinforcement properties of a composite.

The sizing essentially consists of a polymer (film former), an adhesion promoter (silane), slip agents, antistatic compounds, and other additives. Water serves as the dispersion medium and typically a sizing formulation has a solid content between 5 to 10 %. The positioning of the different sizing ingredients can be determined for instance by means of XPS measurements[111] and a basic model of a sizing interphase is shown in Figure 5.87.

The reinforcing properties depend on how well the glass filaments adhere to the plastic matrix. The film former used in the sizing must therefore be compatible with the matrix so as to ensure an adequate interaction. Depending on the application process of the glass fibers and the selected matrix materials, various polymers, e.g. polyvinyl acetates, polyester resins, epoxy resins and polyurethane dispersions, can be used in the sizing formulation.[112] For example, polyurethane-sized glass fibers are preferred for the reinforcing polar thermoplastics such as polyamide.

Polyurethane-based glass fiber sizings are stable to the shear forces that occur during the application process and inert to certain sizing components which could further

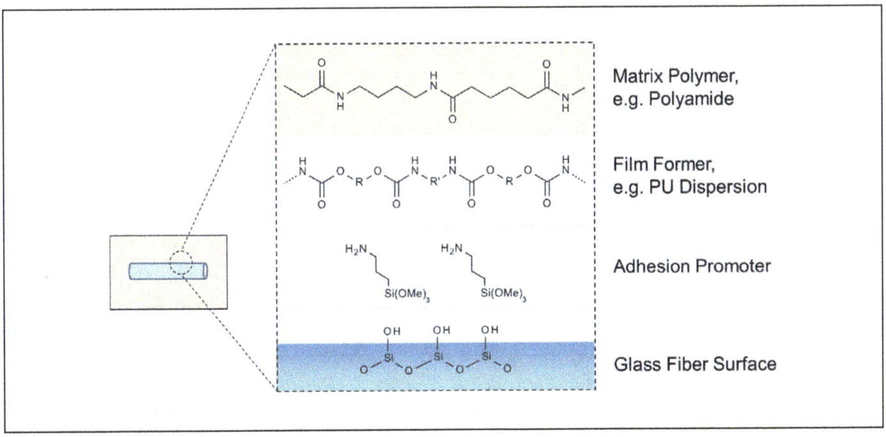

Figure 5.87: Model of sizing interphase

destabilize the film former dispersions. To cover the fibers properly they also need outstanding film-forming properties and very good colorfastness on exposure to heat as most subsequent processes of glass fibers require high temperatures.

Polyether and polyester polyurethane dispersions based on aliphatic diisocyanates are mainly used to formulate polyurethane glass fiber sizings. Both non-functional and crosslinkable dispersions are available.[113, 114] Thanks to their good and tailorable compatibility with the different matrix polymers, polyurethane sizings ensure very good adhesion and thus positively influence the reinforcing properties in the end product, such as impact resistance, tensile strength and flexural strength. Similar film forming systems can also be used to formulate sizings for carbon fibers.

The low weight of glass fiber reinforced materials and the excellent properties are expected to lead to a significant growth of this class of products. A main driver will be the digitalization and the trend to energy saving transportations systems like lighter cars, trains and planes.

5.7 Use on textiles and leather

In 2017 the textile coating and leather industry consumed more than 5,100 kilotons of resins like polyurethane, PVC, polyacrylate, silicones and others globally. Polyurethanes accounted for approximately 50 % of these resins (see Figure 5.88). Annual growth is

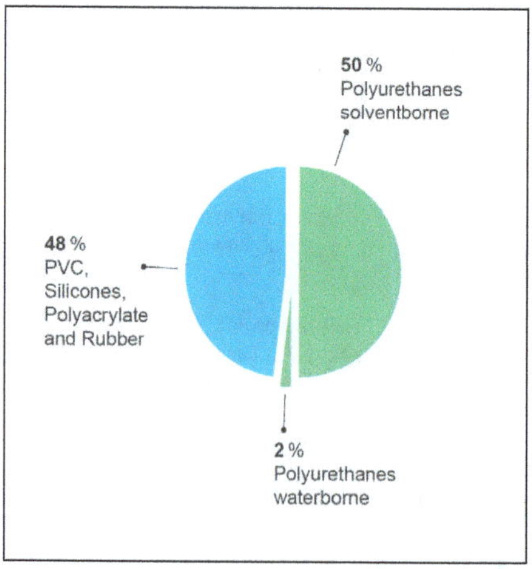

Figure 5.88: Global consumption of coating resins (supply form) in the textile and leather industry by chemical systems 2017, Total: 5,100 kt[115]

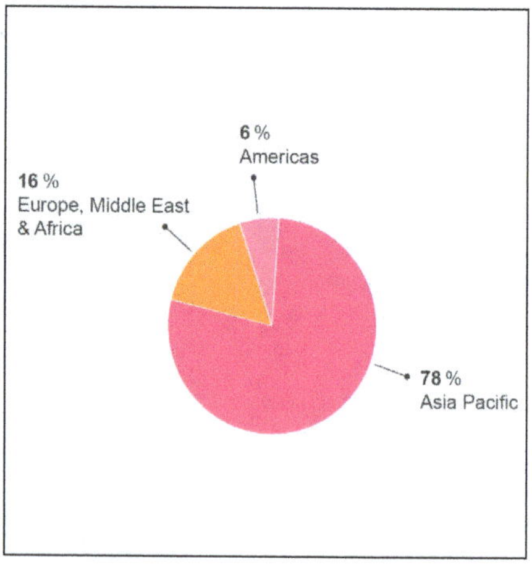

Figure 5.89: Global consumption of coating resins (supply form) in the textile and leather industry by regions 2017, Total: 5,100 kt[115]

estimated to be about 4 %. Most textile coatings (78 %) are used in the Asia Pacific region followed by Europe and the Americas (see Figure 5.89).

The reasons for the significant share and growth of polyurethanes in textiles lies in the excellent physical properties of the finished articles and the versatility of this class of raw materials. Polyurethanes can be used in the form of granulates, solvent-borne systems, high-solid, or water-borne dispersions for the coating of textiles. Over the past 5 years, polyurethane dispersions have made great strides within the textile industry driven in large part to new environmental laws.

Due to the higher material costs of polyurethanes compared to e.g. PVC, they are used only in applications which specifically require their superior technical properties: high abrasion resistance, flexibility, high brittleness temperature, chemical resistance, tensile strength and peel strength. As even lower film thicknesses of polyurethane fulfill their technical function, the cost disadvantage can often be reduced significantly or even offset completely in some cases by a lower dry add on.

5.7.1 Textile coating

The first polyurethane systems for coating textiles, which came onto the market in the 1950s, were two-component systems.[116, 117] These systems were applied by a direct textile coating method (see Figure 5.90) to textile materials used in clothing for men, women and children, for work clothing, accessories and luggage.[118]

Applications

With the introduction of one-component polyurethane systems and the transfer coating process (see Figure 5.91), new applications were developed in the early 1970s, such as production of articles with a leather-like appearance (see Figure 5.92). In the transfer coating process, the topcoat is applied on release paper (silicon- or propylene-coated paper), which can also have a grain, and then transferred to the textile using a laminating adhesive (adhesive coat).

In the 1980s and 1990s, the demand for clothing with improved wear comfort increased. This required high water vapor permeability (WVP) and improved wind resistance. Special hydrophilic microporous modified polyurethanes were developed for this purpose.[119–121]

Polyurethane coated articles are found today in the following areas:

- clothing, e.g. outerwear, shoes, screen printing;
- protective textiles, e.g. work clothing, rainwear, windproof clothing, supported gloves;
- materials for accessories, e.g. bags, luggage, wallets;
- industrial textiles, e.g. filters, seals, awnings, conveyor belts;
- environmental textiles, e.g. drainage mats, erosion nets;

- automotive interiors, e.g. seat covers, door and side panels, headliners, side curtain air bags;
- medical and hygiene articles, e.g. adhesive plasters, bandages, mattress covers;
- sports textiles, e.g. grips, balls, backpacks, sports clothing, and
- contract upholstery, e.g. hotels, marine, hospitals.

Manufacturers worldwide offer a wide range of products for a variety of different applications. Because of the combinability of polyurethane coating raw materials, textiles coated with these products can be adapted easily to meet the needs of fashion. The typical performance of polyurethane coatings includes:
- variable formulation and application,
- textured, leather-like surfaces (see Figure 5.92),
- haptic properties,

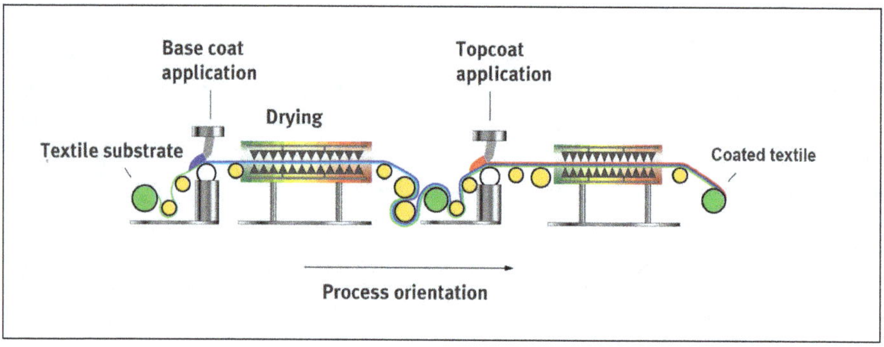

Figure 5.90: Direct textile coating process

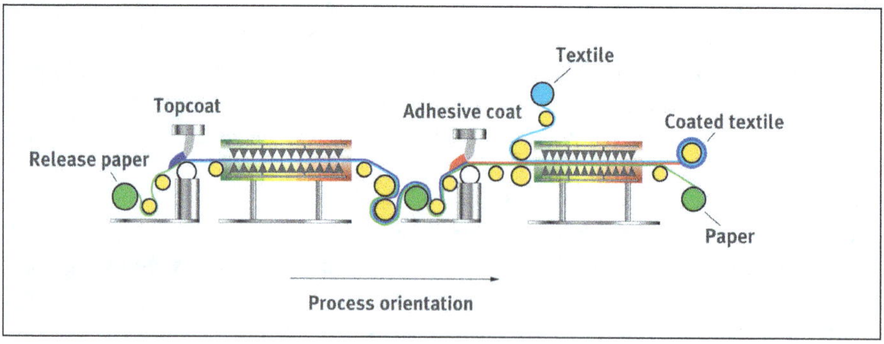

Figure 5.91: Transfer textile coating process

- resistance to washing and dry cleaning,
- adhesion to the various substrate,
- high elongation and elasticity,
- low-temperature flexibility (without plasticizer),
- abrasion resistance,
- scratch resistance,
- high flexibility,
- resistance to water and solvents,
- resistance to oils and grease,
- water vapor permeability and windproof properties, and
- low specific weight.

The type of textile coated, the polyurethane system used, the coating process and the finish products applied determine the quality and physical properties of the end product.

Textile substrates

The importance of polyurethane coatings has increased with the growing variety of textile materials. Today, coatings are applied mainly on woven and knitted goods made from various natural and synthetic fibers and their blends, as well as on non-wovens made from synthetic fibers and micro-fibers.

Formerly, when only direct coating was possible, the napping of a cotton fabric was the decisive factor in the quality of the coated goods. With the introduction of thermoplastic one-component polyurethane systems and suitable release papers, lighter and less expensive fabrics can be coated by the transfer process. It is even possible to produce softer articles in this way. Now, the napping of a cotton fabric can be substituted by a foamed polyurethane intermediate coat.

For many years, it was standard to use PVC, latex and acrylate for this purpose. Polyurethane systems are foamed by the addition of blowing agents (chemical compounds which split off nitrogen on exposure to heat)[122] or, in the case of dispersions, by the mechanical incorporation of air (mechanical foaming). Polyurethane-coated cotton fabrics without napping are used to manufacture articles such as

Figure 5.92: Example of a textured surface produced by the transfer coating process

light coats for men and women. Similarly produced light to heavy polyamide fabrics are used to manufacture work clothing, backpacks and camping equipment. Polyester and polyamide fabrics coated with polyurethanes are suitable for tarpaulins and technical articles.

Polyurethane raw materials and application methods

As any chain branching caused by triols, amines or isocyanates on the polyurethane structure can result in undesirable hardening of the coating film and a reduction in the flexibility and softness of the coated fabric, the polyurethanes developed for textile coating applications have a largely linear chain structure.[123–125] On the other hand, greater chain branching improves the resistance to solvents and water.

The polyurethanes developed for textile coating varies not only in their technical properties, but also in their preparation. Figure 5.93 shows that there are four types of product on the market: granulates, solvent-borne systems, high-solid and water-borne systems.

Granulates

Granulates are normally two-functional and are processed using isocyanates, mostly MDI, and polyols like polyethers, polyesters and/or polycarbonates by the extrusion process. The

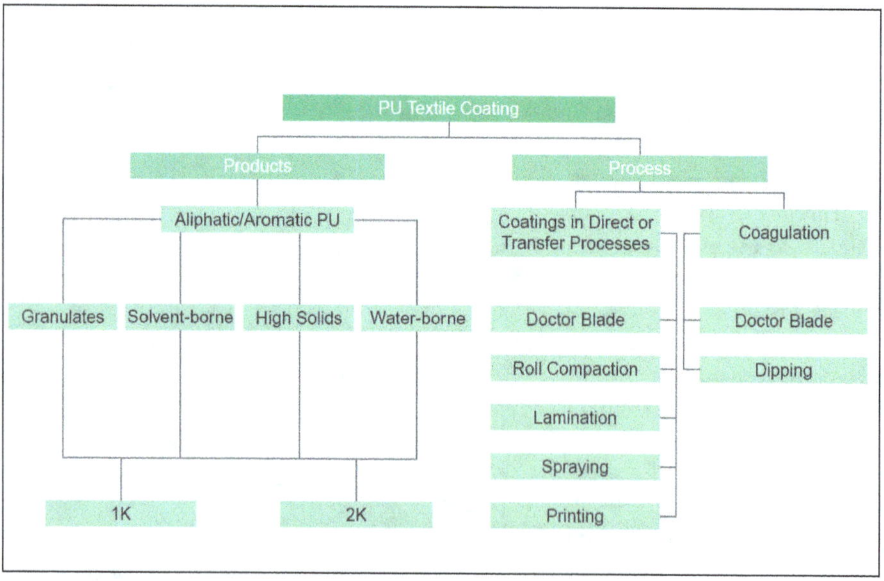

Figue 5.93: Classification of polyurethane textile coatings and processes

advantage of granulates compared to other supply forms are lower transportation costs and greater versatility for formulating tailored made systems. The major application of polyurethane granulates after being dissolved in solvents, is in the wet coagulation process using dimethyl formamide (DMF).

Solvent-borne systems and high-solids

One-component polyurethanes yield films with the desired properties without chemical crosslinking, i.e. solely as the result of interactions such as hydrogen bonding. These films are of medium hardness, highly elastic, relatively difficult to dissolve, and thermoplastic. As they are based on aromatic isocyanates, the one-component systems require the use of highly polar solvents or solvent mixtures such as dimethyl formamide/methyl ethyl ketone.[126, 127]

Using cycloaliphatic isocyanates such as isophorone diisocyanate, lightfast one-component polyurethane ureas were developed in the 1970s. Soft solvents such as alcohols and toluene are normally used with these systems. The aliphatic polyurethanes do not only have good lightfastness, but can be processed easily and safely to yield coatings with outstanding flexibility.

Usually harder polyurethane ureas are used as finishes on polyurethane coatings when necessary. These finish films, normally 5 to 8 µm thick, improve the feel, appearance, and fashion aspects of the coating, as well as its scratch resistance. They are applied by spraying, printing or knifing by a doctor blade.

The high-solid systems mentioned are two-component polyurethanes consisting of thermally activated PU prepolymers reacted with cycloaliphatic diamines (see Figure 5.94). 1-Methoxypropylacetate-2 is normally used as the solvent.[128]

Before application, the diamine is added to the prepolymer at room temperature. These mixtures have a long pot life and the crosslinking reaction only starts at temperatures above 60 °C. Depending on the type of prepolymer used, temperatures of 150 to 175 °C are required for full crosslinking.

A typical multilayer system of polyurethane mostly applied by the transfer coating process is for manufacturing of soccer balls (see Figure 5.95 page 262).

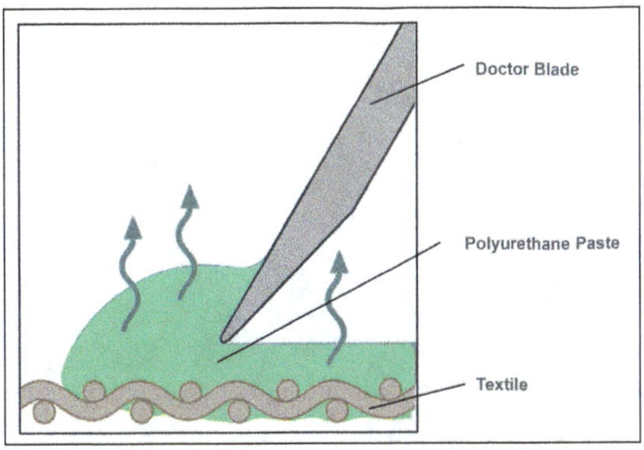

Figure 5.94: Application of a textile coating via knifing by a doctor blade

1. Aliphatic topcoat made from polyurethane solution
   Protected underneath clear coat, prevents abrasion of the printed surface.
2. Aliphatic intermediate coat based on two-component high-solid polyurethane
   Protects the ball from external influences and ensures exceptionally high elasticity.
3. Syntactic coat based on two-component high-solid
   Consists of polyurethane containing millions of gas-filled microspheres. This gives the ball its outstanding resilience, which significantly improves the flight properties.
4. Adhesive coat made from polyurethane solution
   Bonds the various layers to the textile substrate.
5. Special polyester/cotton fabric.
   Serves as the substrate.

Water-borne polyurethane systems

Water-borne dispersions with good storage stability are produced from linear one-component polyurethanes (see Chapter 3.8.1). They contain internal emulsifiers – incorporated into the molecular chain – as well as ionic groups which produce an additional hydrophilic effect, or corresponding non-ionic segments. Depending on the type of ionic groups incor-

Figure 5.95: Selected world cup soccer balls manufactured with multi layers of polyurethane coatings

porated, a distinction can be made between anionic and cationic dispersions, whereby the latter is not as commonly found in textile coating.

An alternative manufacturing process uses external emulsifiers. Most of the polyurethane dispersions are co-solvent-free, just in a few cases, such dispersions contain small amounts of co-solvents, as N-methylpyrrolidone (NMP) or N-ethylpyrrolidone (NEP). They are used as adhesive coats or topcoats, and as foamed intermediate coats applied by the mechanical foam method.

The use of water-borne one-component polyurethane dispersions and products with a low solvent content such as two-component polyurethane high-solid system offers considerable potential for reducing the solvent content. The resulting ecological advantage is the reason that this type of polyurethane coating is being used more frequently.

Direct coated textile for water-borne applications are becoming more common, e.g. for curtain air bags, conveyor belts (see Figure 5.96) and the apparel industry, as new governmental regulations have helped to move the industry from solvent-based systems to water-based systems. These applications were dominated by solvent-based systems for many years.

Effect of the application methods

Special techniques exist for the application of granulates. One is the melt calendaring process using thermoplastic polyurethane granulates. Another is coagulation, in which the coating material is precipitated in water from a one-component solution in dimethylformamide (DMF). A third option is film lamination using the thermal transfer method.[129] In addition to the material used, the type of coating method[130-132] also influences the properties of the finished article. For example, the application rate and the feel of the finished product changes with the type and hardness or flexibility of the application instrument (knife-over air, knife-over roll) when applying a polyurethane solution straight to the nap of the fabric by direct coating.

By contrast, the transfer process yields particularly soft coatings as e.g. synthetics. The coating material is applied as a solution on release paper. It is cured by heating and then laminated onto the textile using

Figure 5.96: Conveyor belt coated with water-borne polyurethane *Source: markobe – stock.adobe.com*

a thin adhesive coat. Even high-solid and dispersions, adjusted to the necessary coating viscosity of about 5,000 mPa s with just a small amount of thickener, can be applied by transfer coating. Dispersions thickened in this way and then foamed mechanically can yield very soft materials for outerwear, even though they contain no emulsifier.

5.7.2 Polyurethane synthetics and microporous coatings

An essential characteristic of leather is its microporosity. It has the ability to adsorb and release water vapor. Efforts to develop a synthetic material with the appearance and properties of leather started some 65 years ago.[133] About ten years later, "Corfam" (DuPont) was the first poromeric synthetics on the market.[134] The manufacture of this product was later licensed to Clarino in Japan where development continued to enhance the product with a micro-fiber non-woven. Brands such as "Alcantara", "Amara" and "Amaretta" are now of importance worldwide.

Synthetics by coagulation process with DMF

One of the first steps in this development was the polyurethane/dimethylformamide (DMF) coagulation process (wet process)[135, 136]. The substrate – a woven fabric, a non-woven, or in special cases a knitted fabric – is coated with a solution of polyurethane that has already started to gel. It is then passed through a number of baths containing DMF/water in decreasing concentrations of DMF and lastly through a bath of pure water. As the result of phase transfer, a polyurethane sponge forms from the polyurethane solution in DMF.

As the result of the heat which is generated by mixing DMF and water, the product is voluminous and porous. Depending on the method of application, only the top side of the product is coated, or it is dipped to coat it on both sides.[137] The DMF can be recovered from the process water (water : DMF = 75 : 25) by distillation. The full coagulation process causes a high energy and water consumption and is less sustainable than using a full water-borne process.

Depending on the intended end-use, the product can be printed, embossed, buffed, transfer-coated, impregnated or tumbled. In this way, materials with a microporous structure can be produced for the manufacture of shoe uppers, shoe linings, upholstery, bags and luggage, clothing, cleaning and polishing cloths and filters.[138]

Synthetics based on polyurethane dispersions

Because DMF is listed on the candidate list of "Substances of Very High Concern" in accordance with the EU REACH regulation there have been various efforts to produce thick, porous textile structures as synthetic material without the use of solvents.[139]

The development of new highly dispersed aliphatic polyurethane dispersions has made it possible to produce a mechanical foam structure and apply it on a textile substrate in a thickness of a few millimeters in just one step via the transfer coating process (see Figures 5.97 and 5.98).[140]

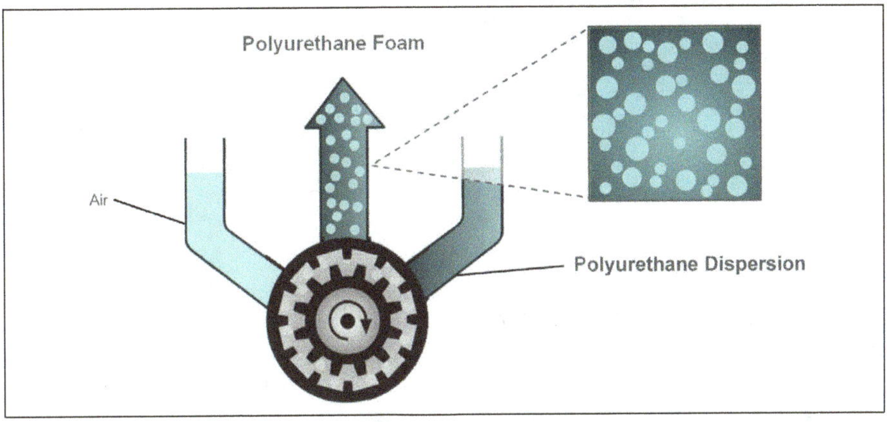

Figure 5.97: Process of manufacture of mechanical foam based on polyurethane dispersions

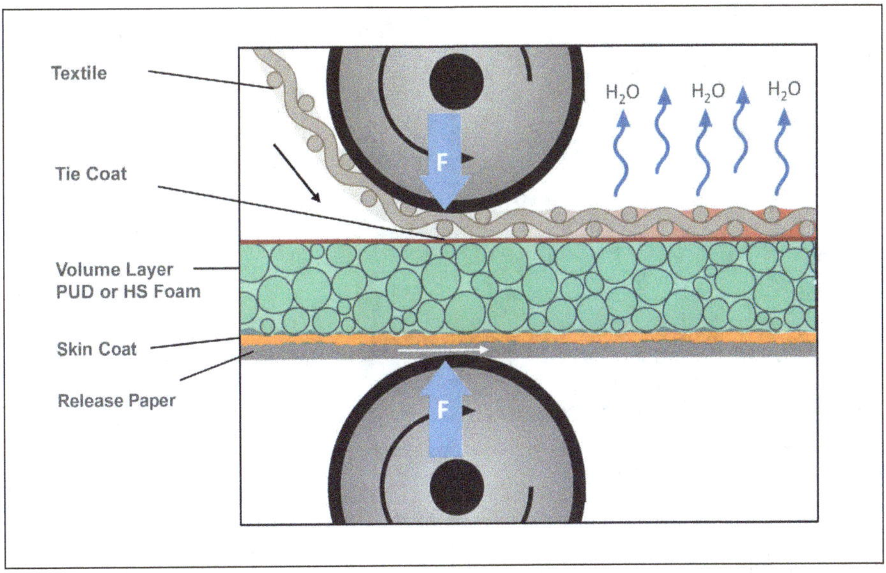

Figure 5.98: Lamination of textile during the transfer coating process with water-borne polyurethane for synthetics

Table 5.29: Typical formulation of a water-borne polyurethane topcoat for furniture synthetics

Component	**Brand name**	**Content [g]**
Aliphatic polyurethane dispersion	e.g. "Impranil" DLC-F (Covestro)	1000
Surface additive	e.g. "BYK" 333 (Byk)	3
Thermally activated polyurethane crosslinker	e.g. "Imprafix" 2794 (Covestro)	30
Matting agent	e.g. "Ceraflor" 1000 (Byk)	20
Water		1000
Pigment	e.g. "Aquaderm" white B (Lanxess)	100
Thickener	e.g. "Borchigel" ALA (Borchers)	2–6

The advantage of this method is that it can be run on conventional textile coating equipment units and, in contrast to the DMF coagulation method, does not require any solvents. Table 5.29 shows a typical water-borne polyurethane topcoat for furniture synthetics.

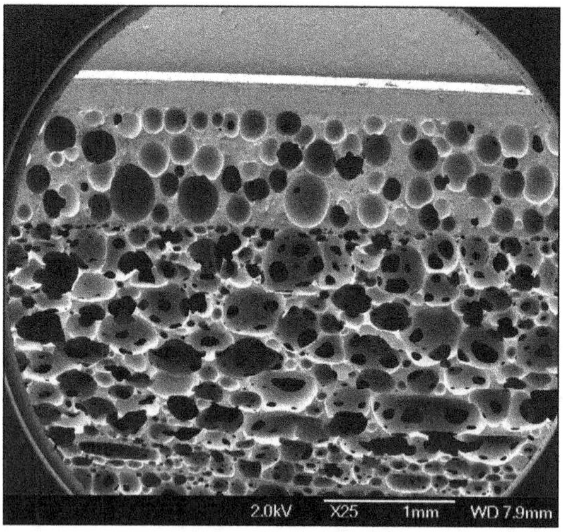

Figure 5.99: Structure of a synthetic material based on a water-borne polyurethane formulation
Four-layer transfer coated material: from bottom: low density mechanical polyurethane dispersion foam with a density of 250 g/l followed by a mechanical polyurethane dispersion foam with a density of 600 g/l both based on a 60 % "Impranil" DLU polyurethane dispersion, top: compact, water-borne topcoat

High-solid polyurethane dispersions with a solid content of approximately 60 % make it possible to economically produce synthetic materials by this process. Due to increased interest in bio-based raw materials, polyurethane dispersions based on pentamethylene diisocyanate (PDI) are under investigation.

If high-quality raw materials such as polycarbonates and polytetramethylene glycols (PTMG) are used as the polyol components, synthetic materials with high hydrolysis resistance can be produced.[141]

With this process by using polyurethane dispersions it is possible to save up to 95 % water and the Global Warming Potential is 45 % less compared to the conventional DMF-coagulation process.[142]

Polyurethane top finishes

To get different types of properties such as haptic, appearance, and mechanical performance, synthetics need a special formulated polyurethane coating as a top finish. For example, a deep matt PU clear coat is commonly used for PVC and PU automotive synthetics; waxes, blowing agents containing pigmented are widely used for pull-up and nubuck type finishes.

Compared to genuine leather, synthetic materials have an extremely large volume, uniform size, and most are in a need of a topcoat finish. Thus, the high efficiency coating process, gravure printing, is designed for synthetics topcoat finish. Different rollers contribute to different add-on amounts, approximately one pass pick-up range is 10 to 50 g/m^2 (wet). Depending on the substrate and requirement, the difference is typically 1 to 3 passes. Because the coating add-on is quite low and there is no limitation for the curing temperature, this type of synthetics finish operation is quite desirable due to its high line speed, up to 15 to 30 m/min with a 15 meter long oven.

5.7.3 Screen Printing

Screen printing is a technique particularly suited for flat or relatively flat surfaces. The process involves a fine mesh or screen that is tightly stretched around a rigid frame. The areas that are not to be printed are masked out on the screen. To create the print, the framed screen is positioned over the substrate to be printed, a high viscosity ink paste is applied at the top of the screen. A squeegee or blade is then used to press the ink through the screen. The masked areas prevent the ink from passing through, but the unmasked areas allow the ink to be imprinted on the material (see Figure 5.100).

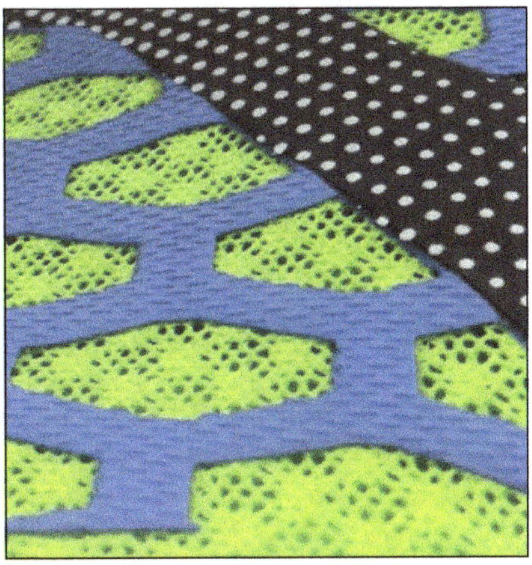

Figure 5.100: Screen printing with waterborne polyurethane

If multiple colors are in the final design, the process is repeated with different screens (one for each color). The screens are usually placed on a rotary press that allows the different color prints to be properly aligned or registered with each other. Some screen-printers have fully automatic presses that do not require any manual labor other than set-up and loading/unloading.

Textile screen printing technology has several application areas, such as clothing, shoes, upholstery, bags, umbrellas etc. Printing on clothing, upholstery, umbrellas, and bags is mainly for decoration. Printing on shoes should simultaneously possess functional and decorative characteristics.

The printing resins commonly used in the market include water-based acrylates, solvent-based polyurethanes, water-borne polyurethanes and silicone polymer systems. Water-based acrylate resins are inexpensive and stable. Limitations of polyacrylate resins are that the physical properties average, the hand feeling is stickier than polyurethanes at similar modulus levels, and they become rigid at low temperatures, which makes it difficult to meet the high-end requirements. Solvent-based polyurethanes are also popular because of its advantages such as simple processing, and excellent physical properties. One issue with solvent-based systems is that they contain organic volatile substances that can have significant impact on employee health and the environment during the production process. In recent years, water-borne polyurethanes in industrial applications show a rapid growth trend. The water-borne polyurethanes contain little or no solvent, applying the technology is simple, and overall performance is excellent.

5.7.4 Micro-fiber dipping

Micro-fiber is a synthetic fiber finer than one denier or decitex/thread, having a diameter of less than 10 µm. After impregnated by PU, the micro-fiber substrate could better simulate the appearance, hand feeling and performance of genuine leather; it's soft, supple with very good physical properties. PU dipped micro-fibers are widely used in high end applications such as sports shoes, upholstery, auto, apparel and suitcase.

The most widely used PU in micro-fiber dipping applications are still solvent-based, but aliphatic water-based PU has drawn more and more attention in recent years. Compare to solvent-borne PU, water-borne PU has the following advantages:

- low VOC and odor,
- does not have issues of DMF solvent,
- safer working environment – elimination of occupational health risk,
- less pollution – elimination of pollution risk to air, water, and
- efficient – can be processed with less water and less energy.

5.7.5 Supported glove dipping

Supported gloves are a type of hand protecting gloves made of a coated fabric. Usually, the coated supported gloves are made by dipping a knitted or woven cloth liner into a liquid compound. The main binders in the compound could be PVC, natural rubber, nitrile butadiene rubber, polyacrylate, styrene butadiene rubber, chlorobutyl rubber, solvent-borne polyurethane or water-borne polyurethane. Among them, NBR (nitrile butadiene rubber) and solvent-borne polyurethane (SB PU) are the most popular coating types used today.

The advantages of a SB PU coating are soft, breathable, light & inexpensive gloves, but the process and the final glove contains DMF (a key component in the formulation), which brings health and environmental issues, e.g. California proposition 65 requires a label with DMF containing products.

Water-borne polyurethane (WB PU) can be used separately or combined with NBR to provide a DMF-free solution for supported glove coating, both foam and compact coatings are applicable (see Figure 5.101).

5.7.6 Leather coating

Leather is a substrate of natural origin that is used in the production of shoes, clothing, and furniture, as well as in other diverse applications. The high wear comfort of leather is based on the excellent adsorption and desorption of moisture. In chemical terms, leather is a collagen, or a polypeptide obtained from animal hides. The soaked hides are fixed by tanning to stabilize the collagen fibers and ensure that the leather remains flexible at the temperatures of use. The tanning quality depends on the temperature at which a leather dries, shrinks and cures. Above 90 °C, most tanned leathers cure to an irreversibly brittleness.

In most cases animal hides are too thick for processing. For this reason, it is separated into two layers. While the top layer still shows the natural grain, the bottom layer or split leather has a rough surface and is often used to produce so called suede. Leather is coated for different purposes. Generally, the surface treatment of dry leather is called finishing. This may result in coloring, altered haptics, water and dirt resistance, and improvements

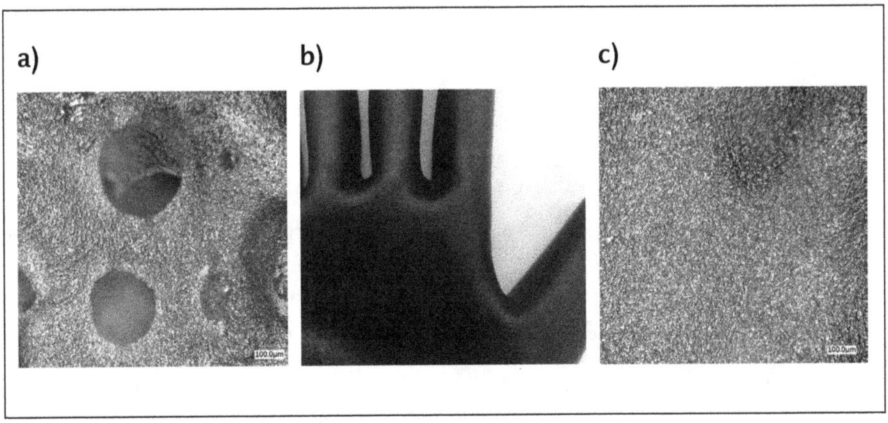

Figure 5.101: Water-borne polyurethane foam glove coating and compact glove coating structure scanning electron microscope (SEM) images, a) breathable coating, b) glove, c) water proofed coating

of abrasion and scratch resistance. Starting from the leather substrate typical leather finishes consist of impregnation, base- and color coats, effect coats and topcoats. The drying temperatures for the individual coats should not be too high to avoid damaging the leather (see above).

In addition to polyacrylates one- and two-component polyurethanes are used when particularly high performance is required. Polyurethane dispersions are usually crosslinked either with water-dispersible polyisocyanates (see Figure 5.102). The highest performance will be achieved with crosslinked water-based polyurethane finishes.

Standard methods of applying the leather finishing products are spraying, curtain coating, transfer coating (analogous to textile coating), and film lamination. The demand for suede depends on prevailing fashion trends. However, since large amounts of split leather are produced, methods have been developed to produce artificial graining on the surface of split leather. One method is direct leather coating followed by embossing.

Alternatively, a grain pattern can be produced by reverse coating with negative graining on the release paper, lamination and removal of the paper. In a third method, a matrix of silicon rubber is used as the negative pattern. The artificial grain produced by one of these methods has the advantage that there are no defects (e.g. scars caused by barbed wire or damage from skin parasites that are found on natural grained leather). Also, it is uniform and entirely reproducible. These artificially grained split leathers are hard to distinguish from natural grained leather.

Figure 5.102: Steering wheel with polyurethane transfer coating on split leather

5.7.7 Outlook

Along with population growth, the demand for textile coatings will increase in the upcoming years. Sustainability aspects will lead to significant changes in this industry.

The high usage of the traditional wet process, which is based on DMF, will change the most. A new kind of chemistry is more likely the solution as e.g. high-solid two-component polyurethane systems or water-borne polyurethane. These changes will be supported by new regulations regarding environmental protection, measures to improve pollution and worker safety, stricter rules and standards, and stronger penalties for non-compliance. Some city governments and brands have already announced their own programs to reduce emissions and to reduce water and energy consumption. Furthermore, raw materials based on renewable resources will gain favor in in the development of new textile coatings.

5.8 Coating and finishing of paper and films

In the coating and finishing of thin substrates such as paper or films, a distinction is made between the following segments:
- papermaking,
- specialty papers,
- coating of printed products,
- production of packaging materials,
- production of decorative films for furniture and interior design, and
- modification and finishing of technical films.

5.8.1 Papermaking

Polyurethane raw materials still play a comparatively subordinate role in papermaking.[143] The chemical crosslinking of paper components – cellulose or starch – with polyisocyanates significantly strengthens the structure of the paper. This results in an improvement in the wet strength. Properties such as printability and tear resistance of the end product – the paper – can also be improved. Hydrophilically-modified polyisocyanates are especially suitable for this purpose. The polyisocyanates are typically added to the pulp but it is also possible to apply them as a surface coating.[144, 145] Depending on the usage of the paper, food safety regulations must be taken into consideration.

Polyurethane dispersions (PUD) can also be used in the production of paper.[146] They are particularly suitable for improving the properties of paper for inkjet and thermal printing[147–150]. In the past, carbonless copy papers and papers for thermal printing used polyurethane raw materials for their manufacture. However, these applications have been outstripped by the digitalization and safety concerns of certain raw materials used in the process.[145, 151]

5.8.2 Paper coating

Printed products are often overcoated with a clear coat both for technical reasons and to enhance their appearance. Such an overprint coating accentuates the print color and increases its brilliance (see Figure 5.103). It also protects the print from mechanical and chemical damage. In response to growing quality requirements, chemically crosslinking systems such as two-component polyurethane (2K PU) coatings are being used increasingly for this application. They are applied by roller coating and by flexographic and rotogravure printing. The coatings are dried in heat and air, e.g. in IR, jet or float dryers. In order to ensure high throughput, the two-component systems are formulated such that physical drying takes place in the drying unit and chemical crosslinking occurs once the coated products are stacked or rolled. This process is used to apply HDI-based aliphatic polyisocyanates combined with physically drying polyacrylate polyols, especially for lightfast coatings. On account of their high curing speed, radiation-curing polyurethane systems in particular are established as binders for overprint coatings. Many of the raw materials used in polyurethane overprint coats are also suitable for the formulation of printing inks.[152]

Coating of printed packaging materials

To a large extent, flexible packaging materials, prefabricated as films, have replaced rigid packagings made from glass or metal. Polyurethanes play an important role in printing, coating, laminating and embossing the film substrates.[152–156] Polyurethane raw materials are often used to formulate the coatings, printing inks or lamination adhesives (see Chapter 6.3.4). Low molecular weight polyurethanes can act as modifiers, for example, they are combined with nitrocellulose in the production of composite packaging.[155] High molecular weight grades can be film forming on their own.[156] Pure polyurethane systems are

Figure 5.103: High-quality applications for paper

discussed to substitute PVC.[157] Two-component systems, e.g. two-component white, are also used in high-quality laminates. Polyurethane dispersions have been investigated for application in water-based ink binders for standard printing applications, as well combinations with lignin are proposed.[158, 159]

In general, the statements made with regard to printed products also apply to the coating of packaging material made from plastic films or aluminum foil. In this segment, the following properties are exploited or optimized:

- high scrub resistance,
- high chemical resistance,
- broad variability of flexibility,
- good adhesion, although polyolefin films should be pretreated, e.g. by corona discharge,
- resistance to sterilization,
- resistance to freezing,
- sealability, and
- barrier properties.

In order to satisfy this wide range of requirements, the broad variety of products, both polyols and polyisocyanates, can be exploited in multiple formulation options. In the design of packaging, haptic properties are increasingly viewed as distinguishing features.[160]

5.8.3 Production of decorative films for furniture and interior design

In the furniture and interior design segments, around half of all wood and wood material surfaces are finished by direct coating. Alternatively, decorative films are applied. For the finishing of MDF materials, polyurethane dispersions are used in the heat activation process (see Chapter 6.5.6).

Suitable substrates used for decorative and finishing films are specially impregnated paper, thin paper and thermoplastic films. The coating is applied through continuous processes like roller coating or gravure rolling. The coated substrate is then dried in float dryers.[161] High requirements are made with regard to the lightfastness and chemical resistance of the coatings used to produce decorative films.[162] As these films are also applied on shaped components such as the fronts of furniture or rounded edges, the coatings must also have a certain flexibility and good blocking resistance. As polyurethane coatings combine these properties, they differ from the acid-curing coating systems which have been widely used until now in the furniture industry. The polyurethane systems used are usually two-component formulations based mainly on polyacrylate polyols and aliphatic HDI-based polyisocyanates. In the case of producing decorative films, the polyurethane coatings are formulated so that they are tack-free when films are wound to prevent sticking, but crosslink on the roll. This is possible because of the rapid physical drying of the coating.

5.8.4 Finishing of technical papers and films

There is a wide range of applications for polyurethane systems in the processing of film-like materials. Significant uses include:

- Production of abrasive paper
  - As a rule, phenolic resins are used as the embedding compound for the abrasive material. However, polyurethane binders yield products with much higher water resistance. They are therefore used primarily in the manufacturing of abrasives for wet sanding. Two-component systems based on polyester polyols and crosslinked with aromatic isocyanates are normally used for the purpose.
- Embossed release paper
  - Embossed release paper is used, for example, as the negative for grain embossing in the manufacture of synthetic leather (see Chapter 5.7). The textured contact side must be chemically inert and must release the coated synthetic leather non-adhesive. The product must also yield good texture reproduction. These requirements are satisfied by crosslinking polyurethane topcoats as well as by other systems such as radiation-curing coatings.
- Scratch-resistant finishes for films used in machine and automotive production
  - The coatings used in this application are expected to fulfil strict requirements like scrub and scratch resistance but must also be suitable for deep-drawn 3-dimensional parts. For this reason, fast-drying polyurethane resins are favored, which can be crosslinked easily with aliphatic polyisocyanates if necessary. Also, in development are systems based on polyurethane dispersions and water-reducible polyisocyanates.
- Stamping foils
  - Stamping foils are widely used to decorate plastics components, e.g. in cosmetics packaging. They originated in the graphics industry and were optimized technically for the aforementioned application. The significance of polyurethane coating raw materials in these applications lies in their use as finishes. They make it possible to satisfy specific requirements in terms of the resistance properties and to achieve adequate blocking resistance and an accurate image during the stamping process. It is therefore necessary to use very hard coatings based on aromatic TDI polyisocyanates.
- The finishing of technical film and foils with polyurethane materials goes beyond the few examples given here. Each application requires a special technology and demands

different property profiles. Frequently, only polyurethane raw materials in all their chemical versatility allow the formulation of such customized coatings.

5.8.5 Outlook

The emerging trend of digitalization of several applications which depended on printed paper and films is expected to impact the sustainability and/or the growth of these applications. For example, airplane boarding passes, entry tickets to events, are being replaced by digitalized QR codes on smart phone applications. Furthermore, efforts are underway to replace typical packaging applications which do not require barrier properties. These trends are also likely to impact the further development of these applications. On the other hand, there might be future growth options as design features will be placed on many packaging items (e.g. perfumes, consumer goods in general) in order to differentiate and to increase optical value.

5.9 Construction applications

Polyurethanes have many uses in the construction sector. They are used as adhesives and sealants, for floor and architectural wall coatings, in waterproofing membranes for roofs, secondary containment, and balconies, as well as for thermal insulation foam.[189] The flooring, architectural, and waterproofing coating segments will be covered in this chapter with the main substrate being concrete but also includes drywall in some architectural applications and metal for some roofing and waterproofing applications.

5.9.1 Floor coatings

The demands made on floors vary greatly depending on the intended use. Vinyl and composite coverings and coating materials are being used increasingly alongside parquet, ceramic tiles and carpets.

Floors are of particular importance in ensuring the functionality of industrial and commercial facilities. The durability and look of a floor is instrumental in the smooth operation of a production facility as well as a decorative commercial application.

The requirements made of floor coatings are therefore correspondingly high and various. Depending on the type of facility and the prevailing conditions, the following characteristics may be required:

- mechanical and dynamic strength (resistance to compression, impact, scratches, and abrasion, crack-free),
- flexibility and toughness,
- chemical resistance,

- electrical resistance,
- thermal resistance (long-term and short-term),
- colorfastness and weather stability,
- smoothness,
- non-slip properties,
- ease of care, cleanability,
- ease of repair,
- service life, cost/benefit ratio,
- VOC and/or solvent content, and
- odor.

Many of these properties are defined in, for example, Germany's DIN standards and technical guidelines and, in sum, make up the service life of a floor. The selection criteria vary by region, service environment, and specifier requirements.

The primary function of floors is to absorb and distribute static and dynamic loads. Irrespective of its actual structure, a floor can be defined as consisting of one or two layers:

- load-bearing layer such as a monolithic concrete slab or a combination of a concrete slab and screed floor overlay;
- wear layer which is adhered to the load-bearing layer. This must withstand the service environment of the floor area such as chemical and mechanical resistance as well as satisfying other user-specific demands, e.g. non-slip properties, ease of cleaning and aesthetic properties as listed in more detail above.

A floor coating is such a wear layer. In order to achieve a successful system, the properties of the individual layers must be compatible. The wear layer is often based on synthetic resins, particularly reactive systems, such as epoxy resin (EP), polyurethane (PU), or others including poly(methyl methacrylate)/PMMA. Epoxy resins hold the largest share at 65 %, followed by polyurethane coatings with 25 % (see Figure 5.104).

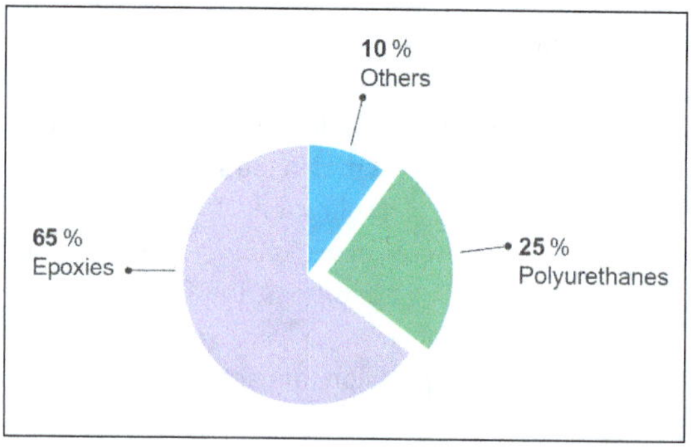

Figure 5.104: World site-applied floor coating market, 2016, total: 550,000 tons finished product[31]

Epoxy-based systems are often used where a floor needs minimal to moderate protection and aesthetics, e.g. in the primer and base layers since

Table 5.30: Variability of the properties of polyurethane floor coatings; Tested according to DIN EN, ASTM, and China GB and GB/T standardized testing methods

Property	**Range**
Pot life [min]	2–120
Tensile strength [N/mm^2]	1–60
Elongation at break [%]	6–1,000
Modulus of elasticity [N/mm^2]	0.1–2,500
Compression strength [N/mm^2]	≤ 100
Tear propagation resistance [N/mm]	1–100
Shore hardness	A 20–D 80
Abrasion resistance [mg]	5–150
Adhesion to iron [N/mm^2]	≤ 30
Glass transition temperature [°C]	-40–100
Water vapor diffusion resistance factor μ	500–100,000

they adhere well to marginally prepared concrete and contribute thickness to the overall system thickness requirements. Most epoxy-based floor coatings will discolor over time when exposed to direct sunlight, sunlight passing through an exterior window, or strong indoor lighting. Epoxy floor coatings are often employed where a good coating at a low cost is needed.

PMMA, or poly(methyl methacrylate), can be used in exterior or well-ventilated indoor spaces where a fast cure is needed. PMMA can be re-applied to PMMA without major additional surface preparation due to the solvating power of monomer contained in the system. Disadvantages of PMMA include a short working time and marginal adhesion, especially in high humidity environments. By far, the largest limiting trait is the very strong odor during application which often precludes its use in indoor applications.

The use of a polyurethane-based floor coating is preferred because of its chemical and abrasion resistance, elimination of seams or grout lines, design aesthetics in colors, and the favorable cost/benefit ratio. Polyurethane and polyaspartic floor coatings are often used over top of an epoxy primer and/or base coat in a multi-coat system due to the better properties mentioned previously. This class of coatings exhibit excellent color and gloss retention which make them suitable in exterior applications as well as indoor applications exposed to sunlight or strong indoor lighting. While more expensive than a typical epoxy coating, they exhibit an extended service life. Polyurethane and polyaspartic coatings typically require better surface preparation than epoxy and are more susceptible to environmental conditions such as high humidity or temperature.

The floor coating may be in the form of a penetrating sealer, a thin coating of up to 0.3 mm or a higher build covering of up to 3 mm thick and may involve multiple functional layers. Other options include synthetic resin mortars or decorative synthetic resin screeds applied in thicknesses of 3 to 10 mm (colored sand mortar). In the latter case, the wear layer additionally has a load-bearing capacity.

Sealers, thin coatings and coatings are formulated with varying levels of solvents and volatile organic compounds (VOC) from solvent-free, solvent-borne, or water-borne.

Synthetic resin mortars are always formulated organic solvent-free. There can be additional components besides the main resins and solvents such as flow and leveling additives, defoaming additives, pigments, and fillers. These minor components vary widely depending on the application conditions.

Technology of polyurethane floor coatings

The wide variability of the polyurethane technology is exploited fully in the floor coating segment.

Table 5.30 (see page 277) provides an overview of the range of main properties tested to a variety of test methods for comparison. This data demonstrates that polyurethane coatings can be customized to yield property profiles that suit the individual flooring application needs.

Polyurethane coatings may range from highly elastic (for membranes, floors in sports facilities) to hard and highly resistant to chemicals (for chemical/process production plants). Typical applications of polyurethanes are:

- industrial, commercial, and decorative floor topcoats over multi-coat systems based on the requirements of the construction or water protection specifications;
- one- and two-component primers and sealers for interior concrete floors as well as exterior brick pavers;
- flexible floor coatings for sports facilities on both concrete and wood substrates;
- membranes, e.g. waterproofing membranes;
- one- and two-component synthetic resin mortars and colored sand mortars.
- Instrumental in the success of polyurethane systems in these applications are the following properties:
- full cure even at low temperatures;
- very good adhesion to a variety of substrates;
- adjustable hardness and flexibility and, as a result:
  - excellent chemical resistance with higher crosslinked coatings;
  - crack-bridging with elastic and tough yet flexible formulations;
  - seamless application;
  - variable water vapor diffusion values;
- good resistance to hydrolysis and low water absorption; very good weather stability and colorfastness in formulations containing selected aliphatic products;
- faster return-to-service in the case of moisture cure polyurethane and polyaspartic coatings.

Polyurethane floor coatings have been in widespread use for more than 50 years and a large number of reference objects are proof of their outstanding quality and long service life.

Coating systems

The most important base products in the formulation of polyurethane floor coatings are isomer blends and prepolymers of diphenylmethane diisocyanate (MDI), oligomers and adducts of hexamethylene diisocyanate (HDI) and isophorone diisocyanate (IPDI), and prepolymers of toluene diisocyanate (TDI) and isophorone diisocyanate.

The preferred co-reactants are compounds such as polyester polyols, polyether polyols, and polyacrylic dispersions. Aromatic or sterically hindered aliphatic amines, such as polyaspartic acid esters[160, 161] and polyether amines, as well as latent hardeners, such bisoxazolidines[162], are also utilized. The latter are converted into reactive products by the effect of moisture.

Classic one-component moisture-cure polyurethane systems are also used in which the polymer is formed by the reaction of the NCO-terminated prepolymer or oligomer with humidity out of the environment. These latter systems are used mainly to produce thin coatings, as foaming occurs in thicker coverings.

Low and high molecular weight products, both linear and branched, with differing chemical structures are available as the polyol compounds. Polyester polyols, for example, ensure good solvent resistance, polyether polyols ensure resistance to water, acid and alkali environments, and derivatives of acrylic and methacrylic acids ensure high weather stability.

It is also possible to use water-borne polyurethane systems based on specially designed hydroxyl-functional polyacrylic dispersions. These formulations can be applied in a wide range of thicknesses from thin traditional concrete coatings to thick cement and/or filler-based overlays. The thick polyurethane-based overlays require the addition of cement and/or hydrated lime (three-component systems) to capture the carbon dioxide produced as a byproduct of the water-isocyanate reaction.[164] These polymer cement overlays are characterized by very high chemical and temperature resistance. For that reason, they are often used in food and beverage applications that require steam cleaning and sterilization.

Aspects of formulation technology

In the formulation of a floor coating, it needs to be considered that these materials will be used at the construction site and therefore must be somewhat forgiving of potential disruptive influences, especially the moisture and humidity which is unavoidable on any construction site. This can result in foaming when applying thick coatings due to the reaction of the isocyanate with moisture. Systems with a short pot life and those containing amines instead of polyols exhibit less issues with respect to the water reaction due to the higher reactivity of the amines compared to water. It is a different matter if systems with a long pot life are used. Here, the hydrophilicity or hydrophobicity of the systems and the associated degree of absorption of ambient moisture play a crucial role.

The coating formulator may select specific polyurethane raw materials in order to address needs. For example, polyether polyols modified with fatty acid to ensure hydrophobicity are among the substances that have been found ideal to resist substrate moisture in a self levelling system. Polyaspartic resins are selected when a faster cure time and a potential faster return-to-service is desired. Molecular sieves can be used to scavenge water from the fillers, pigments and other components often used in a typical polyurethane formulation.

A formulated coating is blended with, for example, a dissolver or butterfly mixer, preferably under vacuum. The extenders, pigments, molecular sieve and surfactants are mixed into the polyol component. Working under vacuum makes it easier to remove air from the formulation, thus reducing the risk of microblistering due to air inclusions during curing. Before filling the coating material into work containers, a suitable deaeration agent is added to ensure a smooth pore-free surface upon field application.

Aspects of application technology

In the different regions, conformity with the appropriate local requirements for floor coatings must be ensured before application. These are contained in the regional standards and guidelines. They include the removal of loose particles, laitance and dirt and application only on a load-bearing substrate which is dry to slightly moist (normally ≤ 4 % water) and has a surface tensile strength of at least 0.8 N/mm^2 (walkable surfaces) or 1.5 N/mm^2 (trafficable surfaces). Most standards include varying requirements related to application temperature range and dew point.

For example, in North America the SSPC (Society for Protective Coatings) has developed the SSPC-SP13 standard which defines the correct observations and preparation of a concrete substrate prior to the application of a topical coating.[165] Additionally, SSPC-Guide 23 defines field methods for measuring concrete substrate moisture content.[166] For guidance of the in-field application, there is the SSPC-PA 7 "Applying Thin Film Coatings To Concrete" specification.[167]

In China, the substrate preparation needs of floor coatings can be found in the national standard GB 50209-2010 "Code for acceptance of construction quality of building ground" and the national standard GB 50212-2002 "Specification for construction and acceptance of anticorrosive engineering of buildings". For example, in GB 50212-2002, fourteen specifications are listed for concrete substrate preparation such as evenness, compression strength of concrete, moisture content in concrete and aging time of new concrete to name a few.

A floor coating system may consist of one or more coats or layers that can include a primer/sealer, self-leveling filler coat, base coat, and/or topcoat. Depending on the service environment, expected service life, and environmental and surface conditions, these layers may represent several different technologies such as epoxy and polyurethane or may be homogenous in technology such as an all polyurethane multi-layer system. Often, pigmented systems in solid color applications will use an epoxy primer and/or base coat that is then topcoated with a higher durability, colorfast polyurethane or polyaspartic coating.

For clear systems, one or two coats of an unpigmented polyurethane or polyaspartic coating is applied that will remain clear and non-yellowing.

The polyol and polyisocyanate components are mixed on site using a cage or blade mixer in the case of a two-component coating. In the case of single component floor coatings such as a moisture cure one-component polyurethane, the material is often stirred or mixed to ensure all components are homogeneous and have not separated or settled.

The floor coating formulations are typically applied using a brush, roller or a notched trowel. The floor coating system can be completed with the application of a decorative, abrasion-resistant topcoat based on e.g. "Desmodur" N and a suitable hydroxyl or amine terminated product type resin.[168]

Specific polyurethane coatings flooring applications

The property profile of polyurethane floor coatings can be varied across a wide range. Depending on the raw materials used, the coatings may be tough but flexible or hard. Typical applications examples are described in the following paragraphs.

Floor coverings for sports facilities

The coatings used on the floors of sports facilities need to exhibit the traditional physical properties such as abrasion and scratch resistance and additionally provide either point or full-surface elasticity in accordance with global and regional standards such as IAAF certificates or Germany's DIN 18 032.[169] Different coating qualities can be achieved thanks to the variable hardness of the topcoat and the substrate, which may consist of polyurethane-bonded rubber granule mats, possibly incorporating glass or textile fabrics. These satisfy the requirements for physiotherapy use but can also be formulated such that multipurpose floors can be produced (Figure 5.105).[170–172]

Decorative floor coatings

Hotels, restaurants, department stores, residential and public buildings are increasingly using coated floors as an aesthetic design element of the architecture. Decorative floor coatings play an important role here. A wide range of colors and looks are in demand such as multicolor or terrazzo effects. Indi-

Figure 5.105: Flexible flooring based on two-component polyurethane[202]

vidually designed floor coatings using two-component polyurethane or polyaspartic systems based on aliphatic isocyanates combined with polyols or polyaspartic resins have gained importance in recent years as an aesthetically pleasing addition.[173]

In the application of decorative floor coatings, a solid-color base coat or primer based on epoxy, polyurethane, or polyaspartic is applied first. Decorative additives such as metalizing pigments, quartz, or color flakes are incorporated into or broadcast on to the base coat to yield the colors, patterns, and effects required. The applicators can walk across the liquid base coat wearing spiked shoes (see Figure 5.106).

The coating may be topped with a highly abrasion-resistant, transparent polyurethane or polyaspartic topcoat (see Figure 5.107).

Figure 5.106: Application of a decorative floor coating

Figure 5.107: Polyaspartic floor coating can be used on concrete floors where higher aesthetics and fast return-to-service are desired traits in addition to excellent durability

Penetrating sealers and sealing compounds

Penetrating sealers and penetrating primers strengthen the surface of a substrate, preventing abrasion of concrete or cement screed. The properties required include exceptional toughness and abrasion resistance, coupled with high resistance to water, chemicals and solvents.

These systems can be formulated with moisture-curing one-component polyisocyanate prepolymers, water-borne one- or two-component polyurethane sealing compounds or two-component polyaspartic coatings, applied at the rate of 100 to 300 g/m^2. Two growing segments are the use of these higher performing sealers on interior polished decorative concrete (worldwide) and on exterior concrete brick pavers (in

North America). In both cases, increased durability, stain resistance, and color enhancement are desirable traits.[174]

Coatings for car park decks

In the past, it was standard to apply only a sealing compound to car park decks for protection. However, compliance with standards and guidelines should be ensured. This requires that steel-reinforced concrete components, which may crack across their entire cross-section if exposed to water with a high salt content (e.g. caused by road salt), must have special protection. This largely applies to statically indeterminate, area-covering structural elements such as car park decks. Given the stresses to which these are exposed, cracking is virtually unavoidable.

A crack-bridging coating is recommended as special protection for steel-reinforced concrete. Specifications for suitable coatings are stipulated in the regional and European standards and guidelines. These guidelines cover in particular, the requirements for durable crack-bridging of existing and newly formed expansion cracks caused by temperature- and stress-related movement of the structure. They specify minimum crack-bridging of 0.3 mm. in a temperature range of -20 to +70 °C. Coatings which meet the requirements of these guidelines are applied at a film thickness of 3 to 4.5 mm.

Both one- and two-coat systems can be used. In the case of two-coat systems, each coat may fulfill different functions. For example, the first coat provides crack-bridging, in particular, at low temperatures, while the topcoat protects against abrasion, chemicals, weathering and mechanical impact. One-coat systems must combine all these functions in a single coat. To ensure crack-bridging at -20 °C and adequate tensile strength, tear propagation resistance and abrasion resistance, the binder used for the system must be selected carefully and fine tuning is required during formulation.

Today, both types of system – one- and two-coat – are primarily polyurethane-based due to the inherent flexibility that can be formulated into the coating. Other technologies, such as epoxy and acrylic (see Figure 5.108), are lacking in either the long-term durability or use plasticizers to achieve flexibility and can become less flexible over time as plasticizers migrate out of the coating. The one-coat variant is normally the more economical of the two.

Figure 5.108: High end floor coating in a private garage setting

Industrial floor coatings

Tough and hard floor coating systems are considered indispensable in factories and warehouses, for example, in the automotive, food processing, pharmaceutical and chemical industries and in wholesale markets. The coatings have to meet a wide range of requirements in terms of mechanical properties, safety and optical properties.[175] Individual customer requirements may also be considered. Thanks to their toughness, polyurethane and polyaspartic systems are suitable for use on cementitious substrates. They can also be customized via the choice of raw materials for use on magnesia and anhydrite screeds compliant with current regulations and guidelines bearing in mind the construction physics and water vapor diffusion resistance. Polyaspartic coatings, with their fast return-to-service, are often used on industrial floors where the area must be put back into service quickly for production or safety reasons.

Secondary containment

The operators of production units in which hazardous substances are used are obliged to implement suitable protection measures to avoid water contamination. There are specific standards at regional, country, and local levels that dictate the requirements of the containment system. For example, in North America, NSF 61 is often specified for concrete and metal containment and pipes. In Germany, § 19 of Germany's Water Resources Act (WHG) is the primary standard. China has several relevant national GB and GB/T standards depending on the service application.

These comprise concrete containment basins that are often coated with systems which have to satisfy particular stringent requirements in terms of crack-bridging, elastic-

Figure 5.109: Exterior concrete walls of this production facility were protected with a two-component pigmented water-borne polyurethane coating[184] Source: Megx [181]

ity and chemical resistance.[176] Applied in a multi-coat system, polyurethane coatings can provide resistance to many chemical classes defined in Germany's construction and testing guidelines. A one-coat system (without primer) already provides resistance to some of these chemical classes. The incorporation of carbon fibers, for example, yields antistatic floor coatings which are required in areas where solvents are stored and handled.

5.9.2 Architectural wall coatings

In the past, architectural coatings were primarily decorative in function, providing minimal physical properties. However, there are several architectural coatings applications where durability and surface protection are key attributes. The prevailing conditions must always be kept in mind when selecting a coating system. Exterior masonry wall conditions differ significantly from interior walls with respect to coating requirements.[177–179]

Exterior architectural wall coatings

Exterior wall coatings must display the following characteristics:

- mechanical strength,
- weather stability,
- resistance to aggressive substances,
- impermeability to driving rain,
- gas diffusion properties, and
- ease of cleaning (lotus effect).[180]

As masonry is often exposed to wide fluctuations in temperature, many masonry coating systems are elastomeric with the ability to stretch with the building movement. Other key criteria for a good coating system are adequate colorfastness and chalking resistance on exposure to weathering and a minimal tendency to microcracking. In large metropolitan and industrial regions, the coating systems come into contact with aggressive substances in the air such as sulfur dioxide (SO_2), nitrous oxides (NO_x), ozone and smog. It is particularly important that the coatings not age and lose their protective function as well as retain their original desired color and sheen. Some exterior applications such as commercial buildings and concrete bridge abutments benefit from the application of a clear or pigmented highly chemical resistant polyurethane coating which can be washed with strong cleaners to remove sprayed-on graffiti without damaging the protective coating or suffer changes in appearance.

Impermeability to driving rain is a fundamental requirement of any masonry coating system. If moisture penetrates a wall, it can cause substantial damage such as frost cracking, efflorescence and reduced insulation properties. The coatings must also have a good barrier effect to acid gases without restricting the water vapor diffusion too much. If the water vapor diffusion is too low, any moisture trapped in the masonry cannot escape. This can easily result in damage inside the building, such as mold growth.

Polyurethane-based exterior coatings can fulfill these requirements due to their excellent chemical resistance while still allowing some permeability. Also, since these coatings systems typically contain an aliphatic hardener, they resist rain, weathering, sheen, and color changes long term unlike some acrylic-based exterior masonry coatings which degrade faster under these conditions (see Figure 5.109 page 284).

Interior architectural wall coatings

Interior wall paints mainly have a decorative function. As a rule, ease of application and a low organic solvent content and odor are key requirements. The resistance properties required depend very much on the area of use. In the case of paints used in private homes, it is normally sufficient to ensure medium resistance to standard household cleaning agents. That is why most typical interior wall paints are formulated with water-based vinyl and acrylic resins.

However, in buildings such as hospitals, laboratories, food-processing plants and other selected process industries, stricter cleaning and disinfection protocols may be specified such as resistance to cleaning agents, steam cleaning, disinfectants and decontamination chemicals. Even though the walls may be exposed to these conditions, they must retain the original color and look.

These strict requirements can be satisfied by using a higher crosslinked coating system such as a polyurethane or polyaspartic coating systems. For example, a two-component water-borne polyurethane exhibits ten times the scrub durability of a high quality acrylic wall coating, much better stain and chemical resistance, and very low odor.[182]

These commercial coating systems are not suitable for do-it-yourself (DIY) applications. They have to be contractor applied due to the added application and safety training needed for a successful painting project.

Coating system types

A large number of wall and masonry paints and coatings covering a broad spectrum of price and quality levels are available on the market. Generally speaking, masonry paints or coatings can be divided into two types:

- systems that dry only physically, such as emulsion paints based on vinyl copolymers and silicates, and
- reactive resin systems based on, e.g. epoxy resins (EP) or polyurethanes.

For most standard architectural applications where low physical and chemical resistance is required, emulsion paints based on polyvinyl acetate (PVA) or acrylates have become the systems of choice.

Water-borne vinyl copolymers

Vinyl copolymers are the dominant binders for the formulation of wall and masonry paints. Pure acrylates represent a particularly high-quality variant. Like the other vinyl copolymers, they dry physically. In other words, film formation results exclusively from the evaporation of the solvent, which is predominantly water. It is the relatively low price of the paint systems that has led to the widespread use of pure acrylates in wall and masonry paints.

This type of system is easy to handle and can be applied in various film thicknesses. Subsequent overcoating with the same system is possible. Adhesion is normally adequate on various mineral substrates and drywall. The dry paint films are colorfast, UV-stable and provide an adequate barrier to atmospheric carbon dioxide. However, pure acrylic paints do have a few disadvantages resulting from the absence of crosslinking:

- low solvent and standing water resistance,
- vulnerability to graffiti,
- thermoplasticity, and
- edium abrasion resistance.

The quality of such coatings can be increased by adding polyurethane dispersions because of the contribution of additional crosslinking as a function of hydrogen bonding from the polyurethane. This imparts higher properties as compared to the pure acrylic-based coating. For example, the addition of 30 to 40 % of polyurethane dispersions can increase abrasion, chemical, and water resistance.

Epoxy (EP) systems

Epoxy (EP) systems yield very high-quality wall coatings. They are two-component systems that form a film and cure as the result of a chemical reaction that yields a three-dimensional network. The films are hard, have outstanding chemical resistance, adhere well on different substrates and act as an effective barrier to carbon dioxide. Solvent-borne, solvent-free and water-borne systems are available.
The following properties of EP systems may be a limitation:

- inadequate through-curing at low ambient temperatures,
- poor recognition of the end of pot life in water-borne systems,
- limited UV stability resulting in yellowing of films over time, and
- yellowing after exposure to certain cleaning and disinfection chemicals.

Polyurethanes

Like EP systems, polyurethane systems cure to form a three-dimensional network yielding thermoset films. However, in contrast to the EP systems, one-component systems that cure by reaction with atmospheric humidity can be formulated. The systems available on the market also include solvent-borne, solvent-free and water-borne two-component

Table 5.31: Typical formulation and characteristic data of a water-borne two-component polyurethane basedcoating for architectural wall application

	Raw material	Parts by weight
Component A		
	"Bayhydrol" A 2542, (Covestro)	22.38
	"Bayhydrol" A 2546, (Covestro)	6.82
	Pigment and thixotropic agent	27.28
	Wetting and dispersing additive	1.02
	Defoaming agent	0.82
	Diluent (DI Water)	13.69
	Coalescent agent	5.12
Component B		
	"Bayhydur" XP 2547, (Covestro)	22.87
		100.00
Characteristic data		
Solid content by weight [%]		63.2
Volatile organic compounds [g/l]		7.2
Drying time to touch (23 °C/50 % rel. humidity) [min]		20

formulations. The raw materials used are polyesters, polyethers and polyacrylates, which are combined with a polyisocyanate crosslinker.

In recent years, water-borne two-component polyurethane systems also achieved greater market acceptance. They are formulated with hydroxyl-bearing polyacrylates such as "Bayhydrol" A 2542 and hydrophilically-modified polyisocyanates such as "Bayhydur" XP 2547 or comparable products. An example of an architectural wall coating guide formula based on two-component water-borne polyurethane is listed in Table 5.31. Additionally, polyaspartic acid ester resins, such as "Desmophen" NH 1420, 1520, and XP 2850, are used as a reactant with aliphatic polyisocyanates as a fast curing, durable finish.

Polyurethane coatings have good resistance to solvents, graffiti and chemicals, as well as very good weather stability and colorfastness. They are hard yet with a distinct degree of elasticity. From the construction physics aspect, they provide mineral substrates with excellent protection, have adequate water vapor permeability and are effective barriers to acid gases such as carbon dioxide. Possible limitations of polyurethane coatings include:

- higher material costs compared with vinyl copolymers;
- a certain sensitivity to high moisture levels in the substrate – varies widely among the polyurethane systems available on the market;
- the need to maintain a maximum film thickness on application to prevent film defects such as blistering.

On account of their higher costs, high-quality polyurethane systems are only used when higher durability requirements are required in terms of the resistance properties that cannot be satisfied by "normal" emulsion paints. This may be of particular interest if, for example, high secondary costs (scaffolding) make long-term protection properties an important requirement[183]. Typical applications of polyurethane coatings are therefore on the exterior of apartment blocks, towers and stacks; on the interior walls of pharmaceutical and food-processing plants with their high demands in terms of cleaning properties; on the interior walls of schools, hospitals and other buildings with a high visitor frequency. There is also growing recognition of the suitability of polyurethane systems as "anti-graffiti coatings".[184] (see Figure 5.110).

Polyurethane wall coatings

To ensure the durability of a coating on interior or exterior walls, the careful selection and application of a multicoat concept is essential. A typical polyurethane coating concept consists of multiple coats based on the specific application and environment. These coats may include a penetration primer, a base coat and/or a topcoat.

The first coat (primer), the pigmented base coat and the topcoat are usually applied with a brush or roller. In the case of exterior applications, if the coating system is adjusted to a suitable viscosity, compressed-air or airless spraying is also possible. For interior applications over drywall or plaster, a two-coat system is employed consisting of a primer and topcoat. In both exterior and interior applications where the coating is spray applied, proper personal protective equipment must be used according to the specific environment and local prevailing requirements.

Figure 5.110: Two-component polyurethane coating with high resistance to graffiti and ease of cleaning

Primer

For masonry or concrete wall construction, a penetration primer may be needed. Due to the alkalinity of many mineral substrates, such as concrete, saponification-resistant binders based on polyether, (see Figure 5.110), or polyacrylate are preferred for the primer. Formulations are adjusted to a particularly low application viscosity to ensure good penetration of the porous substrate. In the case of solvent-borne systems, this often means an elevated organic solvent content.

Water-borne polyurethane systems based on "Bayhydrol" and "Bayhydur" are available for the formulation of virtually odorless and ultra-low VOC primers over masonry and drywall. As pigments and extenders have a negative impact on the penetration properties of the primer, they are generally not used for masonry but are often pigmented for drywall and plaster applications.

The cured polyurethane primer strengthens the substrate and provides a firm base for the subsequent coating layer(s). Depending on the type and porosity of the substrate, the application rate varies from 60 to 100 g/m^2 for compacted concrete surfaces or drywall and 80 to 250 g/m^2 for highly porous substrates. Depending on the coating system and the ambient conditions (temperature, humidity), the subsequent coats should be applied between 8 to 24 hours after the primer.

Base coat

The pigmented base coat must be matched to the roughness of the substrates in terms of its flow properties, body and hiding power. It must also have good adhesion to the primer and to the topcoat. The application rate of the base coat is 150 to 200 g/m^2 for solvent-borne systems, 60 to 120 g/m^2 for solvent-free formulations and 60 to 100 g/m^2 for water-borne coatings. The topcoat can be applied on the base coat after about 4 to 24 hours.

Topcoat

The topcoat is usually a lightfast aliphatic polyurethane system crosslinked with an aliphatic polyisocyanate. It can be a pigmented or clear coating, depending on the primer or base coat that was used. The application rate is 60 to 200 g/m^2 for solvent-borne systems, 60 to 120 g/m^2 for solvent-free systems and 60 to 100 g/m^2 for water-borne formulations. Depending on the system used, the topcoats achieve adequate initial drying (touch dryness) after 5 to 8 hours (solvent-borne), 1 to 8 hours (solvent-free) or 1 to 24 hours (water-borne), provided an adequately high room temperature and air circulation are ensured. The total film is 100 to 250 μm (solvent-borne), 90 to 140 μm (solvent-free) and 70 to 150 μm (water-borne).

Formulation of polyurethane wall coatings

Several different concepts using various systems have proven suitable for use on masonry and drywall. For masonry, a polyether polyol and an aromatic polyisocyanate crosslinkers are used in the penetration primer and the base coat. This saponification-resistant combination is relatively inexpensive but has the disadvantage that it displays a certain sensitivity to moisture during application because of the high reactivity of the polyisocyanate.

The topcoat is based on a polyester polyol crosslinked with a lightfast aliphatic polyisocyanate. It is highly crosslinked and therefore has very good resistance properties. Alternately, a second option uses a hydroxy-functional polyacrylate crosslinked with an aliphatic polyisocyanate in all three coats. This system is far less sensitive to moisture and thus allows the blister-free application of thicker films. It has the additional benefit that it can be overcoated even after long interruptions in the coating operation. When selecting the pigments and extenders, it should be ensured that they have good wetting properties. Poor pigment wetting will result in the relatively rapid occurrence of chalking for exterior applications.

For interior drywall or plaster walls, one-component water-borne polyurethane modified acrylics or two-component water-borne polyurethane coating systems are used. Since these applications are often in high stain and high scrub areas, scrub-ability and stain resistance properties are maximized. For two-component water-borne polyurethane wall coatings, a hydroxy-functional polyacrylate is crosslinked with a water dispersible aliphatic polyisocyanate.[182]

Two additional advantages of this type of coating solution are:

- the ability to formulate a low sheen coating by selecting a resin such as "Bayhydrol" A 2546, and

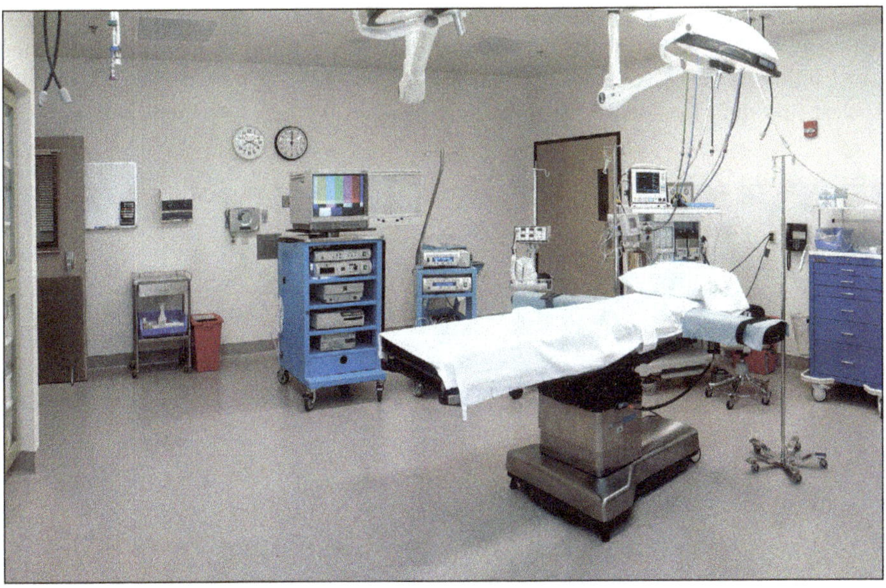

Figure 5.111: Two-component water-borne polyurethane architectural wall coatings provide the durability and low odor needs in areas like this hospital operating room

- the ability to formulate an extremely low odor coating which is important for healthcare and school settings, see Figure 5.111 page 291.

Typically for this type of system, either the primer or the primer and topcoat are pigmented. Low emission, low solvent systems are also suitable for interior wall coatings. The ultra-low solvent systems are based on low viscosity polyisocyanates which react with atmospheric moisture to form urea groups and yield a highly crosslinked film with very good resistance properties. The systems may be pigmented or formulated as clear coats.

Water vapor permeability considerations

Water absorption and the effect on carbon dioxide and water vapor diffusion are the most important properties of wall coatings with respect to construction physics.[185] Generally speaking, polyurethane coatings absorb very little water. They are therefore able to prevent the penetration of water, even on prolonged exposure to moisture, and thus block the growth of microorganisms.

Polyurethane coatings are not particularly permeable to carbon dioxide. Standard polyurethane coating concepts yield carbon dioxide diffusion resistance values ($\mu\,[CO_2]$) of more than 2×10^6 and diffusion-equivalent air layers of more than 200 meters. Therefore, polyurethane coatings provide reliable long-term protection against carbonation.

If properly formulated, polyurethane wall coatings satisfy European requirements in respect to water vapor permeability. Their water diffusion resistance values ($\mu\,[H_2O]$) are between 7,500 to 50,000 µ. The coatings are therefore relatively impermeable to water vapor diffusion. However, as dry film thicknesses of only 100 to 150 µm are normally applied, the diffusion-equivalent air layer is between 0.75 to 3.8 meters – depending

Figure 5.112: Existing roofs can be restored with minimal waste and with higher energy savings using a two-component aliphatic polyurethane coating pigmented white for solar reflectance

on the system – and thus below the maximum value of 4 meters generally required in specifications.
There is no specification to water vaper permeability in the relevant industrial standards in China right now.

In Northern America and some parts of Latin America, ASTM E96/E96M-16 "Standard Test Methods for Water Vapor Transmission of Materials" is often referenced in specifications with varying permeance rate requirements based on the project and specifier preference.

5.9.3 Waterproofing and Roofing

Waterproofing is understood to be the application of a water-impermeable system on construction elements, usually made of concrete. These may be roofs or balconies, but also secondary containment, ponds, canals and bridge decks, for example. The following two sections discuss the sealing of flat roofs and balconies. The other mentioned applications use highly reactive polyurea systems or hybrids of polyurethane and polyurea.

Waterproofing flat roofs with liquid polyurethane membranes

Most flat roofs are waterproofed with sheeting materials based on (polymer-modified) bitumen, polymeric rubber or PVC. However, the use of liquid roofing membranes are an economical alternative with long term durability, especially on roofs with many penetrations such as vent pipes, skylights, gutters, etc. Typical binders for liquid membranes are polyurethanes, unsaturated polyesters and acrylates.[186]

Liquid polyurethane roof membranes are applied in the liquid state over a variety of substrates such as metal, concrete, roofing board, and spray polyurethane foam and cure to form a weather-stable coating with long-term flexibility. They are used primarily in the refurbishment of leaking flat roofs but also in new construction projects. The key properties of these coatings are a long service life and low water vapor diffusion resistance, allowing roofs which have been penetrated by moisture to dry out after application of the membrane (see Figure 5.112).

In contrast to the use of roof sheeting materials, liquid polyurethane membranes yield a seamless skin.

Standards and regulations

In Europe, there are regulations and guidelines relevant to the use of liquid membranes in the refurbishment of flat roofs. These are being updated frequently so it is important to reference the most current version of the standard or regulation.

In North America, several organizations have performance-based specifications that address durability, flammability, and facility certification:

- ASTM International: ASTM has a series of roof coating technology/performance specifications based on the type of roof design and composition to ensure code compliance.
- FM Global: Factory Mutual has multiple specifications that are based on the complete roof build-up system. FM also performs the testing and certification of roof assemblies.
- Underwriters Laboratories (UL): UL 790 (ASTM E 108) "Standard Test Methods for Fire Tests of Roof Coverings" covers the fire resistance performance of roof coverings exposed to simulated fire sources originating from outside a building on which the coverings are installed.

In China, the national standard GB/T 19250-2013 polyurethane waterproofing coating, both non-exposure & exposure applications (for roofing and waterproofing) is involved. For roofing use, performance tests are needed, including the application properties of liquid coating, original mechanical strength, water and chemical resistance, durability and fire resistance. If there is traffic bearing needs for the roof, additional performance requirements are needed, i.e. shore A hardness, abrasion resistance & impact resistance.

Polyurethane coating systems

Four general categories of formulations are typically used in roofing and waterproofing applications. For all, some common properties are shared such as weatherability, elasticity, and adhesion to the substrate. For a summary, see Table 5.32.

- **One-component aromatic coating:** typically, a low NCO TDI polyether prepolymer, it will often contain a variety of other materials such as solvent, bisoxazolidine, pigments, and additives.
  Bisoxazolidine, a latent cure agent (see Chapter 3.7.7), ensures blister-free curing of the formulation in the necessary film thickness of about two millimeters. It reacts with atmospheric moisture to release amino and hydroxyl groups, which react in turn with the TDI-based prepolymers. This suppresses a direct reaction between the isocyanate prepolymer and atmospheric moisture, which would result in the formation of carbon dioxide and cause blistering. At the same time, bisoxazolidine improves the mechanical properties of the coating.
  Aluminum paste as part of the formulation (aluminum leafing pigment in plasticizer or aliphatic hydrocarbon solvent) reflects the sun and thus contributes to the weather stability of the polyurethane system. Additional solvents or thixotropic agents may be used to adjust the viscosity of the system for ease of application on the substrate. Multiple coats may be required in order to create the specified thickness.
- **Two-component aromatic coating:** highly reactive two-component formulations can also be used for this application. They are fast-curing and therefore are typically applied using plural component equipment. These products may consist of, e.g. a TDI

Table 5.32: Types and properties of liquid polyurethane membranes for sealing flat roofs

Binder	Application method	Raw materials	Remarks
One-component, aromatic	Manual or 1K spray equipment	TDI prepolymer, polyether polyols, oxazolidines	Coating typically contains aluminum leafing pigment or titanium oxide for light and weather stability
One-component, aliphatic	Manual or 1K spray equipment	HDI and IPDI prepolymers, polyether and polycarbonate polyols, oxazolidines	Coating can be pigmented a variety of colors
Two-component, aromatic	2K plural component spray equipment	TDI and MDI prepolymers, polyether polyols, amines	Coating typically contains aluminum leafing pigment for light and weather stability, fast rain-ready system
Two-component, aliphatic	Manual or 2K plural component spray equipment	HDI and IPDI prepolymers, polyether and polycarbonate polyols, amines	Typically provided as a white pigmented system for solar reflectance purposes, fast rain-ready system

or MDI prepolymer or MDI polyisocyanate that is cured with a polyol and/or polyamine/amine combination. In this coating, light stabilization is achieved by the addition of carbon black or aluminum powder. The system is normally designed to be low enough viscosity that it can be formulated at 100 % solid with no added VOC or solvents. The increased crosslinking obtained from the two-component polyurethane, adds a higher level of toughness. Typically, one coat is applied to the desired specified thickness.

- **One-component aliphatic coating:** Applied either by spray or manual methods, they can be pigmented in any color since they are based on aliphatic isocyanate prepolymers prepared from diisocyanates such as "Desmodur" I and polycarbonate or polyether polyols. Usually there is some solvent in the system, especially for those systems that are spray-applied. Multiple coats may be required in order to create the specified thickness.
- **Two-component aliphatic coating:** A more recent and growing development is the use of an HDI or IPDI oligomer or prepolymer reacted with a blend of polycarbonate or polyether polyols and amines for a fast curing, tough, weather stable two-component aliphatic coating. Similar to the two-component aromatic coating, the extra crosslinking yields higher properties such as tear, abrasion, and impact resistance. The coating can be pigmented to a variety ofcolors but white is the most common in order to maximize long-term solar reflectance where needed due to the season or latitude. Typically, the desired and specified thickness is achieved in one coat.

Roof coating concepts

If refurbishing a roof with liquid polyurethane membranes, the old roofing materials which have become leaky (possibly bituminous roofing felt, TPO or PVC sheet, or elastomeric acrylic coating) do not have to be removed first if they are dry, sound, and well adhered to the underlying roof deck. Depending on how the liquid membrane is to be applied, there are two alternative concepts:

- One-component systems for manual application:
  - primer (undercoat to strengthen substrate and bind dust);
  - base coat (approx. 1.5 kg/m^2 of liquid polyurethane membrane, applied by roller or spray equipment);
  - polyester matting (approx. 100 g/m^2, pressed into the liquid membrane to provide mechanical reinforcement and ensure achievement of the desired film thickness);
  - topcoat (approx. 1 kg/m^2 of liquid polyurethane membrane, applied after the base coat has dried to cover the polyester matting).
- Two-component systems for plural component machine application:
  - primer;
  - two-component polyurethane spray formulation (film thickness 1 to 5 millimeters);
  - before application of the liquid polyurethane membrane, the following conditions must be satisfied:
    - the substrate must be clean and strong enough;
    - the residual moisture content of the substrate and the underlying insulation materials must not exceed four percent.

The highly reactive two-component systems are applied using plural component spray units with gear or piston pumps.

Table 5.33 summarizes the polyurethane-based primers normally used with liquid polyurethane membranes. The primers and one-component liquid membranes are applied by brushing, rolling, or 1K airless spraying.

Waterproofing balconies with polyurethane membranes

Balconies, terraces and galleries on the outside of buildings are exposed to the weather and mechanical wear. Therefore, the load-bearing concrete substrate requires a high level of protection, particularly against moisture penetration. If this protection is absent, the moisture may migrate into adjacent floors and corrode the rebar of the concrete balcony slab or the balcony railing anchorage embedded in the concrete. Additionally, water penetration into the concrete may cause spalling due to repeated freeze/thaw cycles. The result is a reduction in the load-bearing capacity of the balcony slab with the ultimate effect that the balcony can no longer be used.

Similar to the previous discussion on polyurethane roofing coatings, liquid polyurethane membranes applied to balconies and terraces can provide the necessary waterproofing protection and fulfill a decorative function at the same time.

Table 5.33: Standard primers for liquid polyurethane membranes on various substrates

Substrate	Primer
Concrete	1K pMDI and/or MDI prepolymers, e.g. "Desmodur" E 21, 1K and 2K epoxy
Roof sheeting	1K pMDI and/or MDI prepolymers, asphalt, and modified bitumen

Standards and regulations

There are no special regulations covering the use of liquid membranes for waterproofing balconies and galleries in Europe and China however local standards and guidelines should be followed. The following general standards should be consulted in the North American region:

- For North America, balcony waterproofing in construction specifications is listed under 07 "Thermal and Moisture Protection" within MasterFormat 2004, by the Construction Specifications Institute (CSI)[187]
- ASTM C1127 "Standard Guide for Use of High-solids Content, Cold Liquid-Applied Elastomeric Waterproofing Membrane with an Integral Wearing Surface"[188].

Chemical principles

Liquid membranes used to waterproof balconies must display a high degree of lightfastness and weather stability (topcoat) and remain elastic in fluctuating temperatures (both base coat and topcoat). Polyurethane materials fully satisfy these requirements.

Figure 5.113: Aliphatic-based balcony coating with high resistance to weathering[163]

Formulations based on IPDI/polycarbonate polyol prepolymers and HDI polyisocyanates are adequately lightfast and can be used as transparent waterproofing systems. TDI prepolymers yield highly flexible, two-component membranes that may either be combined

Table 5.34: Polyurethane raw materials for balcony and waterproofing systems

Binder	Raw materials
One-component, aromatic	TDI and MDI prepolymers, polyether polyols
One-component, aliphatic	HDI and IPDI prepolymers, polyether polyols
Two-component, aromatic	TDI and MDI prepolymers, polyether polyols, amines
Two-component, aliphatic	HDI and IPDI prepolymers, polyether and polycarbonate polyols, amines

with topcoats based on IPDI or HDI polyurethane prepolymers that have been stripped to lower monomer content or used as sealing systems underneath ceramic tiles.

As far as the co-reactants are concerned, the use of polycarbonate polyols in the polyurethane prepolymers yields particularly weather-stable, flexible membranes. Prepolymers based on polyether/polyester polyols are less weather-stable and therefore used less in formulating the topcoat that is exposed to weathering. Two-component polyaspartic coatings using flexible "Desmodur" E 2863XP are a new entry into the balcony and terrace waterproofing coating market, providing both flexibility as well as weather stability. In addition to the binder, the formulations usually contain mineral extenders, pigments and many other additives.

Solvent-borne, moisture-curing, one-component polyurethane systems and solvent-free two-component systems have proven ideal as balcony waterproofing compounds. The most important polyurethane raw materials used in the formulation of balcony sealing systems are summarized in Table 5.34.

5.9.4 Outlook

There are several global trends that will continue to positively affect the use of polyurethanes in construction projects. These are considered in the following categories.

Environmental/regulatory

Around the world, the VOC regulations at a national, regional, and local level are expected to continue their downward adjustment. This will require formulators to use raw materials that meet or exceed these VOC restrictions. The adaptability of polyurethane chemistry along with the continuing investment in new polyurethane raw material product development ensures future coatings will meet the requirements.

In addition to VOC regulations, there continues to be a proliferation of green building standards that consider materials, service life, resource conservation, and other factors. Some examples include:

- The International Green Construction Code (IgCC), developed by International Code Council (ICC);

- EDGE (Excellence in Design for Greater Efficiencies), created by the International Finance Corporation;
- U.S. Green Building Council (USGBC) and their rating system LEED, or Leadership in Energy and Environmental Design;
- BREEAM is a leading sustainability assessment method used primarily in Europe;
- The Green Building Program (GBP) developed by the European Commission;
- GBEL (Green Building Evaluation Label – China Three Star), administered by the Ministry of Housing and Urban-Rural Development of the People's Republic of China (MOHURD).

Polyurethane construction coatings are very suited to meet the requirements of these VOC or green buildings standards due to the ability of formulators to design coatings with water-borne building blocks or high-solid technology.

End user demands

When polled, construction coating contractors and facility managers expressed several relatively new desires as they relate to coatings. There is a need for faster curing coating technologies such as polyaspartic coatings to keep projects on schedule. Additionally, odor during or after the coating process can interrupt the normal operation of a facility so lower odor options are preferred in both new and existing buildings.

To support these environmental, regulatory, and end user requirements, the raw material suppliers as well as coatings formulators will continue to focus new developments on water-borne, high-solid, and solvent-free technologies with faster return-to-service and lower odor as they migrate away from traditional solvent-borne coatings that were used in this site-applied construction coatings segment.

5.10 Light-stable, thick film coatings

Spray-applied polyurethane or polyurea elastomeric coatings are typically very fast curing systems which are characterized by their high film build. Over the last 30 years, these coatings have been used for a wide variety of applications to improve resistance properties of numerous substrates including metal, concrete, wood, and plastic. They often get used instead of conventional coatings when enhanced properties such as crack bridging, toughness, corrosion or abrasion resistance are required due to the demanding nature of the end-use application. Typically, these spray polyurethane/polyurea systems are used in applications such as pipes and tanks for wastewater infrastructure, roofing protection, bridge deck waterproofing, and truck bedliners for personal and industrial use.

Most often, these elastomeric coatings are based on aromatic raw materials, but there has been an increase in the development of aliphatic systems. According to the Polyurea Development Association, the NAFTA thick film coatings market is estimated to be approaching 100 kilotons in total volume in 2017. Certainly aromatic-based products (including polyureas, polyurethanes and hybrid products) heavily dominate this volume. Weatherable aliphatic products make up a very small portion of this total volume (estimated at less than 2 % of the total), but they are showing very high growth rates of above 5 %. These aliphatic systems show much of the same advantages of their aromatic counterparts (such as easy application with plural component equipment, fast curing and property development, and high film build > 1000 microns), while displaying much improved gloss, color and weather stability which allow them to be used in applications requiring this high performance level.

5.10.1 Spray-applied, aromatic polyurethane elastomeric coatings

Spray-applied, fast-curing elastomeric polyurea and polyurea hybrids based on aromatic starting materials have been commercially available for over 30 years. As mentioned above, one of the application areas that has found wide acceptance is for the after-market application of spray-in bedliners for pick-up trucks in NAFTA (see Figure 5.114). Applications for these aromatic systems are limited somewhat as they are typically only available in darker colors, due to their susceptibility to UV degradation (e.g. color change and loss of gloss). The darker pigmentation is used to mask the effect of sunlight.

Figure 5.114: Aftermarket-applied aliphatic polyurethane/urea protective liner for pickup truckbed

Mainly for cost considerations, polyureas and polyurea hybrids are most often based on resins consisting of aromatic isocyanate-functional materials such as diphenylmethane diisocyanate (MDI) adducts or MDI prepolymers and polyol/polyamine materials (such as amine- or hydroxyl-terminated polyethers, aromatic diamine chain extenders or aliphatic diamines). Table 5.35 lists some typical properties of a spray-applied, aromatic polyurethane elastomeric coating used in bedliners.

Table 5.35: Typical properties of a spray-applied, aromatic polyurethane/polyurea system[190, 191]

Processing		
Gel time [s]		15
Tack-free time [min]		2
Physical properties	**ASTM method**	
Density [g/cm3]	D792	1.09
Shore A hardness	D2240	85A
Tensile [MPa]	D412	9.2
Elongation [%]	D412	580
Die C [kg/cm]	D624	59
Split tear strength [kg/cm]	D1938	25
Taber abrasion [mg loss/1000 cycles with H-18 wheel]	D4060	190

These aromatic polymer building blocks are prone to oxidation, which is accelerated by exposure to sunlight. Since the gradual color change (yellowing) is retarded and masked by incorporating carbon black into the elastomeric spray coating, the only detectable degradation is loss of gloss. A common method that the coatings industry uses to detect oxidation is to determine the loss of gloss after exposure to sunlight. Figure 5.115 shows the retention of gloss as a function of time when exposed in South Florida. The 60° gloss readings of aromatic-based systems fall from initial values of about 80 to less than 10 within two months when no UV stabilizer is added, and within nine months when a typical UV stabilizer package is added.

Despite the dramatic loss of gloss shown above, it is also very important to note that the physical properties (e.g. tensile strength and elongation) of the same aromatic elastomeric coatings were still greater than 95 % of their original values after exposure.

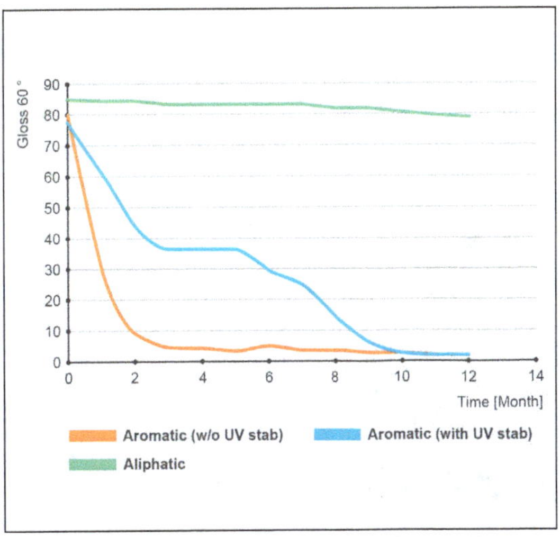

Figure 5.115: Florida weathering of an aliphatic and an aromatic polyurea truck bedliner

5.10.2 Spray-applied, aliphatic polyurethane elastomeric coatings

Because the traditional polyether or aromatic isocyanate polymers are not UV-resistant, other chemical approaches and building blocks must be used to provide non-yellowing spray elastomers. Most of these approaches involve the use of aliphatic isocyanate prepolymers or adducts in place of the aromatic isocyanate prepolymers. These, in combination with light-stable co-reactants (e.g. polyesters or aliphatic di- or polyamines) can yield the necessary weathering performance (see Figure 5.115).

Specific market areas require light- and weather-stable, tough, elastomeric, thick film coatings that can be pigmented to various shades. For example, the automotive industry would like to use this approach to manufacture truck bedliners and other optional accessories (i.e., running boards, rocker panels, fender flares) that match or complement the color of the remainder of the vehicle. For example, Nissan offers a factory installed bedliner for their Titan truck as part of a bed upgrade package which is aliphatic in nature. In order to offer an OEM factory installed bedliner, the material must be much more UV-stable, compared to the aromatic-based products discussed above. Additionally, in the automotive aftermarket, aliphatic-based sprayable elastomers are available in a wide variety of colors which show good gloss and color stability. Other markets with an interest in these color and gloss re-

Figure 5.116: Thick film polyurethane/polyurea coatings can be used to provide light- and color-stable protection for a wide variety of industrial applications (such as amusement parks) where frequent repainting is not an option

tention characteristics could include marine, agricultural and construction machinery and amusement parks, to name a few (see Figure 5.116).

Various approaches have been developed to address this issue[191, 192]. They include the use of light stable spray elastomers-based on aliphatic polyisocyanates, either used alone (i.e., single layer) or in combination with aromatic elastomers (i.e., top layer of aliphatic over an aromatic base elastomeric coating layer in a composite approach).

Low monomer containing prepolymers that have been used to provide light-stable elastomers include those based on poly(caprolactone) or poly(carbonate) diols capped with IPDI or H_{12}MDI. These products, in combination with aliphatic diamino-functional or triamino-functional resins, produce instant-set durable polyurea coatings with excellent physical properties.

Other approaches have been taken which have used aliphatic polyisocyanate dimer or trimer crosslinkers or prepolymers. These products (based on HDI or IPDI) are currently used in standard, high-performance polyurethane coatings in the automotive OEM, refinish, construction, and general industrial metal coating markets (e.g. "Desmodur" N 3400, "Desmodur" N 3900, "Desmodur" XP 2580).

These products are stripped of monomeric diisocyanate yet have relatively low viscosities. They are commonly used in spray coatings within well-established industrial hygiene guidelines. They usually have a NCO-functionality of more than three, which leads to more rigid polymers with plastic-like physical properties. These tough materials display excellent abrasion resistance and energy absorption characteristics (to help prevent dents and other damage), even though they do not display all of the elastomeric characteristics common with aromatic counterparts.

Polyester polyols often serve as co-reactants for aliphatic isocyanate adducts. The molecular design possibilities for these polyesters are very broad and easily customized to compensate for the higher functionality of the aliphatic polyisocyanate adducts used. This is achieved by selecting polyesters with the appropriate molecular weight, functionality, and glass transition temperature (T_g).

Urea linkages are formed when amines are introduced as co-reactants along with the polyester in the formulations. There are several amine choices available. Aromatic amines can be used here with some sacrifice in color stability. The use of primary and secondary aliphatic di- and triamines can help here as well, but the difference in reactivity between these amine (NH_2) reactive groups and the polyester (OH) is very large. Certain faster curing polyaspartic acid ester-based co-reactants are also finding usage as modifiers in the production of these aliphatic sprayable thick film coatings.

If too much of these highly reactive amines are used, they will react preferentially, and the resulting urea can precipitate from the reacting mass, bringing with it unreacted isocyanate. This in turn leaves an excess of polyester in the system that can later exude from the elastomer. Therefore, it may be critical to catalyze the polyol/isocyanate reaction to match

Table 5.36: Yellowing (ΔE) comparison of elastomeric coatings

1,500 hours exposure	Aromatic	Hybrid	Aliphatic
Weather-o-meter	55.6	10.6	3.3
QUV-A (313 bulb)	51.5	22.8	6.2
QUV-B (340 bulb)	45.6	17.4	4.0

that of the amine/isocyanate reaction to obtain complete polymer curing and optimum physical properties.

The resistance to UV degradation of the aliphatic isocyanate-based elastomeric coatings is much improved compared to that of aromatic coatings. Table 5.36 shows the comparison of white pigmented aromatic isocyanate-based spray elastomers (based on MDI prepolymer and polyether polyol) to two types of aliphatic isocyanate-based coatings. The hybrid aliphatic system contains aliphatic polyisocyanate crosslinker based on an HDI allophanate modified trimer, polyester, and an aromatic amine chain extender. It resists yellowing much better than the aromatic isocyanate-based system, but it is not as good as the system based solely on aliphatic materials. The coating based solely aliphatic material consists of the same HDI allophanate-modified trimer and polyester as the hydrid system, but also contains some aliphatic amine as well. The improved gloss and color stability shown by aliphatic polyurethane/polyurea systems opens up a myriad of color and design possibilities, allowing this technology to be used in additional demanding market segments (see Figure 5.117).

5.10.3 Spray application technology

When spray applying an elastomeric polyurethane/polyurea coating, the application equipment selected is a big factor in getting the product to market. Thoroughly understanding the application and the capabilities of your customer will help greatly with this selection. The most successful product introductions usually involve the cooperative support of raw material suppliers, equipment specialists, and end-users.

Figure 5.117: Examples of spray-applied aliphatic polyurethane/polyurea elastomeric coatings showing possible colors and surface textures

Typically, for most applications, multicomponent mixing/spray equipment of some type is used. Typically, these application devices can be divided into two basic categories – low-pressure and high-pressure. They allow a relatively thick coat of the spray elastomer to be applied in a minimum number of passes. Low-pressure equipment (0.7 to 7 MPa) can be used for production and larger scale trials. It can have a broader range of varying mixing ratios and produce a higher output. Low-pressure equipment can be fitted with material heating capabilities that can aid sprayability when using higher viscosity materials. This type of equipment can also use low-pressure guns with disposable plastic static mixer inserts. Some equipment manufacturers will use and/or recommend using an impingement mix gun for fine finish applications.

High-pressure/high-output equipment (7 to 27 MPa) can also be used for production and large trials. It can be equipped with fixed or varying ratios, heating capabilities and generally use impingement-style spray guns. These spray guns can produce a coarse to fine finish. A higher volume of materials is normally required when using this equipment. Due to the much higher pressures, an equipment manufacturer or spray equipment specialist should be consulted when considering high-pressure spray equipment.

Another approach to applying systems like this is to use a cartridge-fed spray gun. A cartridge gun, which is commonly used in the adhesives market, is similar to a caulking gun. It has the capability to handle different size dual-component cartridges and may be used manually or pneumatically actuated (see Figure 5.118). This device uses a piston to drive a plunger into the cartridge, displacing the product through a static mixer. Attached to the end of the static mixer is a spray nozzle that blows the product from the static mixer tip and propels it at the intended substrate[193]. This positive displacement device, in combination with the static mixer and spray tip configuration, works very well when evaluating formulations in the lab, small-scale trials or repair applications. These types of cartridges and guns are relatively inexpensive and can be purchased with a wide variety of mixing ratios (numerous mix ratios available from 1/1 to 4/1 by volume). The disposable tubes and static mixers can come equipped with a spray tip that will allow the applicator to create effects ranging from a textured to a somewhat smooth surface. This cartridge application option has also greatly aided laboratory developments of fast-curing sprayable elastomeric coatings because it is a very simple and cost-effective means of applying experimental systems. The benefits of this

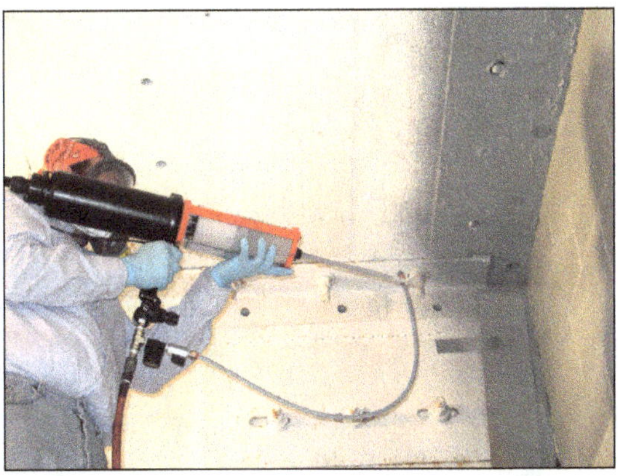

Figure 5.118: Example of high pressure spray equipment for the application of thick film coatings

application method do not end here; applicators are finding niche markets where this application method can work very well.

5.10.4 Outlook

The use of aliphatic, spray-applied truck bedliners are now a commercial reality in the OEM sector and are being applied in a limited number of assembly plants, such as the Nissan facility previously mentioned[194]. This means that pickup truck owners who opt for the coating have a tough, durable and non-fading bedliner protecting their vehicle from the day of delivery. In addition to these OEM manufacturers, suppliers are introducing aliphatic-based polyurethane/polyurea sprayable coatings to the aftermarket for pickup trucks as well as in other industrial applications[195].

Two-component aliphatic elastomeric polyurethane coatings are also being considered for almost any other application that requires resistance to corrosion, chipping or abrasion, in addition to UV durability. Possible additional end uses include floors, bridges, marine applications, roofing, industrial equipment and construction applications.

5.11 References

[1] "Irfab" Global Industrial Coatings Markets (GICM), PRA World Ltd, 2017, 2018
[2] e.g. AgBB (Germany), GB24410 (China)
[3] C. Irle, M. Bayona, Surface Coatings International, Part A, 2005/06, p. 21
[4] C. Irle, R. Roschu, E. Luehmann, S. Feng, PCI Magazine, February 2002, p. 28 ff
[5] M. Almató, E. Tejada, J. M. García, Water-borne wood coatings made easy, PPCJ April 2016
[6] N. Gruber, Market overview: UV/EB curing, Radtech Conference, Prague October 2017
[7] IKEA IOS MAT 066, EN 12 720, GB/T-3324-2017 (China), KCMA/ BIFMA (US)
[8] E. Tejada, Double crosslinking for greater efficiency, ECJ 11/2017
[9] Irfab (Ed.): New Coatings Technology Trends – a VOC-Voice of the Customer – Study 2007–2012 (Brussels)
[10] Floor Forum International 81, p. 43 ff
[11] B. Meuthen, A.-S. Jandel, Coil Coating. Vieweg Verlag. Wiesbaden 2005
[12] Fachhochschule für Technik: Seminar Coil Coating (Esslingen 2006). Esslingen: Fachhochschule für Technik, 2006. – B. Baumbach: 1K-PUR-Systeme für Coil Coating
[13] Fachhochschule für Technik: Seminar Coil Coating (Esslingen 2006). Esslingen: Fachhochschule für Technik, 2006. – K.H. Stellnberger: Flansch-Korrosion – Die gefährlichste Korrosionsart am Automobil?
[14] A. Goldschnidt, H.-J. Streitberger, BASF Handbook: Basics of Coating Technology, BASF Coatings (Hrsg.), Vincentz Network, Hanover 2018, p. 747, p. 763
[15] https://echa.europa.eu
[16] http://impact.nace.org/economic-impact.aspx
[17] ISO 12944 standard
[18] ASM Handbook Volume 5B Protective Organic Coatings www.asminternational.org/bestsellers/-/journal_content/56/10192/23412274/PUBLICATION

[19] C. Hare. "Protective Coatings – Fundamentals of Chemistry and Composition", SSPC 94-17 1994

[20] Olson, A., et al. "Two coat polyaspartic urethane coatings protect Virginia steel bridges for over a decade", Journal of Protective Coatings and Linings January 2016 56-63

[21] Chong, Shuang-Ling, and Yuan, Yao. "Are Two Coats As Effective As Three?" Report No. FHWA-HRT-2006-006. Federal Highway Administration, Sept. 2006. Print

[22] M. O'Donoghue et al., "Windmills Fast production Schedules with Novel Zinc Primers and Polyaspartic Urethane Ester Topcoats," Proceedings of the PACE 2005 Conference, SSPC: The Society for Protective Coatings, Pittsburgh PA, 2005

[23] A. Olson et al. "Reducing the Cost of Shop Painted Steel Bridges through Improving Shop Painting Efficiency." World Steel Bridge Symposium, National Steel Bridge Alliance, Chicago, IL, 2018

[24] B. Castler "Rapid Deployment Technology: A New Concept for Connecticut," International Bridge Conference, Engineers Society of Western Pennsylvania, Pittsburgh, PA, June 10, 2003

[25] ASTM Standard Test Methods for Chatiodic Disbonding of Pipeline Coatings

[26] M. Homann, A. aus der Wieschen, K. Kolping, Solvent-free 2C Polyurethanes – A proven product class for corrosion prevention, Latincorr Congress, Fortaleza, 2006

[27] Covestro, Order-No. COV00081225, 2016

[28] T. J. Murray, Macromol. Mater. Eng. 2008, 293, p. 350–360

[29] H. R. Krichel, (1999) Progress in Polyimide Chemistry II, Chapter 2, Poly(ester-imide)s for Industrial Use, K.-W. Lienert, ISBN 978-3-540-49814-8

[30] H. Kittel, Lehrbuch der Lacke und Beschichtungen, Band I, Teil 2, 581, Verlag W.A. Colomb, 1973

[31] Covestro estimate

[32] Eisenmann Mayer, SurCar, 2011

[33] P. Minko, Axalta Coatings Systems, May 2014 "Niedrigtemperatur-Lacksysteme", DFO Tagung

[34] K. F. Doessel, JOT 9 (2015)

[35] U. Meier-Westhues, M. Mechtel, Th. Klimmasch, J. Tillack, On the road to improved scratch resistance, ECJ 09 (2002), pp. 28

[36] M. Clemmer, W. Kosteniuk, SURCAR Cannes 2017

[37] EP1731582A1, 12.2006

[38] O. Fleck, S Nowak, U. Meier-Westhues, European Coatings Journal 04 (2009), p. 106–111

[39] Dr. M. Groenewolt, BASF Coatings, Karosserielackierung intensiv, Sindelfingen 2016

[40] G. Behnken, A. Hecking, B. Vega Sanchez, Farbe und Lack 08 (2015) GDCh Jahrestagung

[41] Automotive Circle Conference: M. Mechtel, 03.2017, Ingolstadt, Germany

[42] K. F. Dössel, Automotive Circle Conference Bad Nauheim, 11 (2014)

[43] JP5112570B, JP5280580B, US 9,567,485 B2, US 9,802,223 B2: (2017) Covestro

[44] L. Kahl, M. Harsch, Automotive Circle International, Bad Nauheim, 11 (2014)

[45] H. Morita, SurCar Shanghai conference 2017

[46] "Automotive Paints Market by Vehicle Type, by Coat Type (Electro Coat, Primer, Base coat, Clear Coat), by Technology, by Texture – Global Industry Perspective, Comprehensive Analysis and Forecast, 2016–2022", Zion Market Research

[47] J. Krömer, S. Kreiling, Henkel AG & Co. KGaA, Strategies in Car Body Painting, Berlin, 2016

[48] R. Kopp, H. Hess, G. Grögler, Proceedings of the Polyurethanes World Congress (1987) p. 316

[49] V. Mirgel, K. Nachtkamp, Chemische Aspekte vernetzender Einbrennurethanharze (Farbe und Lack, 12, 1983)

[50] H. Casselmann, M. Bock, H. Blum, Progress in Development of Waterhorne Polyurethane Primer for the Motor Vehicle Original-Equipment-Manufacturers Industry, Water-borne, Higher-Solids and Powder Coatings Symposium, New Orleans 1994, Congress Proceedings, p. 735

[51] DuPont Performance Coatings, Wuppertal, Germany
[52] L. Kahl, R. Halpaap, C. Wamprecht, Umweltresistente 2K-PUR-Klarlacke für die Autoerstlackierung, Industrie-Lackierbetrieb 61 (1) (1993) 30
[53] H.-U. Meier-Westhues, Th. Klimmasch, J. Tillack., On the Road to Improved Scratch Resistance, Europ. Coat. Conf., Zürich 2001
[54] K.-F. Dössel, Current Status of the Various ECO Concepts (1. German Automotive Circle No.21), 2004
[55] P. E. Yeske, M. Bock, T. Engbert, L. Schmalstieg, Anorganisch modifizierte Lackrohstoffe für Polyurethan-Systeme 64. Vortragsveranstaltung der GDCh-Fachgruppe APi, Celle, October 1998
[56] M. Bock, T. Engbert, L. Schmalstieg, P. E. Yeske, Inorganic-Modified Raw Material for Polvurethane Coatings, 7i' Annual Advanced Coatings Conference, Detroit/USA, September 1998
[57] US 6 593 417 B1
[58] M. Mechtel, Glänzende Aussichten für 2K-Polyurethan-Klarlacke, JOT 09/ 2005, 540 ff
[59] K.-F. Dössel, DuPont Automotive Systems, 5th Int. Strategy Conf. On Car Body Painting, Berlin, July 2006
[60] T. Engbert, E. König, E. Jürgens, Einsatz einkomponentiger Einbrennurethanbeschichtungen, Farbe und Lack 102 (7) (1996) 51
[61] "Will autonomous vehicles put the brakes on the collision parts business?", 2017, KPMG U.S. Manufacturing Institute Automotive Center
[62] www.arb.ca.gov/drdb/moj/curhtml/R1116.PDF
[63] Dürr Systems AG
[64] M. Sonntag, Les Entretiens De Wolff, Paris, April 1998
[65] L. Kahl, M. Bock, H. J. Laas, Wässrige 2K-PUR-Lacke für die Autoerstlackierung, Teil I: Rohstoffentwicklung, Industrie-Lackierbetrieb 63 (12) (1995) p. 424
[66] H. Liebscher, Technology guidelines for vehicle refinishes, personal note, 1997
[67] F. Richter, H. Mertes, Farbe und Lack 106 (9/2000) pp. 60–72
[68] C. Zwiener, M. Sonntag, L. Schmalstieg, European Coatings Journal (10) (1992) 5/88
[69] V. Mirgel, M. Sonntag, European Coatings Journal (10) (1994) 6/90
[70] M. Melchiors et al., Prog. Org. Coat. 40 (2000) p. 99
[71] M. Melchiors, H. Casselmann, C. Kobusch, K.-L. Noble, M. Sonntag, Polymer Paint Colour Journal 190 (2000), p. 24–28
[72] H.-C. Gruner, K.-W. Reinhart, Materials Engineering (New York, NY, United States) (2003) 21 (Coatings of Polymers and Plastics) p. 317–352
[73] R. Volland, Lackharze im Wandel, Bayer AG, Bestell-Nr. LS 44917, 1995
[74] M. Bock, et al., Globalisierung der Fahrzeugindustrie, Farbe und Lack 102 (9) (1996) p. 132
[75] DD-Lackrohstoffe-Einführung, Bayer AG, Bestell-Nr. LS 44734A, 1992
[76] M. Osterhold, Progress in Colloid and Polymer Science (2006), 132 p. 41–47
[77] C. Kobusch, W. Küttner, U. Klippert, B. Willwerth-Schwarten, 'Comparison of PUR-coatings for plastic substrates: Solvent-borne, high-solids and water-borne' Proceedings of the International Water-borne, High-Solids, and Powder Coatings Symposium (2000), 27th p. 476–489
[78] Plastic Europe
[79] R. Ryntz, Adhesion to Plastics, Global Press, 1998, p. 1
[80] O. Fleck, S. Nowak, U. Meier-Westhues, European Coatings Journal, 04 (2009), p. 106–111
[81] J. Cleuvenbergen, PP and Painting Technologies in the Automotive Industry, CEASP Symposium, Himont/Italien, April 1991, Congress Proceedings, p. 1
[82] D. Briggs, Surface Treatment for Polyolefins in Surface Analysis and Pretreatment of Plastics and Metals, Brewis (Hrsg.). MacMillan Publishing Co., New York 1982, p.1
[83] S. Waddington, D. Briggs, Polymer Comm. 32 (16) (1991) 506

[84] G. Stipan, Zur Lackhaftung auf Kunststoffoberflächen, Dissertation, Universität Stuttgart 1989
[85] M. Schönfelder et al., Plastverarbeiter (1987) 123; W. Cohnen, I-lack 59 (1991) 102; W. Cohnen, Farbe und Lack 98 (1992) p. 102
[86] J. Petzold, Umweltfreundliches Lackieren von Kunststoffen, SKZ-Tagung Würzburg 1993
[87] C. Kobusch, J. Petzoldt, M. Bock, 2K-PUR coatings for plastics in automotive applications, Boya IL International Paint Conference 1997, p. 17
[88] Ninth Adaptation to Technical Progress to CLP Regulation Commission Regulation (EU) 2016/1179
[89] Carbon Blocks for Speciality Applications, Cabot Technical Report p. 136
[90] C. Kobusch, M. Bock, 2K-PUR coating systems for plastic Substrates, 23" Focus Conference, 1998
[91] K. Martin, Conference Proceedings – Polyurethanes Expo, Orlando, FL, United States, Oct. 1–3, (2003), p. 100–105
[92] K. Doessel, Opportunities for Low Bake Coating in Compact Coating Systems, conference proceedings, SurCar, Cannes, 2015
[93] F. Richter, M. Grahl, J. Weikard, Recent Developments on Controlled Polyurethane Formation employing Thermolatent Hardener Technology, conference proceedings, European Coatings Conference, Nuremberg 2015
[94] AutoCreative, publication of Bayer AG, Leverkusen, KU 46 312 d/1 778 177, 1996
[95] St. Jacob, Finishing of plastics with coatings – development and trends, Berichtsband – Deutsche Forschungsgesellschaft für Oberflächenbehandlung (2002), 53 (Kunststofflackierung), p. 9–16
[96] C. Kobusch, J. Petzoldt, M. Bock, Polymers Paint Colour Journal (PPCJ) 188 (4404) (1998) p. 16
[97] Softfeel coatings for plastics, Bayer Inform, publication of Bayer AG, Leverkusen, VS 4564-67e, 1993
[98] C. Kobusch, J. Petzoldt, M. Bock, Aqueous 2K-PUR softfeel coatings for use in car passenger compartments, Surplast VI, 1997
[99] www.paint.org: C. Challener, Formulating soft-touch Coatings, A Balancing Act, Coatings Tech, July 2016, pp 47-51
[100] C. Kobusch, J. Petzoldt, M. Bock, VDI Berichte Nr. 1398, (1998) p. 357
[101] A. Lockhart, L. Gindin, P. Schmitt, P. Lunney, 'Development of UV-curable water-borne polyurethane dispersion for soft feel application' RadTech Report (2004), 18(6), p. 46–54
[102] CCP Polyester Products Application Manual, 9th Edition, (2000)
[103] www.netcomposites.com/education, Guide to Composites, Introduction to Coatings
[104] ww.bayermaterialsciencenafta.com
[105] W. Vogel: Glaschemie. 3. Auflage. Springer, Berlin 1992 (Reprint 2012)
[106] Römpp Chemie Lexikon, J. Falbe, M. Regitz (Hrsg.), 9. Auflage. Georg Thieme Verlag, Stuttgart 1990, p. 1575 ff.
[107] Glasverpackungen und Umwelt, Bundesverband der Glasindustrie und Mineralfaserindustrie e.V., Düsseldorf, (Ed.), Issue 02/93
[108] WO 97 / 05077 (1995) BASF Lacke und Farben AG. WO 90 / 05031 (1988) Brandt Manufacturing Systems Inc.
[109] EP 0 025 992 (1979) Chemische Werke Hüls AG. EP 0 133 090 (1983) Saint-Gobain Vitrage
[110] B. Baumbach, M. Bock, R. Dearth, W. Küttner, Water-borne polyurethane resins for glass coatings, European Coatings Journal (9) (1997) p. 792–794
[111] D.W. Dwight "Glass fiber reinforcements.", p. 231–261, 2000
[112] J. L. Thomason, Glass Fiber Sizing (2015), Published by J. L. Thomason
[113] EP 137 427 (1984) Bayer AG
[114] EP 1 375 550 (2004) Bayer MaterialScience AG
[115] Covestro estimate 2018
[116] H. Träubel et al. in: Kunststoff-Handbuch, G. W. Becker, D. Braun (Hrsg.), Band 7: Polyurethane. Carl Hanser Verlag, München, Wien 1993, pp. 621–641

[117] Druckschrift FB-V Le 1310/1, Bayer AG, Feb. 1973
[118] H. Bergs, R. Quack, DOB + Haka-Praxis (1971)
[119] W. Schröer, D.-L. Schütze, W. Thoma, Wasserdampfdurchlässige, kompakte Textilbeschichtung mit Polyurethanen, Vortrag auf der Textiltagung in Aachen am 26./27.11.91, published in "Coating" 9 (1992) p. 290–296
[120] H. Träubel, D.-1. Schütze, J. Pedain, Bessere Tragehygiene durch wasserdampfdurchlässige Materialien, Textilveredelung 29. Juni 1994, p. 171–177
[121] D.-L. Schütze, H. Träubel, J. Pedain, Improving wear comfort with water-vapour permeable polyurethanes, Vortrag auf dem AATCC Symposium, Coated and Laminated Fabrics: New Processes and Products, in Danvers/Massachusetts, 3.4.95
[122] H. J. Koch, Melliand Textilber. 58 (1977) 319, Bayer Farben Revue 28 (1977) 35
[123] H. J. Koch, Textil-Praxis Int. 3 (1977) 311
[124] W. Thoma, Bayer Farben Revue 25 (1974) 70
[125] J. Gillibrand, J. Coated Fabr. 3 (1974) 153: J.S. Maddams, J. Coated Fabr. 3 (1974) 164; D. A. Atkinson, Eur. Rubber J. (6) (1976)
[126] H. J. Koch, Melliand Textilber. 52 (9) (1971) 1094
[127] H. J. Koch, Melliand Textilber. 53 (11) (1972) 1272; H. Neumaier, J. Coated Fabrics 3 (1974) 181
[128] H. J. Koch, Melliand Textilber. 51 (11) (1970), 1313; J. Coated Fibrous Mater. 1 (1971) 118
[129] DE 2 754 593 (1977) Bayer AG, DE 2 754 603 (1979) Bayer AG
[130] H. Giessmann, Melliand Textilber. 51 (11) (1970) 1348
[131] W. Schröer, Textilveredlung 22 (1987) 459-467
[132] W. Wittke, Coating (1988) 340-345
[133] H. Träubel, Das Leder (1975) 1; T. Hayashi, Chemical Economy and Engineering Review (1974) 25
[134] Brand DuPont: "Corfam"
[135] DE 888 766 (1951) Bayer AG
[136] DE 1 110 607 (1958) Du Pont
[137] B. Zorn, Textil Praxis International 29 (1974) 1706, 1711, 1712
[138] H. Träubel, New Materials Permeable to Water Vapor, Springer Verlag Berlin Heidelberg New York, 1999
[139] https://echa.europa.eu/candidate-list-table/-/dislist/details/0b0236e1807dec94
[140] EP-A 1 669 382 (2004) Bayer MaterialScience AG
[141] US-A 6 642 303, EP-A 1387859 (2000) Bayer AG
[142] Covestro Press release "New data shows INSQIN reduces the carbon footprint of textile coating by half" 5.09.2017
[143] US20150044480A1 (2013) Moulures Transform Inc.
[144] E. Wenderoth, J. Kopp, H. Träubel, H. A. Burkhart, Wochenblatt für Papierfabrikation (9) (1996) p. 379–383
[145] J. Kopp, H. Träubel und E. Wenderoth, Wochenblatt für Papierfabrikation (7) (1996) p. 294–297
[146] B. Thiele, B. Hauschel, Pulp & Paper Europe October/November 2002
[147] PCT Int. Appl. (2010), WO 2010054760 A1 20100520 (Covestro)
[148] US 2005/70 008 794 (2005) Arkwright, Inc.
[149] US 6 866 383 (2005) Avery Dennison Corporation
[150] WO 2010054760 A1 (2010) Covestro
[151] EP 1 116 713 (2001) Asahi Kasei Kabushiki Kaisha
[152] Technik des Flexodrucks, 5. Auflage 2006, St. Gallen: Rek & Thomas Medien AG, K. Heger/F. Reichert, Druckfarben im Flexodruck, Chap. 9, p. 341, 2006
[153] H. Klein, Coating (4) (1989) p. 112–113
[154] EP 0 738 750 (2000) BASF AG
[155] B. Magee, American Ink Maker (79) (2001) p. 70–72
[156] www.flintgrp.com/media/3668/printpur_polyurethane_range.pdf
[157] www.labelsandlabeling.com/news/new-products/hubergroup-develops-pu-based-printing-ink-flexo-demanding-applications
[158] Yang Zhihui, Wu Guangfeng, Zhang Huixuan, Journal of Applied Polymer Science 134 (2017), 45406

[159] PCT/EP2016/063643 Siegwerk Druckfarben AG & Co, KGaA

[160] D. Kummer, "Veredelung von Verpackungen am Beispiel von Faltschachteln", print media messe DRUPA 2004, compass sessions "Verpackungsdruck"

[161] www.valmet.com/globalassets/media/.../white-papers/.../wpd_flotationdrying.pdf

[162] DIN 68861: Möbeloberflächen

[163] Kemper System

[164] H. Gruber, Beschichtungsstoffe auf Polyurethanbasis, Zentralblatt für Industrie 3/75

[165] SSPC: The Society for Protective Coatings. (2003). SSPC SP-13/NACE No. 6, Surface Preparation of Concrete. Retrieved from www.sspc.org/ST-00SP13

[166] SSPC: The Society for Protective Coatings. (2017). SSPC Guide 23, Field Methods for the Determination of Moisture in Concrete and Masonry Walls and Ceilings. Retrieved from www.sspc.org/ST-000G23

[167] SSPC: The Society for Protective Coatings. (2017). SSPC PA-7, Applying Thin Film Coatings to Concrete. Retrieved from www.sspc.org/ST-000PA7

[168] Reinstadtler, S. (2018, February). High Performance Coatings for Decorative Floors. Paper presented at the meeting of the Painting and Decorating Contractors of America, Galveston, TX

[169] DIN 18 032/1-6: Sporthallen

[170] Sportstättenbau und Bäderanlagen. sb, Jg. 20, 1986

[171] Bautenschutz Bausanierung, Jg. 12 (1989), Nr.2, pp. 1–42

[172] AIT. Architektur, Innenarchitektur, Technischer Ausbau, Jg. 104 (1996), Nr.7/8, pp. 8–101

[173] Reinstadtler, S. (2017, January). Case Study: A polyaspartic coating made a commercial bakery floor look really sweet. Paper presented at the meeting of SSPC: The Society for Protective Coatings, Tampa, FL

[174] S. Reinstadtler, (2016, September). High performance coatings for industrial and decorative floors. Course presented at the Concrete Décor Show, San Diego, CA.

[175] AGI-Arbeitsblatt S 10 Teil 3: 1995–11 zu beziehen bei: Arbeitsgemeinschaft Industriebau e.V., Ülsdorfer Straße 106, 51143 Cologne

[176] Bau- und Prüfgrundsätze für den Gewässerschutz, Deutsches Institut für Bautechnik (DIBt), Berlin, July 2005

[177] H. Schuhmann et al., Handbuch Betonschutz durch Beschichtungen, Expert Verlag, Ehningen 1992

[178] K. Weinmann et al., Moderner Bautenschutz, Lexika-Verlag Grafenau, Würzburg 1976

[179] R. Engelfried, Diffusionswiderstandszahlen für Kohlendioxid und Wasser und deren praktische Anwendung, Farbe und Lack 6 (1983) p. 513

[180] Wilhelm Barthlott, Zdenek Cerman, Anne Kathrin Stosch: Der Lotus-Effekt: Selbstreinigende Oberflächen und ihre Übertragung in die Technik. Biologie in unserer Zeit 34 (5), pp. 290–296 (2004), ISSN 0045-205X

[181] Covestro building Dormagen; Artist: Megx

[182] Reinstadtler, S. (2012, January). New architectural coatings technology targeted at stricter hospital infection protocols. Paper presented at the meeting of SSPC: The Society for Protective Coatings, Tampa, FL

[183] DD-Lack – wirtschaftlicher Langzeitschutz im Bauwesen, publication of Bayer AG, Leverkusen, 1985

[184] Reinstadtler, S. Environmentally friendly graffiti resistant coatings. Paper presented at the meeting of SSPC: The Society for Protective Coatings, Phoenix, AZ

[185] W. Kubitza, Polyurethan-Anstriche auf Beton unter Berücksichtigung der bauphysikalischen Eigenschaften, Bautenschutz & Bausanierung 2 (1980) p. 55–58

[186] Abdichtungen mit Flüssigkunststoffen nach ETAG 005, Deutsche Bauchemie, 2005

[187] Construction Specifications Institute. (2004). Masterformat: 07 00 00 Thermal and moisture protection. Retrieved from www.masterformat.com/revisions/

[188] ASTM International. (2015). ASTM C1127/C1127M-15, Standard guide for

use of high solids content, cold liquid-applied elastomeric waterproofing membrane with an integral wearing surface. Retrieved from www.astm.org/cgi-bin/resolver.cgi?C1127C1127M
[189] S. Reinstadtler, PU 104B-Polyurethane Coatings And Elastomers, CPI Polyurethanes Technical Conference, New Orleans, 2017
[190] "Baytec" SPR-085A (Covestro)
[191] D. Zielinski, J. Rosthauser, J. Perry, "New In-roads into Weather Stable Spray Elastomers," Proceeding of the Alliance for the Polyurethanes Industry (API) Conference (October 2003)
[192] D. J. Primeaux, "100% Solids Aliphatic Spray Polyurea Elastomer Systems," Journal of Elastomers and Plastics, Vol. 24, No. 4, 323-336 (1992)
[193] A. E. Sandholm, R. W. Clemens, J. H. Hunter, U.S. Patent 6,601,782 (2003)
[194] D. M. Bratys, "A High Volume Production Multi-Use Polyurea Protective Coating," Proceeding of the SSPC Paint and Coatings Expo (PACE) Conference (January 2006)
[195] B. Naik, "Aliphatic Thick Film Coatings – How You Can Have Your Cake And Eat It Too!!," Proceeding of the SSPC Paint and Coatings Expo (PACE) Conference (January 2006)
[196] Adobe Stock, ValentinValkov
[197] Adobe Stock, Araspixel
[198] Adobe Stock, benoitgrasser
[199] Adobe Stock, Lukasz Z
[200] Adobe Stock, Petair
[201] Adobe Stock, Sailorr
[202] Adobe Stock, golubka57

6 Polyurethane adhesives

6.1 Introduction

Adhesive bonding: comparison with other joining methods

Adhesive bonding is an essential technique to manufacture products economically and with consistently high quality. It is[1, 2], however, only one of many possibilities for joining materials and competes with other assembly methods such as welding or mechanical fastening. The process ultimately used is the one that delivers the highest benefit/cost ratio, after consideration of all costs and risks. Adhesive bonding has several advantages over other joining methods. The parts to be joined are not damaged by thermal stress (welding), nor are they weakened by holes (screwing). Stress forces may be introduced when two substrates are joined and this is more critical when the substrates have a complex shape or are made of different materials. When an adhesive is used the stresses are spread across the bond line instead of being concentrated at the point of a weld or screw. Adhesive use is a preferred joining method but presents a disadvantage from a recycling standpoint. A strong and lasting adhesive bond limits the potential to separate the substrates to allow them to be put into different recycling streams.

Global market of adhesives and sealants

In 2017, the global adhesive and sealants market had a volume of 15 million tons, corresponding to approximately 40 bn. €, with the demand nearly evenly distributed between the regions Europe, Asia-Pacific and Americas. The application areas of converting/packaging, construction and assembly account for 74 % of the market; polyurethane based adhesives play an important role in the transportation and footwear applications as well as packaging and construction[4].

Water-borne adhesives hold a share of about 50 %, while the traditional solvent-based systems have approximately 14 % overall market share. Vinyl and acrylic resins are the dominating resin technologies accounting for approximately 50 % share. Compared to these, polyurethane adhesives and sealants are a specialty with a lower global market share[4] in recent years, however, the more complex application profiles in adhesives and sealants processes have led to a growing demand for high-quality systems. In particular, polyurethane systems are benefiting due to their outstanding property profile compared with other materials.

In the upcoming years (2017–2021) the global market for adhesives and sealants resins in all supply forms is expected to grow at least at 4.0 % per annum, with an above-average growth in Asia-Pacific.[4]

Prominent application examples showing the benefits of adhesive bonding

Adhesive bonding is often not the most inexpensive joining method. Its real advantages are often demonstrated by considering the overall value-added process, for example, by factoring in weight savings or improved construction methods into the overall equation[3]. The possibility of joining different materials yields assemblies that can incorporate the advantageous properties of each substrate. This is especially relevant in the automotive industry. Adhesives are more and more used to join dissimilar materials (e.g. steel/aluminum/polymers) to reduce the weight of car bodies and improve overall fuel efficiency. Adhesives are

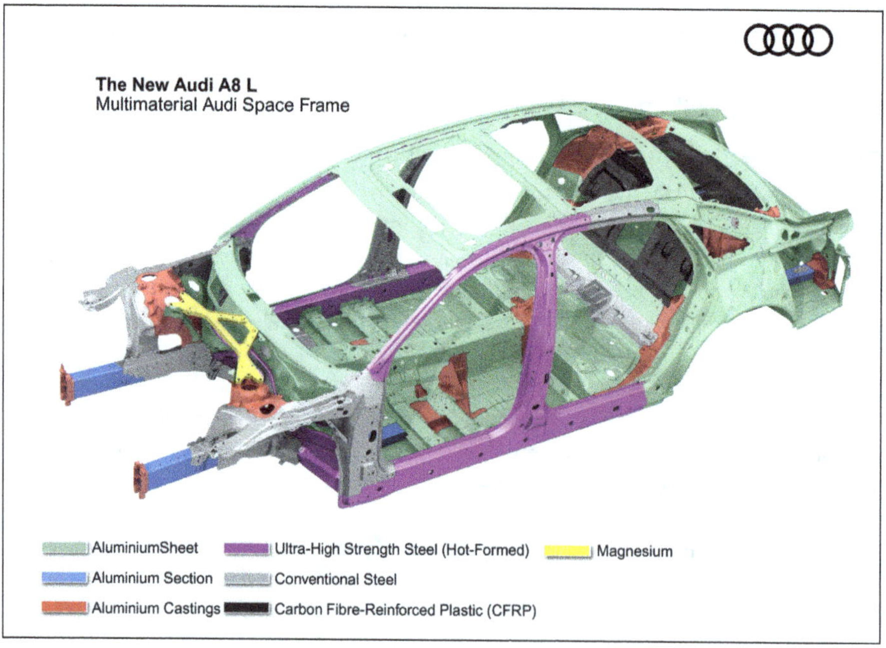

Figure 6.1: Audi A8 spaceframe, consisting of different steel and aluminum grades metals and alloys as well as magnesium and carbon fiber reinforced plastics (CFRP)[5]

essential for manufacturing advanced vehicles constructed from composites and thermoplastics (Figure 6.1). This is a particular strength of polyurethane based adhesives.

In the modern footwear industry, shoes are made from a variety of materials to optimize cost, performance or meet the latest fashion trends. Heat-activated polyurethane adhesives allow shoes to be manufactured in a cost-effective manner and at the same time produce sport shoes with the complex structure and the high performance qualities that are expected by today's consumers.

Another important application is the production of load-bearing engineered wood elements such as beams and joists, where one-component moisture-curing polyurethane adhesives have been used for many years. As an example, cross-laminated timber – prefabricated wooden components of up to 15 meters in length are being made using this adhesive technology. It has recently gained market relevance because it enables the rapid construction and the use of wood as a sustainable raw material in high-rise buildings like residential buildings and even towers for on-shore wind energy plants[6]. The main advantage of moisture-curing polyurethane adhesives in wood bonding is their ease of use without requiring a mixing step and their curing behavior that combines long open times with short pressing times at ambient temperatures.

Properties of polyurethane adhesives

A large number of adhesives of different types are available on the market. They can be separated by different criteria, e.g. into physically setting vs. chemically curing adhesives, into solvent-free, solvent-borne and water-borne adhesives, or they can be separated by the chemical nature of their base polymer. Important types of adhesive base polymers with regard to their chemical nature are vinyl-/polyolefin resins, acrylics including copolymers, natural resins, epoxy resins, silicone resins and polyurethanes.

Polyurethane adhesives[7–10] adhere well to many substrates. This is not solely due to physical bonding forces arising from the close contact between the adhesive film and the substrate. The adhesion can be reinforced by hydrogen bonds, which are formed between the polar groups (urethane, urea) present in the polyurethane backbone and polar groups that may exist on the substrate surface. Moreover, free isocyanate groups present in the adhesive film can react with traces of moisture in or on the substrate as well as with reactive groups present on the substrate, which have been introduced by a plasma or corona pre-treatment to further strengthen adhesion.

Every adhesive must be able to flow for a certain time, so that it can be applied to the parts to be joined. It wets the substrate surfaces and yields initial adhesion, but it hasn't yet developed load bearing strength. The adhesive joint reaches the necessary bond strength (adhesive and cohesive) as the adhesive polymer sets, solidifies and cures (two-component adhesives) to build up molecular weight. Hot melt adhesives develop strength as they transition from the liquid to solid phase; solvent-borne and water-borne adhesives develop strength by evaporation of their solvent and water moieties. This sequence also pertains to polyurethane adhesives. Polyurethane adhesives either already have urethane groups present in the polymer backbone when they are applied or the urethane groups develop as the adhesive cures.

6.2 Classification

Polyurethane adhesives are differentiated as follows:

Polyurethane reactive adhesives

These adhesives can be formulated as one-component or two-component 100 % solids systems without the use of solvents. Polyurethane reactive adhesives have good flow properties since they are prepared from low molecular weight or at most oligomeric raw materials. After the adhesive is applied and the substrates are joined together, high molecular weight polymer is formed by the reaction of free isocyanate groups with water or raw materials containing hydroxyl or amine groups. Solvent-based two-component reactive systems can also be formulated (see Chapter 6.4).

Polyisocyanates used in two-component systems are low molecular weight unmodified polyisocyanates (e.g. monomeric or polymeric MDI), or modified polyisocyanates or prepolymers. The polyols likewise have a relatively low molecular weight. After mixing, the polyaddition reaction takes place and urethane groups are formed. The curing rate is determined by the reactivity of the components used and the processing conditions.

One-component systems typically consist of a liquid-modified polyisocyanate or a prepolymer with an isocyanate content in the range of 3 to 16 %. It cures by the reaction of its free isocyanate groups with atmospheric moisture or with the moisture in the substrate, resulting in the formation of urea groups and carbon dioxide. The curing is dependent on temperature, reactivity of the polyisocyanate components and the amount of water present. Catalysts may also be used to accelerate the reaction.

Solvent-borne polyurethane adhesives based on hydroxyl polyurethanes

Solvent-borne polyurethane adhesives are typically based on a high molecular weight flexible polyurethane polymer dissolved in an organic solvent. In this way, the solid polymer can be applied to the substrate. The solvent promotes the substrate wetting and then evaporates, leaving a thin layer of the solid polyurethane polymer on the surface. The polymers are usually linear hydroxyl-terminated polyurethanes based on crystalline polyester segments and are generally processed as heat-activated adhesives.

Heat activation: The dry, non-tacky film that remains after evaporation of the solvent is heated to a temperature above the melting point of the crystalline polyester segments.

At this point the film exhibits tackiness and the flow properties required for the bonding process. One or both substrates may be coated. The substrates are mated with the application of moderate pressure. A strong bond rapidly develops due to recrystallization of the polymer backbone.

Isocyanate crosslinking: The polyurethane adhesives are often used with a high functional polyisocyanate crosslinker (see Figure 6.2). This two-component formulation shows significantly increased softening temperature of the adhesive bond as well as increased resistance to solvents (incl. water), plasticizers, oil and fat migration. The crosslinker also promotes adhesion to the substrate. The adhesive rapidly develops strength as with the one-component adhesive and the final cure follows over a period of hours or days.

Polyurethane dispersion adhesives

These adhesives are high molecular weight polyurethane polymers dispersed in water. Adhesive films form by the evaporation of water or by migration of water into the substrate. In comparison to solvent-borne adhesives, they have a lower viscosity and higher solid content. They consist of mainly linear polymer chains containing crystallizing polyester segments. These products are predominantly designed for use as a heat activated adhesives.

Heat activation: This is performed as described above for the solvent-borne adhesives. Due to the high molecular weight of the PU dispersion polymers, linear polymers have to be used to ensure the necessary flowability of the activated adhesive film. Particularly exacting demands are placed on the flowability of the adhesive polymer if the activated adhesive film is directly bonded with an adhesive-free substrate surface. The adhesive joint physically sets by recrystallization.

Isocyanate crosslinking: As in the case for the corresponding solvent-borne adhesives, the heat-activated polyurethane dispersion adhesives also develop their full potential only through chemical crosslinking. In this case, water dispersible crosslinking polyisocyanates are used. The isocyanate groups react with reactive groups on the polymer chain. Thus, the adhesive film first sets physically and then crosslinks chemically. The

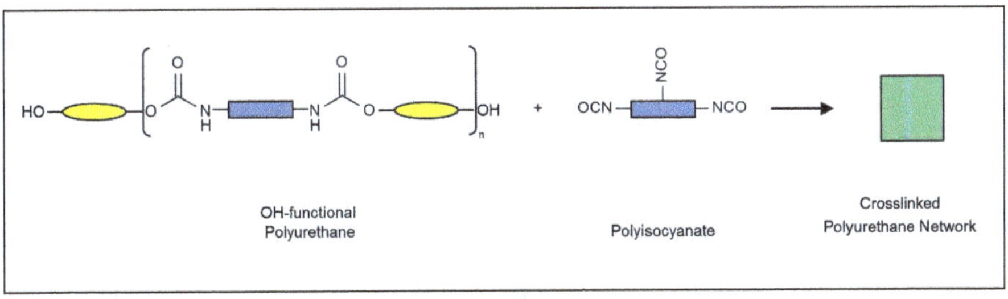

Figure 6.2: Network formation in two-component polyurethane adhesives

softening temperature and resistance properties of the crosslinked adhesive film are significantly increased, comparable to solvent-borne polyurethane adhesives. In the case of polyurethane dispersions other types of crosslinkers, such as polycarbodiimides, are possible (see Chapter 6.5.3).

Polyurethane hot-melt adhesives

These products exhibit very good adhesive properties. They are available as non-reactive (thermoplastic) or reactive types. Non-reactive thermoplastic polyurethane hot-melts are supplied as films or powders based on linear hydroxyl polyurethanes with crystallizing polyester segments. They are applied at temperatures ranging from of 60 to 130 °C, and they set physically by cooling and recrystallization. In contrast, reactive hot-melt adhesives consist of a meltable polyisocyanate polyurethane that is solid at room temperature. Due to their reactive isocyanate groups, these adhesives need to be protected from moisture during storage. They are applied at temperatures between 100 to 140 °C, first set physically by cooling and then chemically by reaction with atmospheric moisture. When processing these adhesives, care must be taken that they are not damaged by excessively high temperatures or prolonged exposure to heat.

6.3 Polyurethane reactive adhesives

Polyurethane reactive adhesives can cure to form elastomers or thermosets, whereby the degree of crosslinking and the strength are dependent on the raw materials used. The mechanical properties and adhesion characteristics of the adhesive films are mainly determined by:
- the type of polyols[11] used, their molecular weight and their functionality,
- the choice of di- or polyisocyanate in terms of type, functionality and NCO content,
- the concentration of urethane and/or urea groups,
- the crosslinking density yielded by the interplay of the molecular weights and functionalities of all the components.

Adhesives with useful properties are only obtained by combining raw materials with a functionality of at least two.

6.3.1 Raw materials[12]

Polyether polyols

Generally, polyether polyols are obtained by the catalytic addition of propylene oxide and ethylene oxide to low molecular weight polyols (e.g. 1,2-propane diol, ethane diol, glycerol or trimethylol propane). Polyether polyols yield polymers with good stability to hydrolysis.

However, their limited resistance to oxidation necessitates suitable stabilization. The very flexible polymer chain does not crystallize and gives the adhesive good low temperature flexibility but poor resistance to plasticizers. Polyether polyols have a lower viscosity than most polyester polyols and are therefore used to adjust (lower) the viscosity of the formulation. In laminating adhesives they promote compatibility with lubricants and slip agents. Last but not least, cost is an important criterion. Polyether polyols are generally less expensive than polyesters.

High molecular weight polyether diols, especially those manufactured by the metal-complex catalyzed process (see Chapter 3.7.3) are suitable to produce adhesive polymers with exceptionally good mechanical and dynamic properties, which are particularly important for sealant applications. In contrast, the level of monohydroxy-functional polyether chains increases significantly using KOH catalyzed technology. As a result, this process is limited to the production of lower molecular weight polyols (molecular weight ~ 2000 g/mol).

Polyester polyols

Polyesters are produced by a condensation reaction of polycarboxylic acids and polyols. Due to the many monomers that can be used, polyester polyols can be amorphous or (semi) crystalline and offer a very broad spectrum of highly varied properties, the most important of which is very good adhesion to a wide range of substrates. Polyester chains are considerably less flexible than polyethers and consequently yield stiffer and stronger adhesive films. However, polyester polyols increase the viscosity and application temperature of the adhesives. The polarity of the polyester polyols results from the high level of ester groups in the polyol backbone. The adhesion properties can be adjusted by the choice of carboxylic acid and polyol. In contrast to polyether polyols, polyesters are more susceptible to hydrolysis.

Polycarbonate diols

Polycarbonate diols offer both superior hydrolysis resistance and oxidative stability. Adhesives with higher level of performance can be prepared by using this type of polyol. A limitation of this raw material is its higher cost.

Polyisocyanates

The polyisocyanate components can be based on a great variety of aromatic (diphenylmethane diisocyanate (MDI) or toluene diisocyanate (TDI)) and aliphatic diisocyanates (hexamethylene diisocyanate (HDI) or isophorone diisocynanate (IPDI)). MDI monomers are the dominate diisocyanate type used in the manufacture of prepolymers which are used for laminating adhesives, reactive hot-melts and structural adhesives (these technologies will be discussed below) as well as for sealants (discussed in Chapter 7), while aliphatic types are being used for special applications, such as lamination adhesives for food packaging (food safety, see Chapter 6.3.2), applications with translucent decors or where the bond line is at least partly visible (UV stability), or to tune the reactivity, pot life or crosslinking speed of the adhesive.

MDI is a solid at room temperature. MDI grades that are liquid at room temperature are available. Liquefaction is possible by blending 4,4'- and 2,4'-isomers of MDI (Figure 6.3), chemical modification of 4,4'-MDI and via prepolymers made from 4,4'-MDI and various polyols.

Polymeric MDIs contain oligomeric fractions of higher functionality in addition to difunctional diphenylmethane diisocyanate (MDI). Polymeric MDI types are available with different viscosities and acidity levels. Low viscosity grades are preferred for prepolymers and acidity adjusted grades are available to control the reactivity of two-component adhesives.

In comparison with MDI, TDI and aliphatic diisocyanates have a high vapor pressure and more stringent safety measures are required. The use of unmodified monomers is avoided, especially in two-component adhesives. Low-monomer derivatives or low-monomer NCO prepolymers are gaining in importance. These diisocyanates are used for special adhesive applications, e.g. as building blocks for oligomeric hydroxyl polyurethanes, as polyfunctional derivatives serving as crosslinkers for adhesive polymers or in low-monomer prepolymer.

Isocyanate-terminated polyurethanes (NCO prepolymers)

NCO prepolymers[13] are low molecular weight polyurethane polymers bearing free NCO groups, and they play a major role in polyurethane chemistry. They occupy an important position between the monomeric isocyanates and the polyurethane polymers. An NCO prepolymer is synthesized by reacting a polyol with an excess molar amount of a diisocyanate. If this excess is very large, the products are referred to as semi-prepolymers. There are many reasons for using prepolymers:

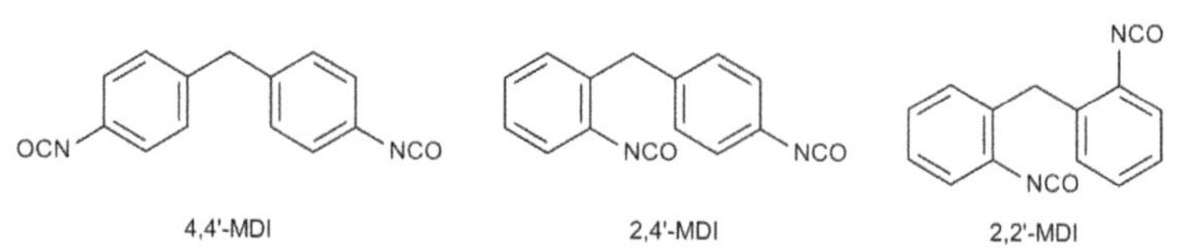

Figure 6.3: Chemical structure of MDI isomers

- A liquid prepolymer is easier to handle than solid monomeric isocyanates, e.g. 4,4'-diphenylmethane diisocyanate which is crystalline at room temperature.
- As described before, prepolymers are a method to safely use polyisocyanates (TDI, HDI) with a high vapor pressure. Prepolymers can be produced with very low-monomer content. The process to prepare prepolymers allows for the formation of a defined polymer structure. This is advantageous, for example, when co-reactants having very different reactivity, such as polypropylene glycol ethers containing primary and secondary hydroxyl groups, are reacted, not simultaneously as a mixture, but consecutively.
- Polyol incompatibility can occasionally be reduced by pre-polymerization.
- Prepolymers can be very beneficial to the processing of two-component adhesives. The exotherm of the crosslinking reaction at the bond line is reduced since a portion of the overall exotherm takes place during the synthesis of the prepolymer. In the same way, the pot life of two-component systems is extended due to the reduced reaction exotherm and to the generally lower reactivity of prepolymers in comparison with monomeric polyisocyanates.
- The isocyanate content of the prepolymer in a two-component system can be adjusted to balance with the OH components to produce a favorable mixing ratio (e.g. 1:1) for processing.

Commercial products: It is possible for an adhesive manufacturer to prepare a prepolymer in-house, but expertise is needed to consistently produce high quality products (e.g. consistent viscosity and NCO content). These can be avoided by working with ready-to-use NCO prepolymers that are supplied as standard products by raw material manufacturers. In the absence of moisture, prepolymers exhibit good storage stability and can be successfully used in many applications.

Low-monomer NCO prepolymers

Low-monomer NCO prepolymers are required for special applications. For example, they are used as polyisocyanate components in laminating adhesives. For this application a key aspect is having a low free monomer content to minimize the potential for migration of any residual monomer or subsequent reaction products. Low-monomer NCO prepolymers are also used as building blocks for low-emission reactive hot-melts. A further application field is the production of low emission one-component polyurethane foams used in construction applications.

Manufacturing low-monomer NCO prepolymers is very complicated and requires an infrastructure for transporting, storing, reacting and distilling monomeric diisocyanates. The producers of these monomers have this expertise in-house. Because of their lower boiling point, HDI and TDI are the dominant base isocyanates in this product group. Recently, however, low-monomer MDI-based products have also become available.

6.3.2 Two-component polyurethane reactive adhesives[14, 15]

Chemical structure and processing properties

Composition: Two-component systems consist of low molecular weight or only slightly pre-reacted polyisocyanate components (or optionally, a modified isocyanate or NCO prepolymer) and likewise relatively low molecular weight polyol components. The latter are often a mixture of different polyols. Short-chain hydroxyl compounds can also be present in the formulation, as well as NCO-reactive amine compounds. Two typical formulations of two-component reactive polyurethane adhesive systems are shown in Table 6.1.

Mixing ratio[13]: The two components have to be processed in a defined mix ratio. This is calculated as the stoichiometric ratio of the NCO equivalents to the equivalents of the NCO-reactive groups (OH and NH groups. Frequently, a two-component adhesive is intentionally formulated with an excess of polyisocyanate groups. This isocyanate excess results in additional crosslinking, compensates for potential problems due to substrate moisture and improves adhesion to suitable substrates. Practical experience suggests the use of 5 to 40 mol% excess of isocyanate above the calculated stoichiometric ratio.

Pot life: After adding and mixing the OH (polyol or prepolymer with terminal hydroxyl groups) and NCO components, the curing reaction commences with the formation of urethane groups. The pot life is essentially determined by the reactivity and functionality of the starting materials, the fillers present in the formulation, and the processing conditions. The temperature of the raw materials, the mixing technique and mass of adhesives, which influences the overall amount of heat generated by the reaction exotherm, are all important variables. Tertiary amine compounds and metal salts are normally used as the catalysts[16]. Metal salts in particular, even in small amounts, may considerably reduce the pot life.

Curing: The same parameters also determine the curing rate. The time to achieve complete cure at room temperature can range from hours to days, depending on the formulation. This curing process can be accelerated by heat and catalysis, and often also results in an increase in the final bond strength.

Additives: Fillers such as chalk, talc, barites, quartz powder or ground shale are often used. They increase the stiffness, but often reduce the elongation at break and the peel resistance. Moreover, they reduce the exotherm and shrinkage on setting – which makes the adhesive less expensive. Fillers are used primarily in adhesives for rigid materials, such as wood, plastics or metals. When bonding highly absorbent materials, any unwanted penetration into the material can be prevented by adding finely dispersed silica as a thixotropic agent. Thixotropic additives ensure the adhesive's stability when applied in beads and prevents run off from vertical surfaces. The additives are usually incorporated into the polyol (OH) component. Nevertheless, they must always be carefully dried. In order to remove

Table 6.1: Typical formulations of two-component reactive polyurethane adhesives

Ingredient	**High hardness [parts by weight]**	**Medium hardness [parts by weight]**
PPG polyether triol	50.0	50.0
PPG polyether diol	20.0	20.0
Castor oil	30.0	30.0
Chalk	100.0	100.0
Fumed silica (hydrophilic)	4.0	4.0
Zeolith (paste)	10.0	10.0
Polymeric MDI Functionality ~ 2.6; NCO content 31.5 %	81.3	-
MDI/polyether prepolymer Functionality ~ 2.5; NCO content ~ 16 %	-	158.4
Pot life (50 g) @ 23 °C [min.]	160.0	180.0
Shore D	80.0	40.0

traces of moisture that may remain even after drying, the addition of a drying agent to the polyol component is recommended, such as molecular sieves. Thixotropy can also be achieved chemically. Small quantities of reactive amine compounds (e.g. isophorone diamine, IPDA) can be included in the polyol component and will react rapidly with polyisocyanates to form highly dispersed polyureas particles that act as in situ or chemical thixotropic agents. This type of thixotropic modification has the advantage that it does not increase the viscosity of the OH component.

End properties of the cured adhesive: Structure, chain length and functionality of the co-reactants and the fillers used determine the properties of the crosslinked polymer. The concentration of urethane groups and possibly urea groups, as well as the chemical crosslinking density are all important variables that can be controlled. Catalysis and processing temperature can also have a considerable influence. The polymer properties are characterized by measuring tensile strength, modulus, elasticity, softening temperature and chemical resistance.

Processing technology

Two-component reactive adhesives are preferably processed using two-component metering and mixing units (Figure 6.4 see page 324).

Filler-free two-component reactive adhesives can be metered with gear pumps as there is no danger of abrasion. Plastic static mixers make processing particularly economical. Gear pumps cannot be used to process systems containing fillers because abrasion (grinding action of the filler on internal parts) would quickly make these pumps unusable. Reciprocating piston pumps have higher durability with formulations containing fillers.

The use of double cartridges with static mixers (Figure 6.5, see page 325) is suitable for trade application of small quantities of adhesives. A pre-condition is the mixing ratio of the components matches the volume specification of the double cartridges. The pot life of the adhesive must be sufficiently long to prevent the adhesive from curing in the static mixer and allow time for the mixed adhesive to be applied. It is very important that the adhesive components are absolutely phase-stable (i.e. that the fillers are sedimentation stable), to insure the contents of the cartridge are thoroughly mixed.

In principle, manual processing is also possible if the two-component reactive adhesives are specially formulated to produce a long open time. As discussed previously, the adhesives cure in an exothermic reaction and the temperature increase depends on the mass of adhesive being mixed. Thus, pot life will vary depending on the size of the mixing vessel and with large volumes the pot life may decrease significantly. Two-component polyurethane systems require high quality mixing to yield homogeneous and reproducible adhesive layers. This may be difficult to accomplish and can lead to variability in performance.

Use of two-component polyurethane reactive adhesives

Two-component polyurethane adhesives are frequently used as structural adhesives. These are adhesives of proven reliability in structural engineering applications in which the bond can be stressed to a high proportion of its maximum failing load for long periods without failure. Structural adhesives are often designed to have a shear modulus of approximately

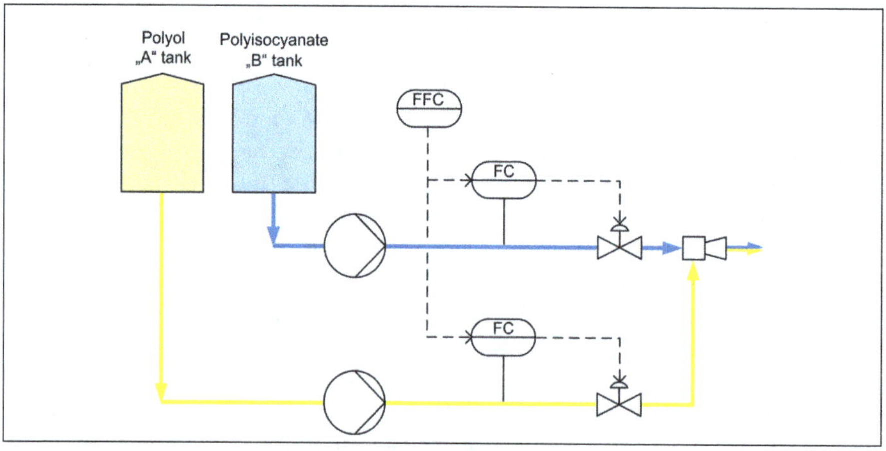

Figure 6.4: Processing unit for two-component reactive systems

FC = Flow control; FFC = Flow ratio control

200 to 600 MPa in fully cured adhesive layers. They will be able to absorb and dissipate stresses to prevent bond line failures (see Figure 6.6).

Two-component polyurethane reactive adhesives are also used for bonding large surfaces, e.g. for manufacturing sandwich panels for refrigerated vehicles, containers and superstructures (caravans, trucks), or for installing large areas of flooring with a rapid return to service. Further applications are as encapsulating compounds for filter elements or electrical components.

Another important application for two-component polyurethane adhesives includes laminating adhesives for flexible packaging application, both for food as well as for non-food applications. This will be discussed in the following chapter.

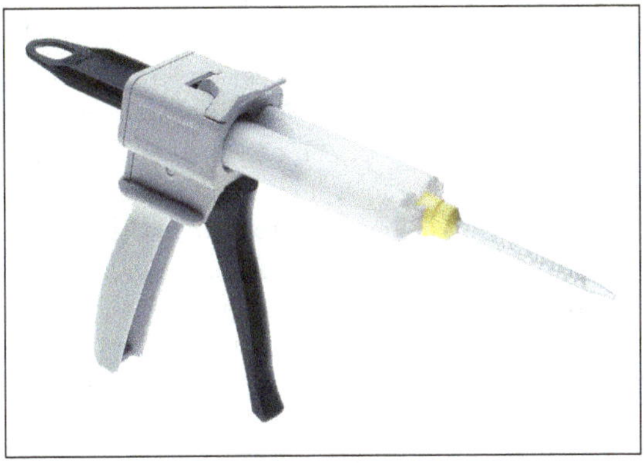

Figure 6.5: Double cartridge static mixer[17]

Polyurethane laminating adhesives[18]

Flexible film laminates for food packaging have increased in importance because they are lightweight, space-saving and cost-efficient compared to other food packing options such as glass containers or cans.

Polyurethane adhesives are the dominate technology used for preparing film/film and film/foil laminates. These adhesives are primarily two-component solvent-based or two-component solvent-free. One-component moisture-curing solvent-based and solvent-free adhesives are also used, but with a limited application scope.

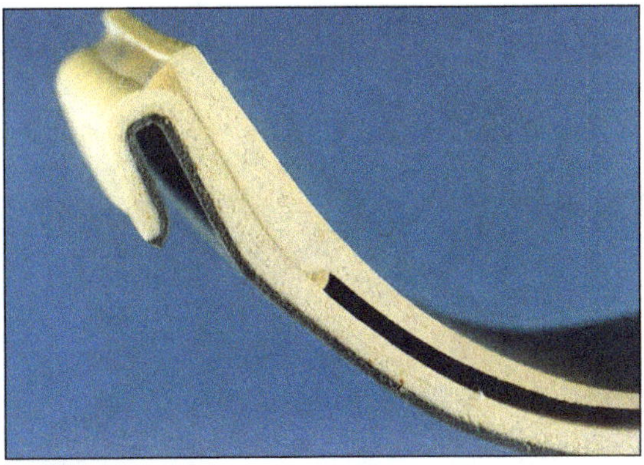

Figure 6.6: Example of a polyurethane integral skin foam part bonded with a two-component polyurethane reactive adhesive

Technical requirements

Laminates for food packaging must fulfill very demanding performance requirements. The list can be extensive and is dependent on the specific application.

For example, laminates may be required to protect the foodstuff from exposure to light, be impermeable to gases, especially oxygen, and moisture. They should also preserve aromas present in the food. In addition for certain applications, laminates are required to withstand steam sterilization processes which run at temperatures up to 134 °C and must also be able to cope with aggressive fillings or fatty food. They have to convey transport nutritional or food preparation information about the product and also function as an avenue to promote the brand. This information is reverse-printed on the inner side of an outer film layer. Finally, the inner or food contact layer of a laminate must be heat-sealable. This is generally achieved by using a polyethylene or polypropylene film.

There is no single film material available to meet this catalogue of requirements. Consequently, different films or film/foil layers have to be combined. Duplex (two-layer laminates) or triplex (three-layer laminates) are standard in food packaging applications.

The laminates are produced in a roll-to-roll process. A typical lamination line is shown in Figure 6.7. The converting industry uses reels of films and/or foils to produce the laminates in a semi-continues processes. The adhesive is applied in a thin layer on one web material. In case of a solvent-based adhesive, the adhesive coated web has to run through a drying tunnel to remove the solvent. In the case of solvent-free adhesives a drying tunnel is not necessary this is an advantage with regard to the line speed as well as to economy of the laminating process. The processing lines for solvent-free adhesives require less space due to the absence of the drying tunnel. Thus, there is a lower investment cost for this type of line. The adhesive coated web then runs through a nip roller where the second film is laminated to the adhesive coated web. This process is called dry-lamination.

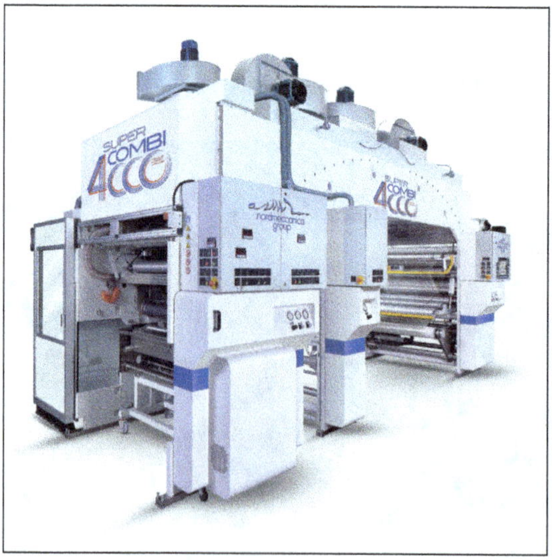

Figure 6.7: Typical lamination line for production of multilayered packaging films[19]

Modern laminating lines allow three layers to be bonded in one process. After the laminating process the web is wound into rolls and the adhesive curing process takes place over a period of 1 to 5 days. Special solvent-based adhesives using aliphatic crosslinkers require elevated temperatures up to 45 °C to fully cure the adhesive.

Table 6.2: Performance classes and requirements for laminated films

Performance classes/global market share	Laminate/application	Adhesive chemistry
High performance/ 15 %	1) All laminates (film/film & film/aluminum) for a) Aggressive products: (e.g. fruit juice, detergents) b) High-strength laminates with high tear strength c) With thermal exposure > 100 °C for sterilizability (e.g. animal feed) boiling resistance (microwave) 2) Aluminum (film/aluminum) laminates with thermal exposure < 100 °C (pasteurizable)	Exclusively solvent adhesives two-component systems: I: – OH-functional prepolymer with high molecular weight Isocyanate crosslinker with low molecular weight II: – NCO prepolymer with high molecular weight OH crosslinker with low molecular weight
Medium performance/ 35 %	1) Film/film laminates with a thermal exposure of < 100 °C (pasteurizable)	Up to 50 % solvent adhesives two-component systems: – OH-functional prepolymer with high molecular weight – Isocyanate crosslinker with low molecular weight
	2) "Simple" aluminum composite with a thermal exposure of < 40 °C (example: coffee packaging)	Up to 50 % solvent-free adhesives two-component systems: – NCO prepolymer with high molecular weight – OH crosslinker with low molecular weight
General performance/ 50 %	1) Film/film laminates for a thermal exposure of < ~ 40 °C (cheese packaging) for low mechanical stress 2) Film/paper laminates	Two-component systems: the proportion of solvent-borne systems is much lower than of solvent-free systems Solvent-free: two-component systems One-component systems: NCO-functional prepolymer with high molecular weight (moisture-curing)

Three performance classes for laminated films

Laminates for food packaging are categorized into three performance classes based on the requirements and the performance level for the application.

Table 6.2 shows the typical requirements for these three performance classes and the formulation principles used for adhesives that fulfill these requirements.

Chemistry of laminating adhesives

Laminating adhesives are available in solvent-based and solvent-free technologies. The isocyanate component of a two-component solvent-free polyurethane adhesive consists typically of a NCO-terminated prepolymer which is cured with OH-components ranging from low molecular weight polyfunctional alcohols to higher molecular weight OH-terminated prepolymers. The aim of the adhesive manufacturer is to keep the content of monomeric diisocyanate in the two-component solvent-free adhesives as low as possible.

Solvent-based adhesives can be either solvated NCO-terminated prepolymers which are cured with low molecular weight polyfunctional alcohols or OH-terminated prepolymers which are cured with isocyanate crosslinkers having functionality > 3.

The isocyanate component is used in an excess (20 to 40 mol%) in all two-component adhesives. The excess is necessary to compensate for side reactions (e.g. with moisture) that consume isocyanate groups and thus inhibit complete curing of the adhesive polymer. Ethyl acetate is the standard solvent used.

Although two-component solvent-free technology continues to grow, solvent-based two-component adhesives still have a strong position in the manufacturing of laminates. This comes from the flexibility and performance they provide. Solvent-based adhesives can be used with all film/film or film/foil combinations and solvent-based adhesives allow the production of laminates for all performance classes. This makes the solvent-based adhesives the most universal adhesive systems for laminates used in food packaging applications.

Laminating adhesives are applied in thin layers onto the web. The thickness of the adhesive layer used is influenced by the performance class of the laminate. The application of the solvent-based adhesive requires the use of an anilox roller applicator, while the application of a solvent-free laminating adhesive is typically done with a smooth roller application unit. Solvent-based adhesives are applied at a coating weight up to 5 g/m^2 and solvent-free adhesives are applied at a maximum weight of 2 g/m^2.

The viscosity of an adhesive formulation is very much influenced by the molecular weight of the adhesive components. Due to the diluting effect of the solvent, solvent-based adhesive polymers can have a much higher "starting" molecular weight comparted to the solvent-free counterparts. The higher molecular weight of the adhesive polymer guarantees higher initial bond strength immediately after the nip roller and should reduce the tendency for slipping between the film or film/foil layer(s) of the laminate. This is of particular importance when thicker adhesive layers are needed e.g. on printed films to fill the topo-graphy of the printing ink layer. Another advantage of solvent-based adhesives is that they form a smooth level film layer which contributes to the excellent appearance of printed as well as transparent laminates.

The lower molecular weight of solvent-free adhesives components requires using a low coat weight for a number of reasons. Higher coating weights increase the risk that the film layers of the freshly rolled laminates will shift. This produces the so-called telescoping effect and makes the laminate unsuitable for further processing. High coat weights may also reduce optical properties of the laminated films which is described as the "orange peel" effect. Despite these potential drawbacks, the solvent-free technology continues to grow faster than the market. The lamination process runs at very high line speeds (up to 600 m/min) using cost-effective lamination equipment.

Water-based adhesives based on polyurethane dispersions are used for special applications for example in paper laminates. An advantage of polyurethane dispersion polymers is that the molecular weight of the polymer has no influence on the processing viscosity of the adhesive. This allows the "starting" molecular weight of a polyurethane dispersion adhesive to be much higher than with the solvent-based or solvent-free laminating adhesives. The bond strength immediately after the laminating process in the nip-roller is high enough, that slitting can take place directly after the lamination process. Polyurethane dispersions are typically used in combination with a hydrophilic modified isocyanate crosslinker in order to obtain the highest level of performance.

The adhesive industry offers a broad range of products for manufacturing laminates for food packaging applications. The details of the adhesive formulations are part of the manufacturer's confidential proprietary know-how. But if a view is taken across all systems and performance categories, laminating adhesives polymers generally consist of ~ 30 % polyether polyols, ~ 40 % polyester polyols and ~ 30 % polyisocyanates.

Raw material selection

Selection criteria for raw materials are:

- molecular weight/functionality/chemical nature,
- viscosity,
- compatibility of the raw materials,
- hydrolytic stability.

Other requirements for raw materials are described below but each application will have its own specific needs.

The proper selection of raw materials will lead to an adhesive with the required polymer characteristics (viscoelastic properties/glass transition temperature) and specific adhesion qualities.

Laminating adhesives used for the high performance retort application must exhibit high hydrolytic stability and elevated temperature performance. The adhesive polymer must be highly crosslinked, and the building blocks should be as hydrophobic as possible. Hexane diol adipate or phthalate esters are recommended.

Diisocyanates, in particular diphenylmethane-4,4'-diisocyanate (4,4'-MDI) or isomer blends of 4,4'-MDI and 2,4'-MDI are the primary raw materials used to produce prepolymers utilized as the polyisocyanate component in two-component solvent-free adhesives. Toluene diisocyanate (TDI), hexamethylene diisocyanate (HDI) and isophorone diisocyanate (IPDI) are typically used to chain extend polyether polyols or polyester polyols or their blends. The chain extension of the polyols component may be necessary to adjust the OH number of the polyol component to achieve a desirable mix ratio of polyisocyanate- and polyol components. It could also be done to improve the compatibility of the polyols used in the adhesive or to adjust the viscosity of the polyol.

The adduct of TDI-TMP as well as TDI-trimer and the hybrid trimer of HDI/TDI are typical aromatic polyisocyanate crosslinkers for solvent-based adhesives.

HDI-trimer is used as an isocyanate crosslinker for both solvent-free and for solvent-based adhesives. Adhesives based on purely aliphatic polyisocyanates have become more important. One advantage of these aliphatic adhesives is non-yellowing which is essential for laminates with white printing.

Another advantage of purely aliphatic adhesives relates to food safety. Adhesives which are exposed to high temperatures (e.g. in a steam sterilization process) may experience some degradation in the polymer backbone. This may lead to the generation and potential migration of harmful primary aromatic amines (PAA) for polymers based on aromatic polyisocyanates[20]. Adhesives with a purely aliphatic backbone cannot produce PAA so they are preferably used for these demanding applications. Other aliphatic polyisocyanate crosslinkers are available but are limited to the use in solvent-based adhesives since they are solids at room temperature. These products include the adduct of xylylene diisocyanate (XDI) and trimethylolpropane (TMP) and the trimer of isophorone diisocyanate (IPDI-trimer).

The universally applicable rules of polyurethane chemistry with regard to structure/property relationships are also valid here. Raw materials with higher functionality yield polymers with a higher crosslink density. This results in a higher durability laminate, but it can also lead to an unwanted stiffening and increased brittleness of the adhesive film. Higher functionality can reduce the pot life of the two-component adhesive but at the same time can produce the positive result of faster curing.

Further processing of the laminated films

Slitting of the laminates requires sufficient crosslinking of the adhesive polymer to prevent the delamination of the polymer films during this handling process. Heat-sealing requires a fully cured adhesive layer since only a high molecular weight polymer is able to withstand the thermal stress caused by the heat-seal process. Food is only allowed to come in contact with the laminate when the adhesive polymer is fully cured thus preventing any migration of low molecular weight parts of the not fully cured adhesive polymer, including residual unreacted isocyanate. This requires waiting periods for the laminated films prior to further processing (see Table 6.3).

Table 6.3: Typical waiting periods for the further processing of laminated films

Typical waiting periods for

	Cutting the roll[d]	Until it can be sealed[d]	Until the residual isocyanate monomer has disappeared [d]	Filling the packaging [d]
Solvent-borne adhesives	0.5–1	3–5	7–14	7–14
Solvent-free adhesives	1–2	5–7	1–7	5–7

Food safety

The EU regulations for materials which comes in contact with food are very clear "Materials [...] shall be manufactured in compliance with good manufacturing practice so that, under normal or foreseeable conditions of use, they do not transfer their constituents to food in quantities which could: (a) endanger human health; or bring about an unacceptable change in the composition of the food".[21]

All participants along the value chain of flexible laminates from raw material suppliers to film producers, adhesive manufacturers, converters and finally the food packers and brand owners have to take responsibility for their part of the value chain. The polyurethane raw material supplier e.g. has to inform the adhesive manufacturer fully about the co-use of auxiliaries such as stabilizers (antioxidants), catalysts or by-products which are formed during the production process of the raw material.

A polymer used in food packaging applications – in either direct or indirect food contact –must comply with the various regional regulations where the flexible packaging laminate is used, for example:

- Regulation (EU) No. 10/2011
- BfR recommendation XXVIII (cross linked polyurethanes as adhesive layers for food packaging materials)
- FDA regulations in 21 CFR 175.105 (adhesives) and 21 CFR 177.1390 (laminate structures for use at temperatures of 250°F and above)
- Chinese hygienic standard GB 9685-2008

Due to their versatility, broad adhesion profile and fast curing characteristics, two-component polyurethane adhesives are expected to remain the dominating adhesive technology for flexible packaging for the coming years.

6.3.3 Moisture-curing one-component reactive adhesives

One-component polyurethane systems

These liquid adhesives, are processed at room temperature or at slightly elevated temperatures. They consist primarily of oligomeric polyisocyanate-terminated polyurethanes (NCO prepolymers) with a range of viscosities. In certain applications lower molecular weight products are suitable. They cure by reaction of the free isocyanate groups with atmospheric moisture or with moisture contained in the substrate to form polyurea groups. Curing of the adhesive layer with atmospheric moisture proceeds from the outside to the inside at a rate determined by the reactivity of the isocyanate groups, the catalyst and the diffusion rate of water into the adhesive layer. Small amounts of carbon dioxide are formed during this reaction, but it does not interfere with the adhesion process provided the free isocyanate content of the adhesive is not too high (< 4 % NCO). Moisture-curing is a slow process and requires that the pieces to be joined should be fixed mechanically until the curing reaction is close to completion.

It should also be noted that, with the onset of curing, a skin may form on the surface of the adhesive where it has contact with moist ambient air. This may prevent the wetting and bonding to the second substrate. These adhesives cure to yield layers with a shear modulus in the range of about 1 to 10 MPa.

In principle, NCO-reactive hot-melts, one-component laminating adhesives and one-component moisture-curing polyurethane sealants could be described in this section since they are based on the same curing process. However, the first two product groups are so specific in their formulation and processing that they are discussed in more detail in separate sections (see Chapter 6.6.2 as well as 6.3.2 and Literature[18]). One-component

Figure 6.8: Application of adhesive in the manufacture of structural wood[22]

Figure 6.9: Hall construction using structural wood[22]

moisture-curing polyurethane sealants are also described in a dedicated Chapter 7. The manufacturing aspect of NCO prepolymers is described in Chapter 3.4.

Applications

The room temperature viscosity of liquid moisture-curing one-component adhesives typically ranges from 2,000 to 10,000 mPas. There are also products in paste form. The adhesives can be applied manually with a spatula or using cartridge guns or mechanically with a robot. They can be used to bond a wide variety of substrates and substrate combinations, e.g. in the manufacture of windows, doors and sandwich elements.

For a number of years now, moisture-curing one-component polyurethane adhesives have been used in an interesting field of application: structural wood construction (see Figures 6.8 and 6.9). This technology has achieved great significance in Europe because it offers considerable advantages over conventional formaldehyde-based adhesives (short curing times without additional thermal acceleration and without costly pressure equipment).

One-component silane-terminated polyurethanes[24]

One-component silane-terminated polyurethanes also cure by reaction with ambient moisture. In contrast to the one-component polyurethane systems described previously, they do not contain terminal NCO groups but are capped with terminal alkoxysilane groups. When they react with water, methanol or ethanol is formed instead of carbon dioxide. Moreover, the substrate adhesion (e.g. to glass) is significantly improved.

6.4 Solvent-borne adhesives based on hydroxyl polyurethanes[25]

6.4.1 Hydroxyl polyurethanes

Production of hydroxyl polyurethanes

Hydroxyl polyurethanes are prepared by reacting polyester diols with diisocyanates. Preparation of a linear hydroxyl polyurethane with high molecular weight and optimal performance (bond strength, flexibility and solubility) requires the precisely controlled reaction of the polyisocyanate with a very slight molar excess of the diol components. In addition to using the optimum diisocyanate/diol ratio, the purity of the raw materials is a deciding factor for the quality of the hydroxyl polyurethanes.

Hydroxyl-terminated polyurethanes are soluble polymers with a low hydroxyl content (approximately 0.1 %). Products are available which differ in their application-relevant parameters of solution viscosity, thermoplasticity and crystallization. These three parameters depend on the structure of the hydroxyl polyurethane. They are essentially determined by the molecular weight, the choice of polyester diol on which the hydroxyl polyurethane is based, and the diisocyanate used. The term "thermoplasticity" refers to the varying degrees of softening of the different products at increased temperature. The strength of the adhesive bond above the softening temperature can be significantly increased by the addition of polyisocyanate crosslinkers (see Chapter 6.4.2).

The crystallinity of the individual hydroxyl polyurethanes fundamentally determines the rate of bond strength development, the ultimate bond strength and the flexibility of the adhesive. Below a temperature of approximately 40 °C, the hydroxyl polyurethanes are partially crystalline while they exist in an amorphous state at higher temperatures. Upon cooling from the amorphous state or after evaporation of the solvent from the adhesive layer, the individual products demonstrate different crystallization properties (as can be shown by measuring the Shore hardness).

Product selection and applications

Adhesives based on hydroxyl polyurethanes show excellent adhesion to numerous synthetic and natural materials. Products based on highly crystalline polyesters yield durable bonds even on highly flexibilized PVC, thus explaining their use in the shoe, automotive and furniture industries. Products containing less crystalline polyesters have a lower resistance to plasticizers. They are characterized by their higher flexibility and better adhesion to non-polar substrates[26]. Examples of use are film lamination applications, footwear sole bonding, and manufacturing abrasive belts (see Figure 6.10).

Manufacturing solution adhesives

The solubility of the hydroxyl polyurethane types varies considerably due to the significant differences in their chemical structure.

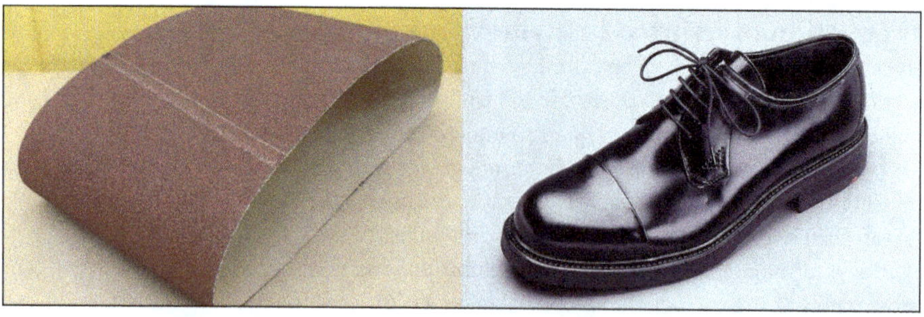

Figure 6.10: Application examples for solvent-borne polyurethane adhesives

Methyl ethyl ketone (2-butanone) is an excellent solvent for hydroxyl polyurethanes and is therefore also used for specifying the solution viscosity. These types are also soluble in acetone. Some grades are soluble in low-boiling esters such as methyl acetate and ethyl acetate, too. Only a limited number of these polymers dissolve in aromatic hydrocarbons and all types are insoluble in aliphatic hydrocarbons. In addition to mixtures of dissolving solvents, certain mixtures of solvents can be used that consist of a dissolving and a non-dissolving solvent. It should further be noted that the solubility is also dependent on the selected solution viscosity specification, which relates to molecular weight of the hydroxyl polyurethane used.

The effect of the solvent on the viscosity of the hydroxyl polyurethane solution is considerable. The solvents also affect the storage stability of the adhesive and its tendency to gel at low temperatures. However, the latter is a purely physical process, which is reversible. Gentle warming restores the original viscosity. In addition, substrate wetting, the drying time and the activation behavior of the adhesive film are affected.

The adhesives are prepared by simply dissolving the raw materials supplied as granules in suitable organic solvents. The granules are gradually stirred into the solvent, as opposed to adding them in one portion, to avoid any agglomeration of the pellets.

With regard to the flammability of the solvent used, suitable measures must be taken to prevent ignition caused by electrostatic charging and the subsequent electrostatic discharge of the solid polyurethane pellets when transferring them from packaging containers. The flammability of the solvents themselves and the danger that solvent vapors form explosive mixtures with air must be kept in mind and the corresponding equipment and procedures must be used to minimize this danger. Moreover, the physiological action of the solvents must be considered. The use of many solvents is restricted or is completely forbidden. This type of information is described on solvent labels. All relevant regional safety legislation must be obeyed.

To ensure optimal storage stability of the adhesive, the solvents used must be of high quality. Their water and alcohol content must not be higher than 0.1 %. These contaminants may react with the polymer backbone causing a drop in the molecular weight of the polymer and a subsequent reduction in initial bond strength. Ketones and esters, in particular, can contain peroxides. An increased peroxide concentration causes oxidation reactions that are evident from a discoloration of the adhesive solution.

Formulation of adhesives

Solutions of hydroxyl polyurethanes are frequently used directly as adhesives. Unique performance requirements can be met when mixing different types of hydroxyl polyurethanes, as well as by including additives in the formulation. The compatibility of the different grades should be considered and tested prior to use. For example, the products with low crystallinity are incompatible with the grades based on highly crystalline polyesters.

The adhesive's property profile can be varied by blending different classes of products. In general, hydroxyl polyurethane adhesives are manufactured without fillers. However, a filler can be added to increase the initial bond strength of the adhesive. This may also have the negative impact of increasing the activation temperature and may also impair the viscosity stability of the adhesive formulation. Highly dispersed silicas have proven to be useful as fillers since they provide e a thixotropic effect on the adhesive solutions.

Processing the adhesives

The adhesive solution is usually applied to both substrates (double-sided bonding). However, there are other applications in which only one substrate surface is coated with adhesive (single-sided bonding). After drying, the adhesive film is partially crystalline at room temperature and non-blocking (not tacky). Bonding takes place in a heat activation process, in the same way as is described in Chapter 6.5.5 for water-borne polyurethane adhesives. Before joining the substrates, the adhesive film is made amorphous by heating it to a temperature above the minimum activation temperature (MAT). The MAT often corresponds

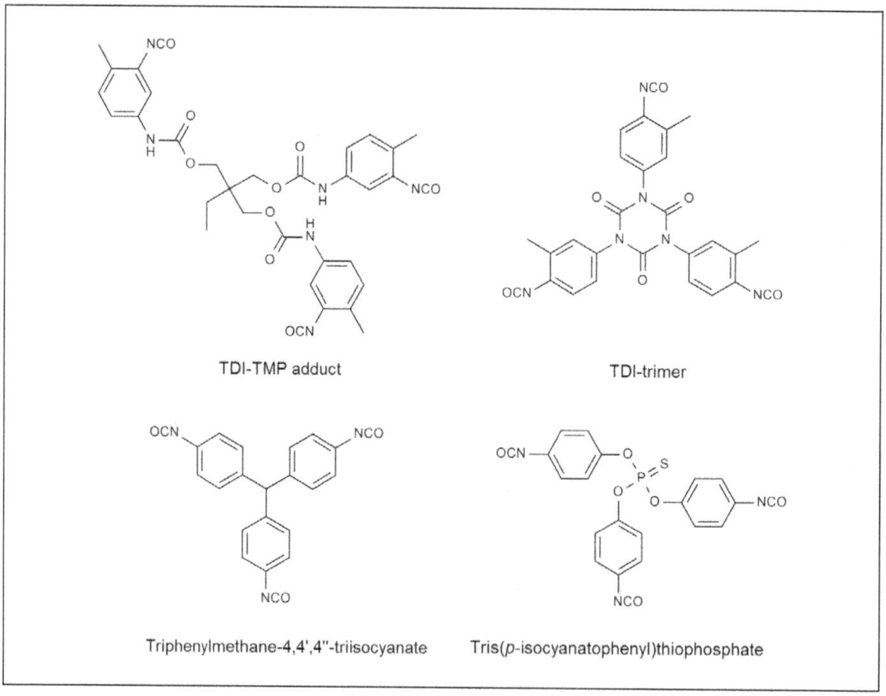

Figure 6.11: Polyisocyanate crosslinkers for solvent-borne polyurethane adhesives

to the melting point of the crystalline segment of the polymer backbone. The resulting tacky film can be joined during the period its hot tack life.

6.4.2 Isocyanate crosslinkers for solvent-borne adhesives

Although adhesives based on hydroxyl polyurethanes, especially those with low thermoplasticity, are often processed as one-component systems, chemical crosslinking using a polyisocyanate crosslinker is recommended. The isocyanate crosslinkers for solvent-borne adhesives are mostly solutions of tri-functional aromatic isocyanates, e.g. TDI-trimer, TDI-TMP adduct, tris(p-isocyanatophenyl) thiophosphate or triphenylmethane-4,4',4''-triisocyanate, supplied in ethyl acetate (Figure 6.11). They are not only suitable for hydroxyl polyurethane adhesives but also for solvent-borne adhesives[27] based on natural and synthetic rubber.

Crosslinking greatly improves the thermal stability of the adhesive bond. It also enhances substrate adhesion and the resistance to hydrolysis, solvents, plasticizers, oils and fats. The products differ in reactivity, color and resistance to discoloration of the adhesive film.

6.5 Water-borne polyurethane adhesives

Dispersions based on polyurethanes or urethane urea polymers containing ionic groups (preferably anionic) as the hydrophilic centers in the polymer chain[28] are of particular interest for the manufacture of adhesives. The incorporation of hydrophilic centers results in a number of advantages, especially high shear stability and good film-forming properties. The solid contents of these dispersions manufactured by the acetone process or by the melt dispersion process are in the range 40 to 50 % by weight. In comparison with solvent-based systems, polyurethane dispersion adhesives[29, 30] yield low application viscosities and high-solid contents. Branched or crosslinked polyurethanes can also be used, which, however, can then only be activated at high temperatures and are primarily suitable for wet-on-wet bonding.

Polyurethane dispersions are an important step in reducing solvent emissions in the field of adhesive processing. A further aspect is also occupational safety because, in contrast to solvent-based systems, there are no fire and explosion risks and consequently costly ventilation, safety and solvent recycling investments can be avoided.

6.5.1 Products

These aqueous products closely correspond in their properties to the conventional hydroxyl polyurethanes used in solvent-based applications that are gradually being displaced for reason of environmental impact and occupational safety whenever possible. Since most of these dispersions are based on aliphatic isocyanates, the adhesive films are widely resistant to UV yellowing. The heat-activated types produce very strong adhesive bonds on most natural and synthetic materials, also on substrates containing plasticizer. They differ in their degree of crystallinity, rate of crystallization and activation temperature. Products that crystallize faster after heat activation are able to absorb the resulting stresses in the adhesive joint more quickly after bonding. This property is of interest, for example, in the lamination of furniture, which often involves three-dimensional structures rather than flat surfaces. After removal from the lamination press, the adhesive joint must absorb the recovery forces of the decorative film that has been severely stretched in some areas.

Products that crystallize less rapidly after heat activation have a somewhat longer open time (hot tack life) compared with the fast-crystallizing types, with the result that somewhat more time is available for positioning and joining the components being bonded. This is a useful property e.g. for shoe sole bonding.

Polyurethane dispersions with low activation temperatures (approximately 40 to 50 °C) are used to bond heat-sensitive substrates, such as natural leather. Products with a higher activation temperature (80 to 100 °C) offer better heat resistance also when no crosslinker is used.

6.5.2 Formulation

Water-borne polyurethane adhesives are widely used as supplied or can be blended with other polymer dispersions. They can be easily applied by spraying, blade or roller coating or brushing. Their low viscosity and high-solid content provide the adhesive manufacturer with a high degree of freedom in formulation.

The rheological properties of aqueous adhesives are highly important for their application behavior. The adhesive should be easy to apply, e.g. by spraying. It should also wet the substrate surfaces thoroughly which may include partial penetration of porous substrates. The viscosity depends on the shear stress. Thickeners based on hydrophilically-modified polyurethanes have proven to be suitable for adjusting the application viscosity. They can be easily incorporated into the dispersion and are highly effective.

Blending with other polymer dispersions

Dispersions with different polymer structures can be combined. However, several points must be taken into account. Dispersions containing different stabilizers (e.g. anionic/cationic) may be incompatible and could coagulate on mixing. The stability of the polymer dispersion can be highly dependent on the pH. Therefore, the pH of the components should be matched to each other before mixing. As the PU adhesive polymers often consist of polyester building blocks, the pH of the adhesive formulation should be between 6 and 9 to avoid hydrolysis of the polymers. The long-term stability of the formulations must be carefully tested because the quality of the formulated adhesive (formation of particulates, coagulation, phase separation) and its application properties may change.

Preservation

In principle, aqueous dispersions are at risk of microbial contamination. Consequently, polyurethane dispersions for adhesive applications contain an in-can-preservative that prevents the reproduction of microbes. In Europe, the Biocidal Product regulation rules the use of such in-can-perservatives[31]. Depending on the type and amount of the active ingredient used, labelling requirements for the formulated dispersion can apply[32]. Additional stabilization may be required when formulating the finished adhesive.

6.5.3 Crosslinkers for water-borne polyurethane adhesives

An effective bonding process requires that the substrate is sufficiently wetted. To meet this requirement in a bonding process, only linear adhesive polymers with the ability to flow can be used. As the finished bond generally has to meet further specifications such as high resistance to heat, the adhesive must crosslink during or after the bonding process. As a result, the polymer loses its ability to flow.

In a crosslinked polymer, the majority of the polymer chains are interconnected by at least one chemical linkage. The chemical links among the polymer chains influence the tensile modulus and "flow" of the polymer. As the number of linkages between the polymer chains increase, the polymer network tightens and the crosslink density increases. This leads to a higher tensile modulus and lower flow properties. Crosslinking increases the softening temperature of the adhesive film and improves the resistance to solvents,

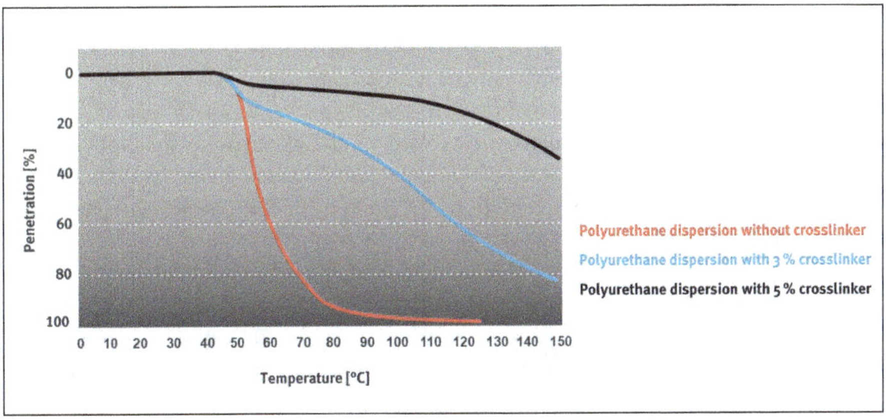

Figure 6.12: TMA of adhesive films of a polyurethane dispersion with and without polyisocyanate crosslinker

Table 6.4: Crosslinking technologies for water-borne polyurethane adhesives

	Water-borne 2K technology	**Latent-reactive technology (1K water-borne)**	**Carbodiimide crosslinking (1K water-borne)**
Type of dispersion	All dispersion types	Semi-crystalline polyester based polyurethane	Carboxyl group containing polyurethane
Crosslinker	Liquid hydrophilic modified polyisocyanate	Surface deactivated solid polyisocyanate	Aqueous dispersions of polyfunctional carbodiimides
Pot life[1]	6–8 hours	3–6 months	3–6 months
Open time[2]	6–8 hours	> 6 months	~ 1 hour
Curing speed	Hours to days	Hours to days	Minutes to hours
Crosslinking reaction	Room temperature curing	Heat-activation required	Room temperature curing

[1] The period of time a reacting composition remains suitable for its intended application.
[2] The period of time after an adhesive has been applied and allowed to dry, during which an effective bond can be achieved by joining the two surfaces.

plasticizers, oil and fat migration, and hydrolysis. Chemical crosslinking is possible with different electrophiles such as polyisocyanates and polycarbodiimides, as described in detail later. Using polyisocyanates has an additional advantage as the isocyanate groups are also able to promote adhesion to some surfaces. The influence of crosslinking on the softening behavior of an adhesive film, as determined by thermomechanical analysis (TMA), is shown in Figure 6.12, see page 339. A lower amount of penetration of the instrument's probe into the surface of the film being tested correlates with a higher softening temperature.

Two main classes of polyisocyanate crosslinkers are available. Water-dispersible hydrophilically-modified polyisocyanates have a long history and are the most commercially important type. An alternative one-component technology is based on a latent reactive curing process using surface deactivated solid polyisocyanate particles (see Chapter 6.5.7). Table 6.4 gives an overview about the processing characteristics of the different crosslinkers and recommendations for the type of dispersion that is appropriate for each crosslinker.

Polyisocyanate-based crosslinkers

Liquid water dispersible polyisocyanates are used in two-component adhesives systems and are dispersed shortly before the adhesive is applied to the substrate. The NCO groups of the liquid crosslinker are not physically separated from the water molecules of the dispersion. A portion of the isocyanate groups and water react with the formation of insoluble polyureas. This reaction limits the pot life of the two-component adhesive dispersions. The unreacted isocyanate groups in the dried adhesive film will cure with reactive groups on the backbone of the polyurethane polymer. The crosslinking reaction can occur at ambient temperature. In contrast, the latent reactive crosslinking technology (see Chapter 6.5.7) requires a heat activation step to initiate the crosslinking reaction.

Dispersible polyisocyanates will crosslink all polyurethane dispersions as well as other polymer types (e.g. vinyl acetate ethylene copolymers). This versatility makes them valuable co-reactants for many adhesive applications. The performance level of the finished adhesive can be adjusted to a certain extent by increasing the amount of crosslinker used. Any excess of NCO groups will react with moisture and the resulting polyureas will ultimately be integrated into the polymer network.

Hydrophilically-modified isocyanate crosslinkers are usually processed without the need for any additives, e.g. solvents to reduce viscosity. Any materials that may be blended with these crosslinkers must not contain any isocyanate reactive groups.

The force needed to achieve a homogeneous blend of the hydrophilically-modified polyisocyanates in water is higher than that required to mix solvent-based adhesives and crosslinkers. These modified polyisocyanates do not just dissolve in water but energy is required to prepare a finely dispersed stable dispersion of the crosslinker. Hand mixing is in most cases not sufficient, mechanical mixing is usually required. After addition of the isocyanate crosslinker, the adhesive dispersion has to be processed within a certain period of time (pot life). The dried adhesive layer also must be bonded to the other substrate before the curing reaction takes place (open time), which would interfere with the ability of the adhesive to wet-out the substrate and produce a strong bond. The pot life as well as the open time is not solely determined by the polyurethane dispersion used. It can be influenced by other components in the formulation which may contain isocyanate reactive groups as well as by the overall dispersion properties such as pH. The pot life and the open time of the adhesive formulations must be optimized in order to meet the requirements of each application.

Emulsifiable polyisocyanates differ in their viscosity and hydrophilic properties. Understanding the performance attributes of the polyisocyanate crosslinker is important in order to formulate an adhesive with optimal performance. Low viscosity, as well as a high hydrophilic content, facilitates blending the crosslinker with the polymer dispersion. In contrast, products with a low degree of hydrophilic modification are preferred in the case when it is advantageous to minimize water swelling in the final application. Upon emulsification, the more highly hydrophilic polyisocyanates yield finer droplets. Consequently, the adhesive has better shear stability during its application and dries to yield a smoother film. The composition of some products allows for their use in food contact applications.

The hydrophilic polyisocyanate crosslinkers are generally manufactured from aliphatic polyisocyanates, so they will not contribute to yellowing of the crosslinked adhesive film.

Polycarbodiimide crosslinker for polyurethane adhesive dispersions

Aqueous dispersions of polyfunctional carbodiimides are effective crosslinkers for polyurethane dispersions containing carboxyl groups in the polymer backbone.

Polycarbodiimides can be synthesized from polyisocyanates (e.g. dicyclohexylmethane-4,4'-diisocyanate) in the presence of a catalyst, often a phospholine oxide derivative, and heat. Two isocyanate groups react to generate a carbodiimide segment and carbon dioxide. The reaction occurs at temperatures higher than 100 °C. After reaching the desired number of carbodiimide segments, the reaction mixture is cooled and the carbodiimidisation reaction stops. This carbodiimide modified polyisocyanate is made water dispersible by the reaction of the remaining NCO groups with co-reactants containing hydrophilic groups. This modified polycarbodiimide is then dispersed in water[33].

Polycarbodiimides are stable in aqueous formulations, provided the pH of the dispersion is at the appropriate value. A dispersed polycarbodiimide based on dicyclohexylmethane-4,4'-diisocyanate requires a pH > 9 to be stable[34]. At pH values below 7 the hydrolysis reaction of carbodiimide with water is accelerated. The hydrolysis of the carbodiimide segments results in urea formation and a drop in the active crosslinker functionality (Figure 6.13).

Polycarbodiimides can be used to prepare storage stable reactive adhesive dispersions with a shelf life of more than 6 months at ambient temperature. The reactive groups in the polymer droplets (-COOH) and in the polycarbodiimide droplets (-N=C=N-) are separated from each other by the water phase of the adhesive dispersion. The crosslinking reaction proceeds rapidly once the water evaporates and the carbodiimide and polyurethane reactants coalesce. The crosslinking reaction results in the adhesive polymer chains being chemically joined via N-acylurea links (Figure. 6.14).

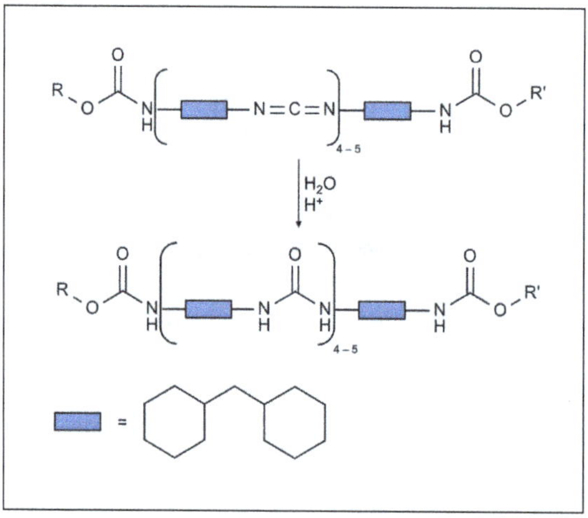

Figure 6.13: Hydrolysis of polycarbodiimide with water at low pH values

The amount of polycarbodiimide needed to achieve the optimal adhesives performance needs to be determined empirically. It has been found that a molar ratio -N=C=N- : -COOH of approximately 2:1 results in the maximum crosslinking density[35].

The use of an aqueous dispersion of polyfunctional carbodiimide should be considered when a one-component crosslinking adhesive with a fast curing speed is desired. This type of adhesive has been used in automated, for example roll-to-roll (R2R, see Figure 6.24 in Chapter 6.5.7), bonding processes, as well as semi-automated bonding processes. Examples of end-use applications include adhesives in the production of film laminates and shoes.

6.5.4 Drying

After evaporation of the water, the dispersion yields a solid and homogeneous adhesive film on the substrate. If the drying equipment is properly designed, drying rates comparable to those for solvent-based systems can be achieved. The drying units can also be utilized for heat activation of the adhesive layers. It is important that the drying operation is well controlled, which can be difficult in high humidity conditions. The process has to

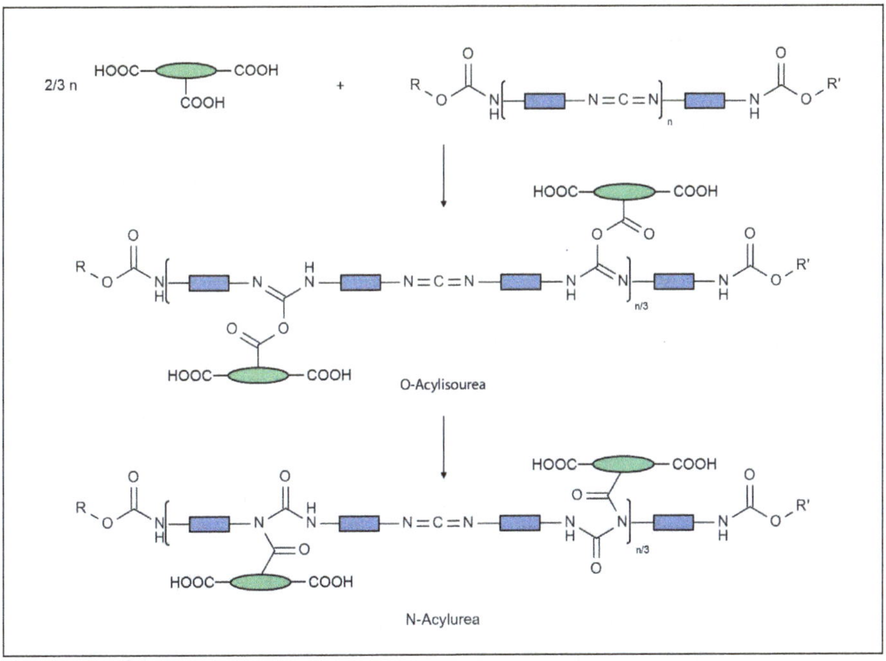

Figure 6.14: Crosslinking of COOH-functional polymers with polycarbodiimid

generate a reproducible film temperature. If it is too low, the film may not achieve the tack level required and if the temperature is too high the film will cool slower and the initial bond strength will be low. The drying process also must insure that there is no residual water in the adhesive film. Water will interfere with bond formation and any moisture trapped between impermeable substrates could lead to hydrolysis of the adhesive over time.

Open time of the dried adhesive film

The NCO groups of the water-dispersible polyisocyanate slowly crosslinks the polymer in the dried adhesive film. This results in a slow progressive decrease in tack and flowability of the adhesive film, with the result that the temperature required for activation slowly increases during storage. In general, the adhesive layer should be heat-activated and bonded within eight hours (end of the open time) after application of the adhesive dispersion.

6.5.5 The principle of heat-activated adhesive bonding

Water-borne polyurethane adhesives based on crystallizing polyester polyurethanes are generally used in a heat-activation bonding process[36]. The activation step is illustrated in Figure 6.15, plotting TMA data as well as tack data as a function of temperature.

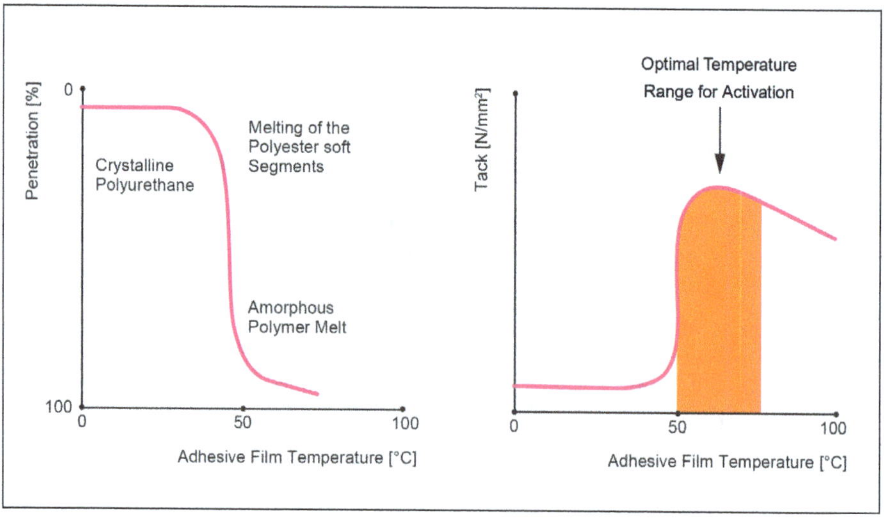

Figure 6.15: Heat activation of polyurethane adhesives containing crystalline polyester soft segments

- **Step 1 Adhesive application:** The adhesive is applied to a substrate and the water evaporates, producing an opaque, non-tacky film. The TMA curve indicates a high modulus polymer.
- **Step 2 Adhesive activation:** The adhesive is heated above the melting temperature (T_m) of the polyester segments and a clear, tacky film results. Activation temperatures range from 45 to 80 °C depending on the product. The temperature required is termed the "minimum activation temperature". The polymer has a lower modulus and is flowable at this point in the process.

Heat activation may take place e.g. in a heated drying tunnel, or by a short exposure to infrared radiation. Flash IR activation is preferred for many applications. The heat is applied for a brief time to warm the adhesive film but the substrate temperature is virtually unchanged. This allows the adhesive layer to cool rapidly and develop strength after the substrates are joined. The hot tack life is the period of time after heat activation in which the film still exhibits sufficient tack and flowability for bonding. The substrates must be joined during this time, which may range from seconds to minutes depending on the polymer structure.

The process of bond strength development and final curing is illustrated in Figure 6.16. The increase in modulus of the adhesive polymer indicates bond strength development and can be measured to track the progression of this process.

- Phase 1: Cooling of the adhesive film after heat activation; within seconds there is a development of initial bond strength as the bond line cools.

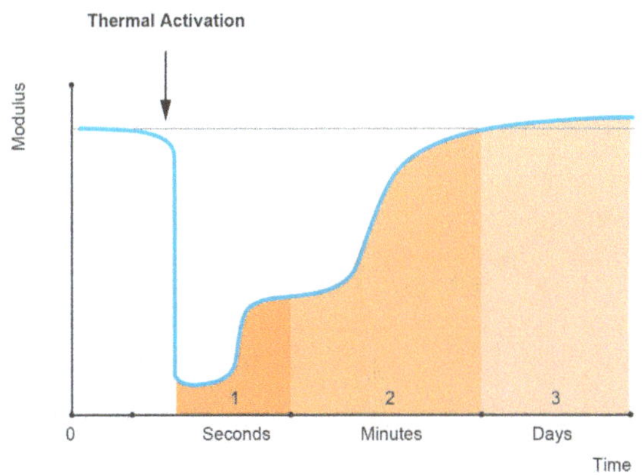

Figure 6.16: Three phases of modulus development after heat activation and cooling of the adhesive film

- Phase 2: Crystallization of the polyester soft segments; within minutes there is another jump in bond strength as the polymer backbone recrystallizes.
- Phase 3: Crosslinking of the polyurethane polymer with a polyisocyanate; the crosslinking reaction continues over a period of days and the final properties are obtained.

6.5.6 Applications and application technology

Water-borne polyurethane adhesives have found their way into a multitude of industrial applications[37], e.g. in the furniture industry, the automotive industry, shoe manufacturing or in the textile industry for manufacturing laminated fabrics. The heat activation process is easily implemented and suitable for many materials as long as they are not heat-sensitive. Depending on the type of adhesive used, the activation temperatures are in the range of 50 to 80 °C.

A typical application of heat activation bonding is in the furniture industry. It is used for the production of three-dimensional furniture front panels by laminating decorative films onto medium density fiberboard (MDF). In this lamination process (Figures 6.17 and 6.18), the adhesive is applied onto the MDF and dried. It is activated through the

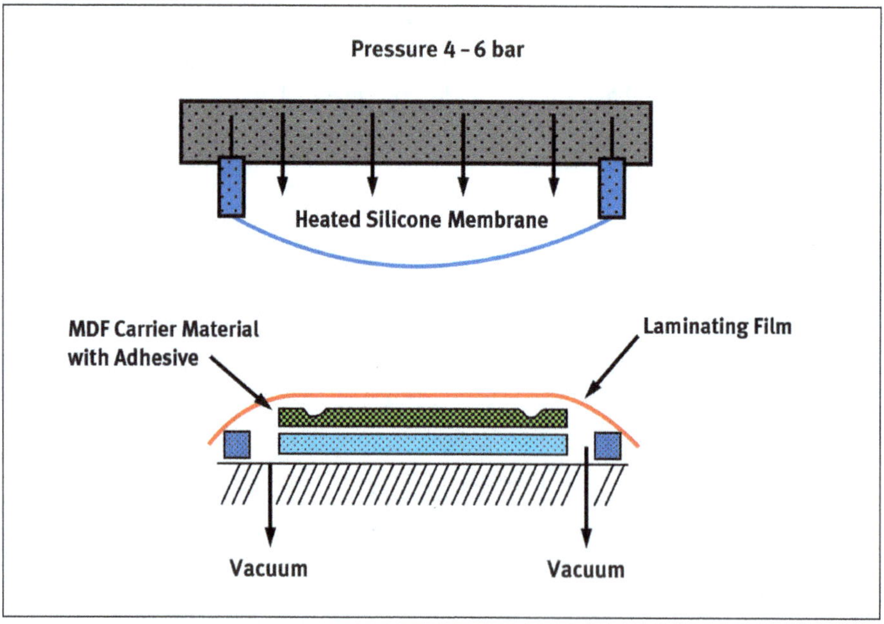

Figure 6.17: Process for the 3D lamination of decorative PVC films onto MDF

pre-heated decorative film by a heated membrane press. The pressure of the heated membrane and the application of a slight vacuum mould the decorative film to the shape of the furniture component and produce a strong bond.

In the automotive industry, the vacuum deep-drawing process is used to manufacture components for automobile interiors. A preheated laminating film is pulled by a vacuum and bonded onto a three-dimensional molded component that is coated with adhesive (Figure 6.19 and Figure 6.20). Adhesives with high initial bond strength are required to ensure short cycle times. As soon as the laminated component is removed from the mould, the adhesive must be able to withstand the stresses that result between the decorative film and the rigid substrate.

Adhesives emitting a minimum of VOC and releasing as low as possible odor are of particular interest to automobile manufacturers.

Water-borne polyurethane dispersions essentially consist of high molecular weight polymers and use water instead of organic solvent in their supply form. Thus, using adhesive dispersions based on these

Figure 6.18: 3D membrane press[38]

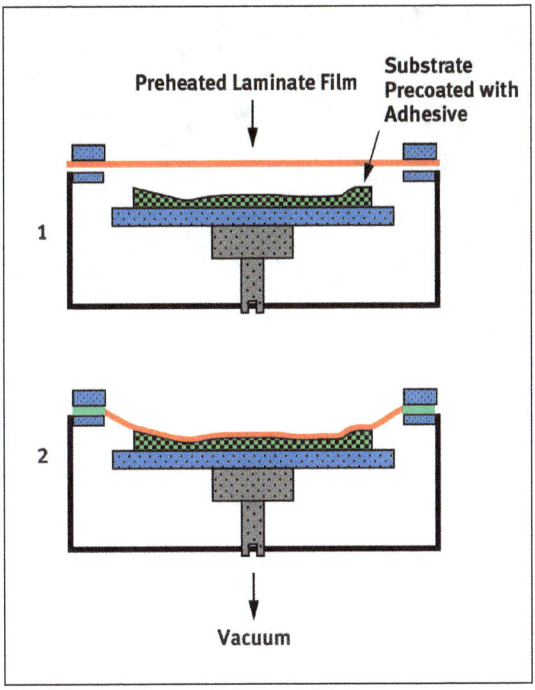

Figure 6.19: Vacuum deep-drawing process for laminating components for automotive interiors with decorative films

polymers can help automotive manufacturers to improve the indoor air quality of the passenger vehicles, allowing compliance with the increasingly stringent VOC requirements. The VOC content can be quantified by standard methods such as VDA 278[39].

Flat surfaces can also be manufactured by hot lamination. The flat carrier material is coated with adhesive and bonded in a continuous process. Heat activation occurs through the pre-heated film or with the help of heated pressure rollers.

In the shoe industry, polyurethane dispersion adhesives are used especially for sole bonding (Figures 6.21 and 6.22). Increasingly complex composite sole structures are the norm in the sports shoe segment and require high-performance adhesives. Polyurethane dispersions are now the state of the art in this application. Dispersion adhesives are also being used increasingly in street shoes and fashion footwear as an alternative to solvent-based adhesives, not least to ensure compliance with current VOC regulations and enhance industrial hygiene.

Figure 6.20: Door panel of BMW 6 series[40]

Some water-borne polyurethane adhesives are also applied in the wet bonding process. The material to be bonded is placed on the wet adhesive layer, which is then dried to produce the

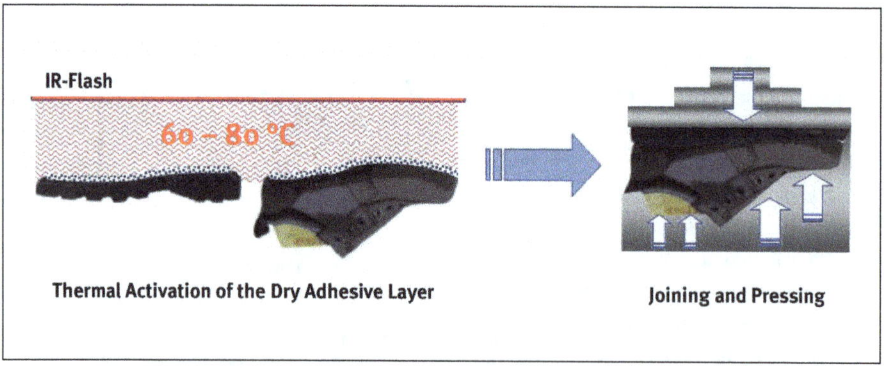

Figure 6.21: Principle of shoe sole bonding with polyurethane dispersion adhesives

bond. The process is ideal for textiles or substrate combinations in which at least one substrate is permeable to water vapor.

6.5.7 Latent reactive polyurethane dispersion adhesives

Latent reactive polyurethane dispersion adhesives[41–43] offer new perspectives for heat-activated adhesives. These are formulations of aqueous dispersions containing semi-crystalline polyurethane polymers and highly dispersed surface-deactivated solid polyisocyanates particles.

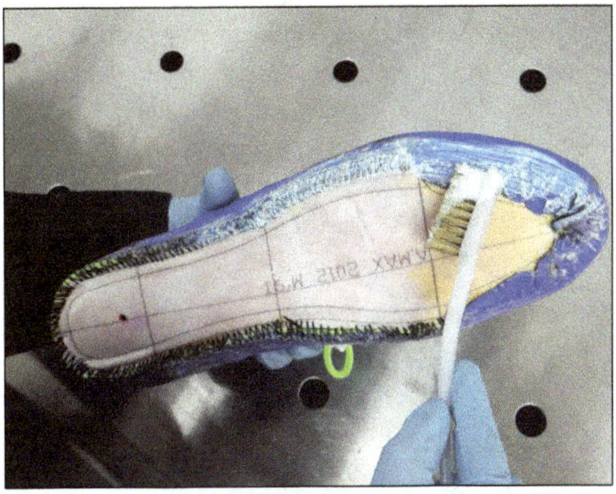

Figure 6.22: Applying dispersion adhesive to a shoe sole

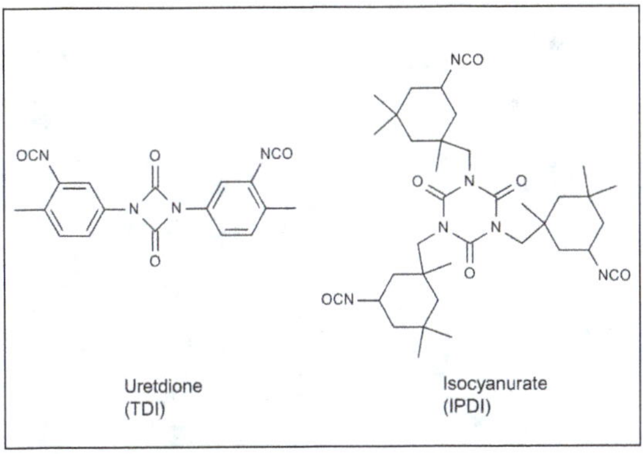

Figure 6.23: Polyisocyanate crosslinkers for latent reactive adhesives

Polyurethane dispersions containing semi-crystalline polyurethane polymers can be used to formulate latent reactive polyurethane dispersion adhesives. TDI-dimer and micronized IPDI-trimer are suitable polyisocyanate raw materials to prepare the dispersed solid polyisocyanate particles (Figure 6.23). The deactivation is achieved by eliminating the NCO groups on the particle's surface through reaction with an amine. The NCO groups inside the particles remain as active NCO groups. This distinguishes the surface-deactivated solid isocyanate particles from a thermally activated PU hardener (see Chapter 3.5).

The stability of the latent-reactive adhesive dispersion as well as of the dried adhesive polymer can be influenced by the amount of amine used to deactivate the solid polyisocyanate particles and the amount of amine added to the adhesive dispersion.

Compared with two-component dispersion adhesives using a liquid polyisocyanate crosslinker, the latent reactive adhesive dispersions have the advantage that they can be stored for long periods of time at 25 °C without decrease in adhesion performance. The latent reactive polyurethane adhesive dispersion can be applied to a substrate to make an adhesive coated material. Under the proper storage conditions, this pre-applied adhesive film is storage stable for several months. In a similar manner, the dispersion can be deposited onto a release liner (Figure 6.24). After subsequent drying under mild conditions, the crystallized polyurethane polymer film containing the deactivated solid isocyanate particles can be removed from liner. This adhesive film can be stored for later use.

The crosslinking reaction begins by heating the adhesive to a temperature above the melting point of the crystalline segments in the polyurethane polymer. Melting of the polyurethane polymer is necessary to allow the polymer to wet-out the surface of the substrate to be bonded. Also, at this time the polyisocyanate particles begin to dissolve (TDI-dimer) or liquefy (IPDI-trimer) in the molten polymer. The isocyanate groups inside the solid polyisocyanate particles are released and become available to crosslink the polyurethane polymer.

Figure 6.24: Roll-to-roll production of latent reactive adhesiv films[44]

The temperature and the length of the heat exposure influence the amount of deactivated isocyanate particles which are dissolved during the heat-activation process. The crosslinking density and the durability of the adhesive layer depend on the amount of isocyanate dissolved during the heat-activation process.

The latent reactive technology used as an adhesive dispersion, a pre-applied adhesive layer or a separate latent reactive adhesive film allows the separation of the adhesive application from the bonding step in time and location.

The adhesion properties obtained with latent reactive polyurethane adhesive polymers match and, in some instances, exceed the performance of the corresponding two-component water-borne polyurethane adhesives.

Latent reactive adhesive dispersions are used for example for manufacturing MDF/ PVC laminates for furniture front panels and for bonding shoe soles.

Latent reactive adhesive films are used in diverse applications[45]. The advantages of latent reactive adhesive films are:

- no handling of liquid adhesives;
- no drying process needed;
- non-blocking adhesive film;
- crosslinking reaction is initiated by moderate heat exposure;
- good adhesion to many synthetic substrates;
- high durability after crosslinking;
- short cycle times of bonding processes.

6.6 Hot melt adhesives[46]

6.6.1 Non-reactive hydroxyl polyurethane hot melt adhesives

Compared with co-polyamides and co-polyesters, hydroxyl-terminated polyurethanes offer substantially longer open times without compromising the bond strength. They can be used to particular advantage in complex manufacturing processes, such as lamination of a second web without additional activation.

Granular and powder types are available. The granular types for extrusion processing must be essentially free of gel particles. Powder types with a particle size of < 600 µm are manufactured by low-temperature grinding and are applied by scatter coating.

Chemically speaking, they are strictly linear hydroxyl polyurethanes based on crystallizing polyester diols and aromatic diisocyanates (i.e. products with structures that correspond to the hydroxyl polyurethanes used for heat-activatable solvent-borne adhesives).

The following technical advantages are significant for their use:
- low activation temperature in comparison with normal hot melt adhesives, therefore they are suitable for bonding heat-sensitive substrates;
- very good adhesion, which is demonstrated by all heat-activatable hydroxyl polyurethanes;
- no penetration of the adhesive into porous substrates because of its viscoelastic properties (textile laminates retain their soft hand);
- high initial bond strength;
- extended shelf life of the adhesive coated substrates.

One of the most important applications is the lamination of textiles (Figure 6.25). The adhesion is achieved by heat activation of scatter-coated powders, extruded adhesive film or the adhesive-coated nonwovens and subsequent joining of the substrates under pressure.

Granules

The granules are used for the extrusion of adhesive films and adhesive-coated nonwoven or for direct coating of other substrates. These adhesive films or coated substrates can be processed with typical hot melt bonding technology, such as heated calendars, membrane presses or continuous presses.

Grades with low thermoplasticity and fast to very fast recrystallization are available. These adhesive polymers heat activate at relatively low temperatures (< 70 °C) and transition from a crystalline to an amorphous polymer. In the amorphous state the polymers have enough flow to wet-out textile substrates, but it is not so excessive that the adhesive penetrates deeply into the fabric.

The scanning electron micrograph (SEM) below (Figure 6.26) shows that the adhesive film only surrounds directly adjacent fibers. The soft hand of the textile is thus retained.

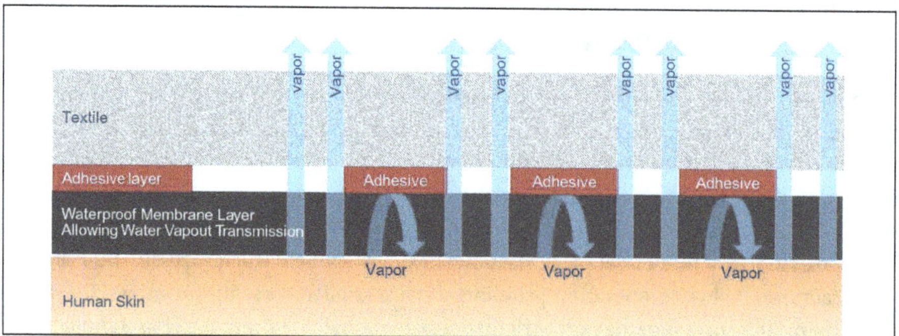

Figure 6.25: Application examples for polyurethane hot melt adhesives in textile lamination (waterproof functional outdoor clothing)

The applications of polyurethane hot melt adhesive films or adhesive-coated nonwovens primarily involve bonding of flat objects, which are mainly textiles. The basic requirement for a usable textile bond is that is must show good resistance to washing and chemical cleaning. This level of performance is achieved by the extrusion grade film adhesives. These adhesives are also suitable for bonding other substrates including paper and flexible films, leather and wood products. Further applications of hot melt films can be found in the automotive industry, for example in the finishing of textile outer fabrics for back molding processes (that prevent the penetration or strike-through of the expanding polyurethane foam), or skin coating of insulating foams. Examples of direct coating are the production of shoe cap materials or the pre-coating of surface films with adhesives for case furniture.

Powders

Hot melt adhesives in powder form are an interesting alternative to adhesive films or for direct coating, especially for textile lamination. These powder adhesives show the same performance profile as the films discussed previously. Polyurethane powders with different adhesive characteristics can be blended if they have similar particle size distributions and morphologies, in order to obtain a formulation with unique performance properties. Powder adhesives can be spray applied, e.g. for bonding porous substrates in the manufacture of breathable textile composites. Spraying is a simple application method and is particularly appropriate for large-scale applications.

Heat-activatable hydroxyl polyurethane powders become tacky at very low temperatures, so they are especially suitable for the gentle lamination of heat-sensitive substrates, e.g. high-quality textiles. Due to their good flow properties in the activated state, a strong bond is produced even when bonding process takes place at low pressure and with short pressing times. These bonding conditions minimize the risk of strike-through into the fabric. The contact properties of the adhesive are not solely determined by the bond line temperature; joining pressure and compression times are also useful control parameters. By balancing these three parameters, the process can be adjusted to match the characteristics of a given substrate. For example, if

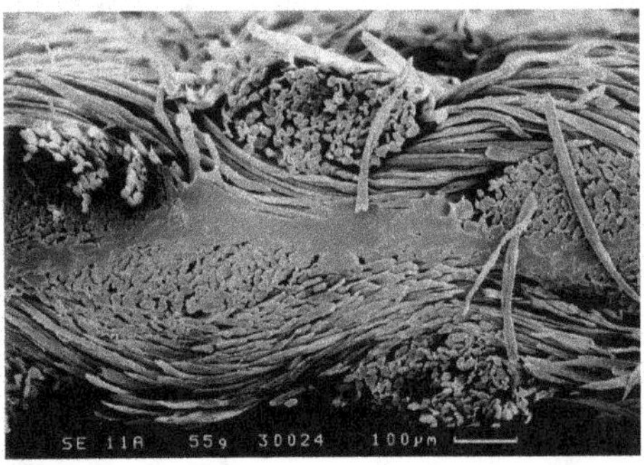

Figure 6.26: Scanning electron micrograph (SEM) of textile layers bonded with hydroxyl polyurethane hot melt adhesive film

only low pressures can be applied, the bonding process can be modified by increasing the pressing time or the pressing temperature.

Scatter coating of hot melt adhesive powders

Powders with particle sizes of less than 600 µm can be applied easily onto the substrate using scatter coating equipment. The principle is shown in Figure 6.27. The powder is contained in a hopper, from which a needle roller conveys a defined powder stream. An oscillating brush causes the powder from the needle roller to fall onto the continuously moving web below. The web then passes through an infrared tunnel where the powder is melted and fixed onto the substrate. In general, the substrate lamination step follows immediately. Examples of possible applications are textile lamination, leather processing and the production of shoe-stiffening materials.

6.6.2 Reactive polyurethane hot melt adhesives

Reactive polyurethane hot melt adhesives are a fast-growing class of adhesives[47–50]. They consist of a polyisocyanate polyurethane (NCO prepolymer) which is solid at room temperature but readily melts at a moderate temperature. They set both physically by cooling and chemically by reaction with atmospheric moisture. Compared with non-reactive hot

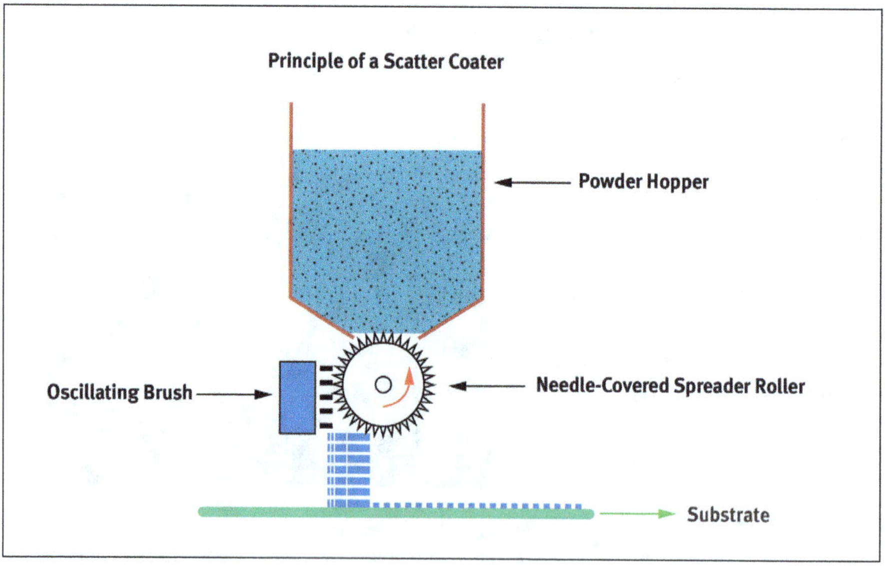

Figure 6.27: Scatter coating of a hot melt adhesive powder

Table 6.5: The function of the starting materials for prepolymers and their significance for the properties of the reactive hot melt adhesive

Starting material		**Function/effect on the properties of the prepolymer**	
Polyester diol	Crystalline	Melting point: > 50 °C	• Low melt viscosity • High initial cohesion, depending on the crystallization rate • Significant effect on the open time
Polyester diol	Amorphous	Glass transition temperature: > 20 °C	• High melt viscosity • High tackiness • Short open time/high initial tack
Polyester diol	Liquid	Glass transition temperature: < 30 °C	• Low melt viscosity • Lower initial cohesion • Longer open time
Polyether diol	Liquid	-	
Diisocyanate		-	Reaction with the diol components in molar excess of the isocyanate to give an isocyanate content in the range of 1–4 %

melt adhesives these reactive types have a lower molecular weight and a much lower melt viscosity. The adhesive develops bond strength as the adhesive layer cools and the prepolymer begins to recrystallize. This process occurs rapidly, and the adhesive bond very quickly reaches an adequate initial strength so that further processing of the joined parts can be done. The final strength and high thermal stability of the adhesive bond result from subsequent crosslinking of the free isocyanate groups with moisture to form polyureas. The moisture is available from the ambient air or from the substrate. Depending on the formulation, reactive polyurethane hot melt adhesives cure to form elastomers with flexible to hard and tough adhesive layers.

Structure and properties

Reactive polyurethane hot melt adhesives are reaction products of polyester diols, sometimes containing a small amount of polyether diols, with an excess of diisocyanate to form an NCO prepolymer. Generally, 4,4'-diphenylmethane diisocyanate (MDI) is used as the diisocyanate. The processing characteristics and final properties are adjusted by varying the NCO/OH ratio, the proportions of crystalline, amorphous and liquid starting materials, and their selection in terms of melting point (T_m), glass transition temperature (T_g) and melt viscosity. The formulations can also contain non-reactive polymers as modifiers, e.g. polyacrylates. The function of the starting materials and their significance for the adhesive properties are shown in Table 6.5.

The NCO/OH ratio is selected to ensure that the NCO prepolymer is solid or highly viscous at temperatures below 50 °C. It must be liquid and have an adequately low viscosity

Table 6.6: Polyester diols for reactive hot melt adhesives

	A	B	C
Chemical structure	Polyester diol	Polyester diol	Polyester diol
	Hexane diol	Hexane diol	Diol mixture
	Adipic acid	Dodecane dicarboxylic acid	Dicarboxylic acids
	Crystalline	Crystalline	Amorphous
Crystallization rate	Rapid	Very rapid	-
Melting point [°C]	~ 60	~ 70	-

at the processing temperatures of 100 to 150 °C. These conditions yield a product with a NCO content in the range of 1 to 4 %. The stability of the product during storage and processing is very important. Since the polyurethane contains reactive isocyanate groups, an increase in viscosity under thermal stress from overly high temperatures or excessively long exposure to high temperatures must be expected. Accordingly, processing should be designed such that the limit of 150 °C is never exceeded. The processing equipment must also ensure that the total thermal stress during melting, conveying and application is minimized.

Reactive polyurethane hot melt adhesives are now available with a significantly reduced level of free MDI monomer[51]. This progress in technology is made possible by the use of low-monomer MDI-based NCO prepolymers as the isocyanate building block. The lower content of free (i.e. volatile) monomeric diisocyanates is a key requirement in lowering emission values. These new hot melt types should be still processed at temperatures < 150 °C even though their monomer content is low. At temperatures > 150 °C the urethane groups present can degrade and volatile monomeric diisocyanates may be released.

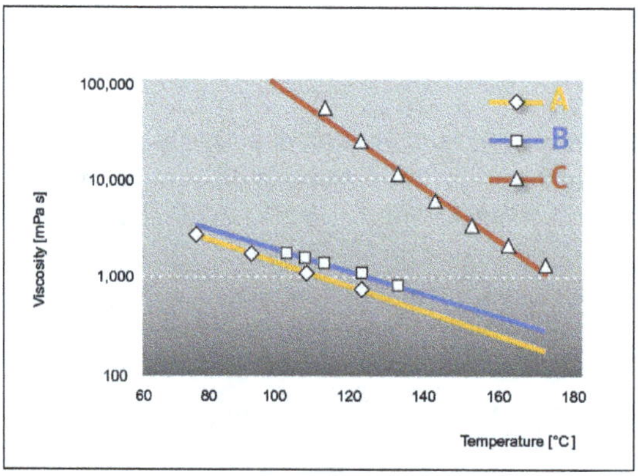

Figure 6.28: Temperature-viscosity curves of polyester diols for reactive hot melt adhesives

Polyester diol components

Several polyester diols used to manufacture reactive hot melt adhesives are summarized in Table 6.6.

The temperature-viscosity curves of these products are

Table 6.7: Sample formulations

	Formulation 1	**Formulation 2**	
Polyester diol	A	C	A
Proportions [%]	100	75	25
	Crystalline	Amorphous	Crystalline
Crystallization rate	Rapid	-	Rapid
Melting point [°C]	~ 60	-	~ 60
Glass transition temperature [°C]	-	~ 30	-
Reaction with 4,4'-MDI to NCO value		3.0 % NCO	

shown in Figure 6.28. In addition to the crystallinity, the viscosity is an important control parameter for the applicability, joint behavior and adhesion properties of the adhesive. The crystallinity and viscosity of the polyester are important parameters to control that greatly influence the adhesion, setting and application characteristics of the adhesive. The significance of these property parameters is demonstrated by the two sample formulations in Table 6.7, differing in the ratio of crystalline to amorphous polyols. The adhesives are produced by reacting these polyester diols with MDI resulting in prepolymers with an NCO content of 3 %. The level of amorphous polyol influences the thermal (T_m and T_g) and tack properties of the adhesive.

Their tack as a function of temperature is shown in Figure 6.29. Formulation 2 contains a high content of an amorphous polyester and exhibits a much higher tack than the adhesive based on the pure crystalline polyester diol (Formulation 1). However, this result should not lead to the conclusion that the highest possible amorphous polyester diol content should be used. Tack alone is not the sole predictor of the overall quality and performance of a reactive hot melt adhesive. The amorphous product strongly increases the viscosity of the adhesive, making processing and joining difficult. The amorphous character can be an advantage for extending open time, but it inhibits the fast development of green strength, which is a highly valued property for hot melt adhesives.

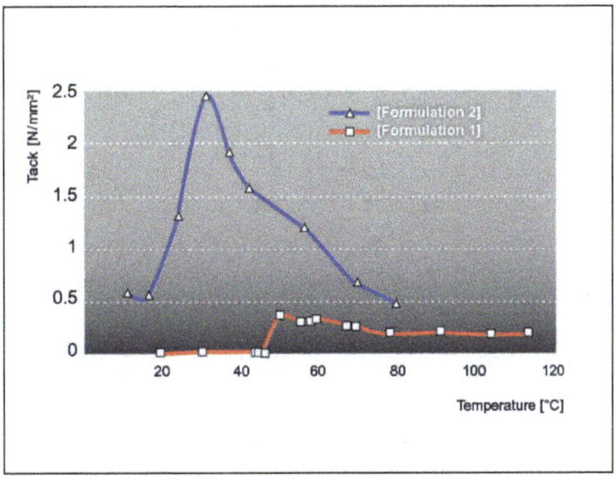

Figure 6.29: Influence of polyester diols on tack

Formulation development thus requires a careful balance

between the adhesive properties and the specific processing conditions and final properties for the given application, with the result that customized solutions are frequently needed.

Processing

When processing reactive polyurethane hot melt adhesives, it is absolutely essential that appropriate application equipment is available, and the user of the adhesive knows how to safely handle these materials. In the melt, the vapor pressure of any residual monomeric diisocyanate will increase. All necessary measures must be taken to prevent emission of these monomers. Each processing step must be reviewed to identify and minimize all potential occupational safety risks. NCO-reactive hot-melts must be delivered in absolutely moisture-proof containers (drums, cans or moisture-proof film pouches, or cartridges) to prevent pre-curing of the prepolymer and pressure build up in the respective container.

The application equipment used with these adhesives should preferably operate in a continuous process that only melts the amount of adhesive that can be used immediately and can accurately control the heating temperature in the range of 100 to 150 °C. As mentioned previously, these adhesives have limited thermal stability which restricts the processing temperature and also limits the time the adhesive can tolerate exposure to thermal stress. The principle of such a processing unit is shown in Figure 6.30. The melt is applied to the substrate by spraying, roller coating or extrusion through a slot die, as shown in Figure 6.31.

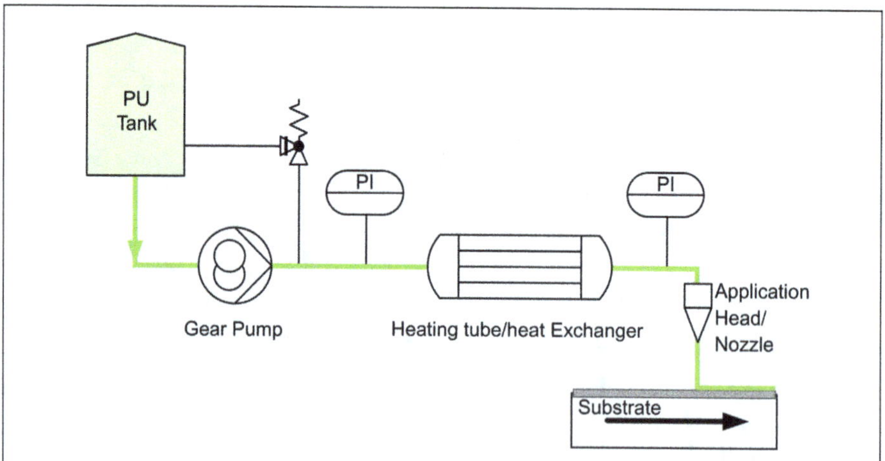

Figure 6.30: Principle of a processing unit for reactive polyurethane hot-melts[52], PI = Pressure indicator

The strength of a hot melt adhesive develops as the bond line cools and the adhesive layer increases in viscosity followed by recrystallization of the polymer. The initial bond strength is strongly influenced by the temperature of the substrate. If the substrate is too warm, the cooling rate will be inhibited, and the initial strength will develop slowly. If the substrate is too cold or rapidly conducts away heat, then the adhesive may solidify too quickly, and it will not wet-out the part being bonded.

The presence of moisture is required to develop the final strength. The curing reaction depends on humidity, any moisture that is present on or in the substrate, as well as on the rate of diffusion of the moisture into the adhesive film. Chemical post-curing therefore occurs significantly more slowly than the physical setting process; depending on the level of moisture available, three to seven days are needed for complete curing. The crosslinking reaction with moisture ultimately yields

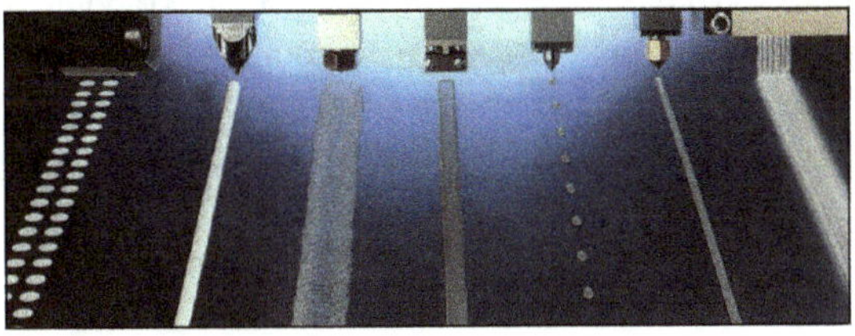

Figure 6.31: Application variants for reactive polyurethane hot-melts[52]

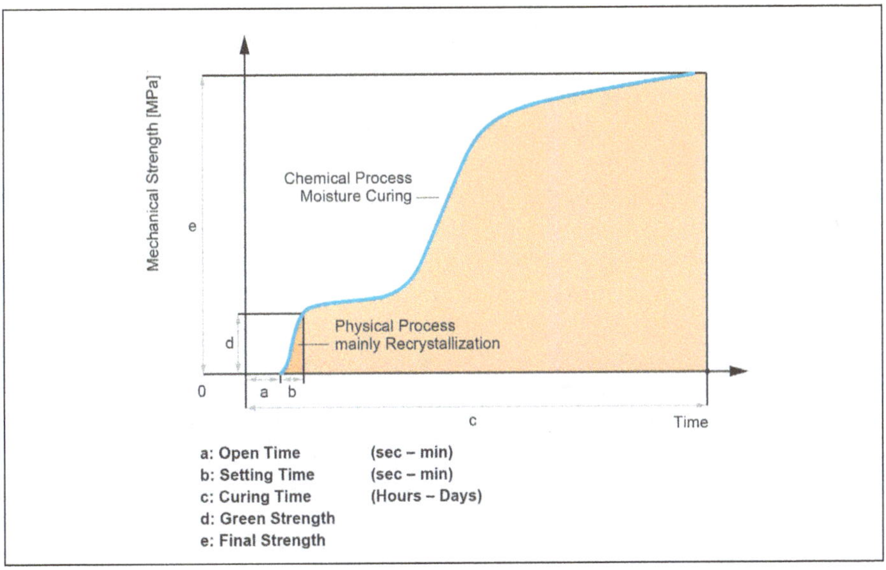

Figure 6.32: Schematic illustration of the increase in adhesive bond strength of a reactive hot melt adhesive

a heat-stable bond with good chemical resistance, e.g. resistance to washing or chemical cleaning in the case of textile bonding.

Figure 6.32 shows a schematic illustration of the increase in adhesive bond strength of a reactive hot melt adhesive. The rapid increase (b) of strength is produced by the increase in viscosity and/or recrystallization on cooling. Final chemical curing by the isocyanate/water reaction follows with some delay (c).

Applications

Customized reactive polyurethane hot melt adhesives are used in many different applications: bookbinding, assembly and profile wrapping in furniture production, textile composites and laminates, footwear components, construction of metal doors and partition walls for offices, manufacture of sandwich panel assemblies for camping vehicles, fabrication of window frames, manufacture of automotive lighting elements.

6.7 Outlook

Adhesive bonding is seen as a key enabling joining technology to develop innovative solutions to address the global megatrends of urbanization, mobility and sustainability. Due to their superior performance level and their high versatility, polyurethane adhesives are expected to show further strong growth in sports and leisure, construction as well as industrial applications. They possess an excellent blend of high bond strength, durability and flexibility and the ability to reliably bond to a broad range of substrates. This combination of properties is a strong argument for using polyurethane adhesives.

Especially important is the current trend to lightweight structures in e.g. automotive, electronics and construction industry, together with the need to join dissimilar materials, that offer the opportunity for continued growth of polyurethane adhesives.

As joining technologies continue to be developed in various applications, new opportunities as well as challenges are expected for adhesives. Examples are progressing automation as well as the fast developing 3D printing technology, which might influence the use of adhesives in some special applications.

Upcoming initiatives such as the new European Plastics Strategy developed by the European Commission to facilitate the target of a more circular economy [23], can create obstacles as well as new opportunities to polyurethane adhesives. This will require further development of this technology to achieve new innovative solutions for the future.

6.8 References

[1] W. Brockmann, P. L. Geiss, J. Klingen, B. Mikhail et al., Adhesive bonding: materials, applications and technology, Wiley-VCH, Weinheim, 2009; S. Ebnesajjad, A. Landrock, Adhesives Technology Handbook, 3rd Edition, Elsevier, Oxford, 2014; C. A. Dahlquist, An Overview of Adhesive Technology, Kautschuk + Gummi Kunststoffe, 38, p. 617, 1985

[2] G. Habenicht, Kleben, Grundlagen, Technologien, Anwendung, 4th Edition, Springer Verlag Berlin Heidelberg New York, 2002

[3] J. Sauer, Composite-Kleben im Sportgerätebau. Adhäsion Kleben & Dichten 44, 7, 2000, p. 16

[4] Chem Research 2018

[5] Courtesy of Audi AG, Ingolstadt

[6] Holzkurier 27, 04.07.2013, 24

[7] K. C. Frisch Jr., Chemistry and Technology of Polyurethane Adhesives, in Adhesive Science and Engineering Vol. 2, p. 759 ff, 2002

[8] D. G. Lay, P. Cranley, Polyurethane Adhesives, in Handbook of Adhesive technology 2nd edition, p. 695, 2003

[9] G. Oertel (Hrsg.), Polyurethane, Becker/Braun Kunststoffhandbuch Vol. 7, 3rd Edition, Hanser Verlag, München, 1993, see there: M. Dollhausen, Polyurethan-Klebstoffe, p. 643 ff

[10] H. Stepanski, M. Leimenstoll, Polyurethan-Klebstoffe, Unterschiede und Gemeinsamkeiten, Springer Berlin, 2016

[11] S. L. Reegen, Adhesion of Urethanes from Oxypropylen Polyols in Advances in Urethane Science and Technology, Vol. 2, Technomic Publishing Co., Inc., 1973, p. 56 ff

[12] D. Dieterich, K. Uhlig, Polyurethanes, p. 665–716 in Ullmann's Encyclopedia of Industrial Chemistry, VCH Publishers Inc., Vol. A 21, 1992

[13] D. Dieterich, K. Uhlig, Polyurethanes, in Ullmann's Encyclopedia of Industrial Chemistry, Wiley-VCH 2002, see p. 592 (polyurethane adhesives)

[14] Z. Aggias, R. Karrer, L. Thiele, 25 Jahre PUR-Klebstoffe für Schiffbau und Kälteisolierung (Vol. 2). Adhäsion Kleben & Dichten, 40, 7, 1996, p. 26

[15] H. Schenkel, Von der Großserie bis zur Reparatur Plastverarbeiter, 47, 6, 1996, p. 99

[16] S. L. Reegen and K. C. Frisch, Catalysis in Isocyanate Reactions in Advances in Urethane Science and Technology, Vol. 1, Technomic Publishing Co., Inc., p. 1 ff., 1971

[17] Mixpac Systems AG; www.mixpac.com

[18] B. A. Morris, The science and technology of flexible packaging, 1st Edition, Elsevier, 2017; Flexible packaging-adhesives, coatings and processes, Rapra Review Reports, 11, p. 3–34, 2000

[19] Courtesy of Nordmeccanica group, www.nordmeccanica.com

[20] K. Ellend, B. Gutsche, G. Steiner, Analysis of laminates-determination of isocyanate residues and primary aromatic amine migration, Deutsche Lebensmittel-Rundschau, 99, 2003, p. 131; W. Frede (Ed.), Handbuch für Lebensmittelchemiker, Lebensmittel – Bedarfsgegenstände – Kosmetika – Futtermittel, 3rd Edition, Springer, 2010

[21] Regulation (EC) No 1935/2004 of the European Parliament and of the Council of 27 October 2004 on materials and articles intended to come into contact with food and repealing Directives 80/590/EEC and 89/109/EEC

[22] Courtesy of Purbond AG; www.purbond.com

[23] A European Strategy for Plastics in a Circular Economy; Communication from the Commission to the European Parliament, the Council, the European Economic and Social Committee and the Committee of the Region, Brussels, 16.01.2018, COM (2018) 28

[24] M. Proebster, Elastisch kleben, Springer Vieweg, Wiesbaden, 2013
[25] C. S. Schollenberger, Polyurethane- and Isocyanate-Based Adhesives, in: I. Skeist (Ed.), Handbook of Adhesives, Springer US, 1990, p. 366
[26] O. Cada, N. Smela, Die Verklebung von Polyolefinen, Adhäsion 18, 1974, p. 198
[27] R. Eftimova, P. Zwettkoff, Über die Rolle der Isocyanate bei der Entstehung der Adhäsionsbindung zwischen Ledermaterialien und 2-Komponeten-Polychloropren-Klebstoffen; Adhäsion 20, 1976, p. 100
[28] H. Honarkar, Waterborne polyurethanes – a review, J. Disp. Sci. Technol. 39, 4 (2018), 507
[29] J. W. Rosthauser, K. Nachtkamp, Waterborne Polyurethanes, in Advances in Urethane Science and Technology. Vol. 10, Technomic Publishing Co, 1987, p. 121; J. F. Dormish, Tack Measurement of Heat-Activated Polyurethane Adhesives, Polyurethanes 2004, Oct 18–20 2004, Conf. Proceedings, 467
[30] M. Pérez-Limiñana, F. Arán-Aís, A. Torró-Palau, C. Orgilés-Barceló, J. Martín-Martínez, Structure and properties of waterborne polyurethane adhesives obtained by different methods, J. Adhesion Sci. Technol., Vol. 20, No. 6, pp. 519–536 (2006)
[31] Regulation (EU) No 528/2012 of the European parliament and of the Council of 22 May 2012 concerning the making available on the market and use of biocidal products
[32] Regulation (EC) No 1272/2008 of the European Parliament and of the Council of 16 December 2008 on classification, labelling and packaging of substances and mixtures
[33] EP2552982B1
[34] EP1644428B1: Stahl International BV 2003, Hesselmans et al., Process for preparation of stable polycarbodiimide dispersions in water, which are free of organic solvents and may be used as crosslinking agents
[35] J. Büchner, Producing film laminates more efficiently, Adhäsion Adhesives Sealants 1-2015 p. 12–16
[36] H. W. Lucas, G. Festel, J. Ramthun, R. Witkowski, J. Dormish, Hot Tack Measurements: An Efficient Development Tool for Water-Based Polyurethanes; Adhesive Age, Vol. 40, 1997
[37] O. Ganster, W. Henning, R. Musch, M. Matner, Neue Rohstoffe für lösemittelfreie Kleb- und Dichtstoffe, Adhäsion Kleben & Dichten, 47, 3, 2003, S. 31
[38] Courtesy of Wemhöner Gmbh & Co. KG
[39] VDA 278, Thermal Desorption Analysis of Organic Emissions for the Characterization of Non-Metallic Materials for Automobiles, October 2011, VDA Verband der Automobilindustrie, Germany, www.vda.de
[40] Courtesy of BMW AG, München
[41] T. P. Abend, Verfahren zur Herstellung und Verwendung lagerstabiler latent-reaktiver Schichten und Pulver, Patent Application, EP 0 922 720 A1, Bayer AG 1997
[42] O. Ganster, J. Büchner, H. W. Lucas, W. Henning, Zubereitungen feinteilig dispergierter oberflächendesaktivierter Feststoffisocyanate mit wässrigen, isocyanatreaktiven Polymerdispersionen, Patent Application, EP 1 600 485 A1, Covestro Deutschland AG 2004
[43] J. Buechner, B. Raffel, Latent-reaktive Polyurethan-Dispersionsklebstoffe, Tagung der Fachhochschule Aachen: Wässrige Kunststoffdispersionen und ihre Anwendung, 2005
[44] Courtesy of Covestro Deutschland AG, Leverkusen
[45] J. Büchner, W. Henning, H. Stepanski, B. Raffel, Lagerstabile latentreaktive Klebfolien und ihre Einsatzchancen, Adhäsion Kleb- und Dichtstoffe, 7–8 2005, p. 23
[46] W. Brockmann, P. L. Geiss, J. Klingen, B. Mikhail et al., Adhesive bonding: materials, applications and technology, Wiley-VCH, Weinheim, 2009, 49; Schmelzklebstoffe, TKH-Merkblatt 4, Industrieverband Klebstoffe, März 2015
[47] W. Li, L. Bouzidi, S. Narine, Current Research and Development Status and

Prospect of Hot-Melt Adhesives: A Review, Ind. Eng. Chem. Res., 47 (20) 2008, 7524; Q. Tang, J. He, R. Yang, Q. Ai, J. Appl. Polym. Sci., Wiley Online Library 38415 (2013)

[48] D. Green, A new beginning in the fabric laminating industry-reactive PUR adhesives, Journal of Coated Fabrics, 28 Oct 1998, p. 116; K. Albers, H. de Jong, S. Katzenmayer, Advanced reactive polyurethane hot melt adhesives for high-performance textile laminates and industrial composites. Melliand International, 9, 4, 2003, p. 321

[49] C. Meckel-Jonas, J. Fett-Schudnagis, High performance polyurethane adhesives for textile lamination of technical fabrics and composites, Technische Textilien/ Technical Textiles, 42, p. 286–288, 1999

[50] M. Krebs, U. Franken, Th. Moeller, Th. Morgeneyer; H. Primke, (Henkel KGAA). Hochelastische Polyurethanschmelzklebstoffe, Patent Application, EP 1 548 042 A2

[51] M. Krebs, Neue monomerreduzierte Polyurethan-Hotmelts Adhaesion Kleben & Dichten, 48, 1, 2004, p. 15; B. Kraemer, K. Paschkowski, M. Schmider, Adhäsion 1–2 2017, 38; E. Cuno, B. Brandt, H. Assenmacher-Maiworm, K. E. Buchwald, J. U. Hahn, T. Hensel, Emissionsverhalten von reaktiven Polyurethanschmelzklebstoffen, Gefahrstoffe – Reinhaltung der Luft 75, 11/12 (2015), 457; M. Krebs (Henkel KGAA), Reactive polyurethanes having a reduced diisocyanate monomers content, WO 2003/006521; M. Wintermantel, P. Reichert et al., (Covestro Deutschland AG), Low-viscosity reactive polyurethane compounds, EP 2794707 B1; D. Vinci, T. Schmidt (DOW Global technologies LLC), Low monomer laminating adhesive, WO2016/060977

[52] Courtesy of Nordson Deutschland GmbH; www.nordson.de

7 Polyurethane sealants

7.1 Terms and definitions

In sealant applications, polyurethanes compete with a variety of other chemistries such as silicones, acrylates, butyl rubber and polysulfides.[1–6] Polyurethane sealants are used in construction, industrial, DIY, and automotive applications because of their superior physical properties, the wide variety of physical properties that can be achieved, and the productivity enhancements they provide to the application process. In most industrialized countries silicone, polyurethane and acrylic sealants have the biggest market share. Other chemistries prevail in special applications, i.e. if high resistance to solvents (polysulfides) or low transmission of gases (butyl rubbers) are required. Both the DIN EN ISO 6927standard and the ASTM C717 standard are in agreement that a sealant is a material that has adhesive and cohesive properties to form a seal.
Sealants are classified in two groups on the basis of their physical properties:

- Plastic sealants (US: thermoplastic): This type of sealant retains plastic characteristics after curing. This means that it has virtually no elastic recovery after application of tensile or compressive stress. Butyl rubber is a typical representative of this class of sealants.
- Elastic sealants (US: thermosetting): These exhibit elastic recovery. Elastic sealants primarily include products based on polyurethane (see Figure 7.1), polysulfide and silicone polymers.

However, many sealants in the market display a mixture of plastic and elastic type of deformation when under physical stress.

There is a large group of reactive elastic sealants that can be used as soft reactive adhesives as well.

7.2 Chemical structure

7.2.1 Polyisocyanate crosslinking systems

Polyisocyanate based sealants are available on the market as two-component and one-component systems.

Two-component polyurethane sealants

The two-component systems can be further classified as polyurethane, polyurea, or polyurethane-polyurea hybrid sealants. The resin components of two-component polyurethane sealants comprise polyols, plasticizers, fillers, thickeners and additives. The resin components of two-component polyurea sealants are the same except amines are substituted for the polyols in the formulation. The hardener component for both systems contains NCO-terminated prepolymers as the co-reactants for the polyol or amine component. These polyisocyanates are often MDI or TDI based prepolymers but may be aliphatic if UV stability or slow reactivity is required. In this case HDI based prepolymers are used.

Two-component polyurethane systems with relatively low application viscosities can be used as self-levelling sealing compounds and permit fast processing because the smoothing step is not required. Since the curing of two-component systems is not dependent on the inward diffusion of atmospheric moisture, deep joints can also be poured without curing problems. Two-component polyurethane and polyurea sealants are used as sealing compounds for horizontal floor joints, particularly when rapid return to service is required. Floor joints in industrial facilities and big box stores are examples. Polyurea sealants are often preferred over polyurethane sealants due to their ability to rapidly cure without foaming and their ability to cure in low temperatures where other chemistry types do not cure. Polyurea sealants are often applied and cut flush with the surface within 20 minutes. The floor can be returned to service quickly.

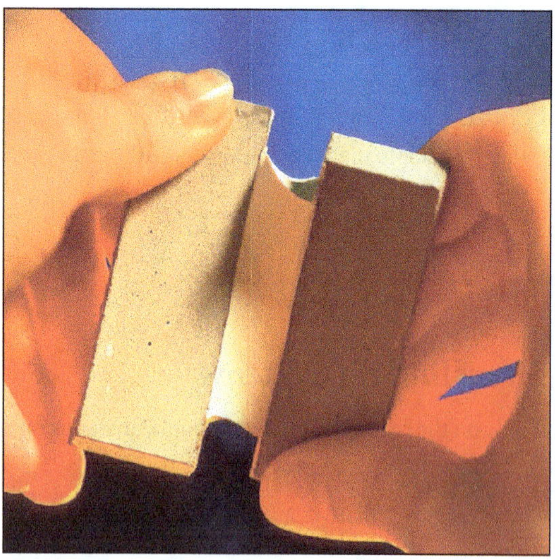

Figure 7.1: Elastic polyurethane sealant

One-component polyurethane sealants

Moisture curing one-component polyurethane sealants are used because of their excellent physical properties included the ability to meet the highest movement capabilities. Thanks to their safe and easy handling, one-component polyurethane sealants have gained broad market acceptance. They usually contain MDI based prepolymers such as "Desmoseal" M that have a relatively low NCO content of approximately 2 %. The NCO content of the sealant formulation has to be limited because carbon dioxide is produced through contact with atmospheric moisture or moisture in the substrate during the hardening reaction. The fewer NCO groups there are in the formulation, the less gas is generated and the lower the risk of bubbles forming as the sealant cures. Additional strategies to limit bubble formation will be covered in the formulation section.

The prepolymer chains are formed using linear polyether polyols, whereby a carefully calculated quantity of trifunctional polyether polyol is used to optimize elastic recovery. The reaction of the NCO-terminated prepolymers with atmospheric humidity, ambient moisture or water in the substrate yields an essentially linear chain extension through urea segments, without the formation of additional branching sites (see Figure 7.2). This type of system is employed in building, automotive construction and shipbuilding.

7.2.2 Silane-modified polymers

Sealants based on silane-modified polymers cure by a consecutive sequence of hydrolysis and condensation reactions of their alkoxy silane end groups. Upon reaction with water, siloxane bonds are formed and alcohol (in most cases methanol) is released. The alcohols are liquid under normal conditions, and there is no danger of blistering or foaming like in one-component polyurethane systems. These products are characterized in particular by a broad adhesion spectrum. Silane-modified polyurethanes have excellent adhesion to glass and other inorganic materials. Whereas the isocyanate groups in polyurethane sealants can react with OH and NH groups on substrates to form covalent bonds leading to excellent adhesion, the adhesion of sealants based on silane-modified polymers is dependent on the combination of alkoxy silane end groups of the polymers and added alkoxy silane based adhesion promoters.

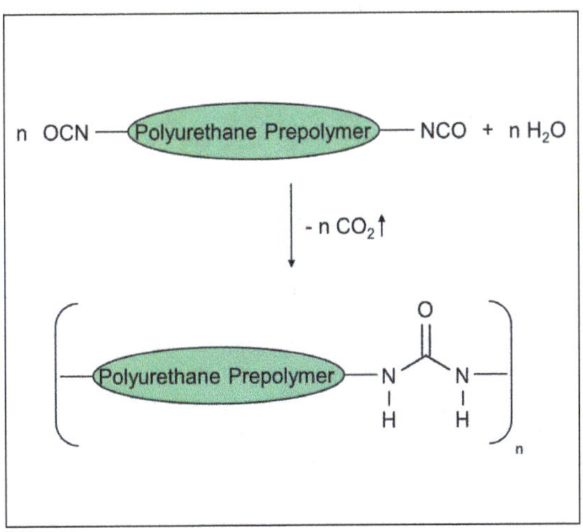

Figure 7.2: Curing of a one-component polyurethane sealant by chain extension of NCO prepolymers upon reaction with water

Silane-modified polymers are used in flexible sealants, as well as increasingly in flexible adhesives, and find applications in automotive manufacturing, machine assembly, in the container industry, electronics and in the construction industry. They can be painted with many common types of paints, exhibit good UV stability, weathering stability, and high elasticity even at low temperatures. Depending on the type of polymer backbone, the silane-modified polymers are subdivided into two structurally different systems:

- Silane-modified polyethers with polyoxypropylene-backbone, and
- Silane-terminated polyurethanes with polyurethane-backbone.

This structural distinction is reflected in the application and end properties of the sealants formulated with these products.

Silane-modified polyethers

The polymer backbone of silane-modified polyethers (see Figure 7.3) is polyoxypropylene, which is also an important building block for polyurethanes. Silane-modified polyethers

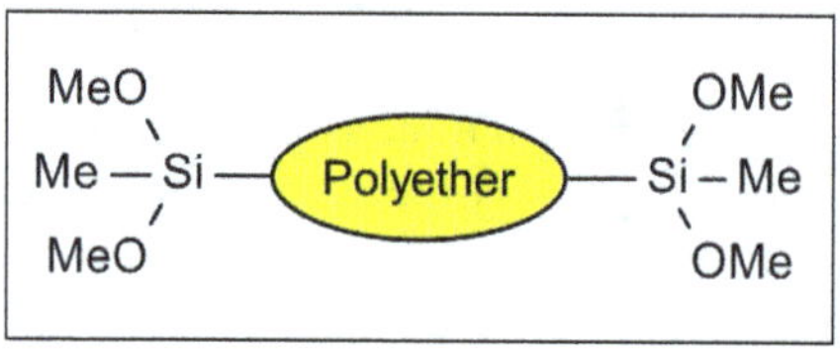

Figure 7.3: Fundamental structure of a silane-terminated polyether

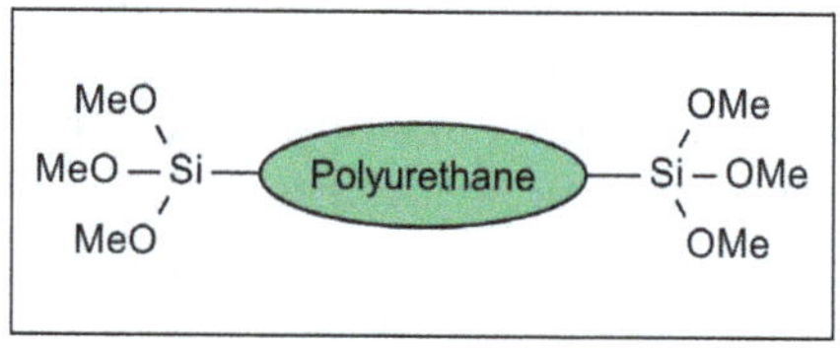

Figure 7.4: Basic structure of a silane-terminated polyurethane

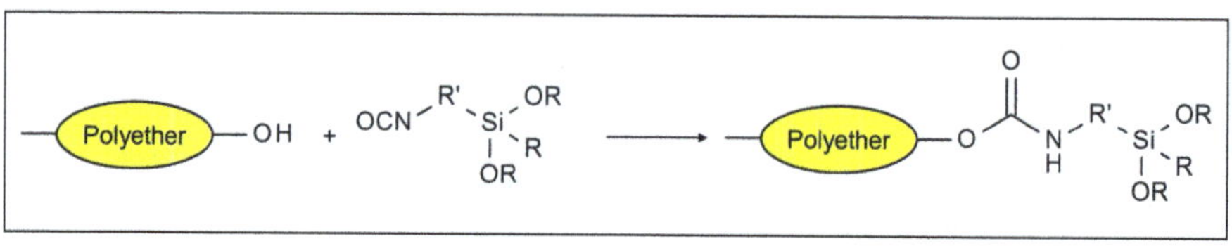

Figure 7.5: Silane-terminated polyurethane by reaction of OH-terminated polyether with alkoxy isocyanato silane

R= Methyl, Ethyl, R'=CH_2 (alpha) or $(CH_2)_3$ (gamma)

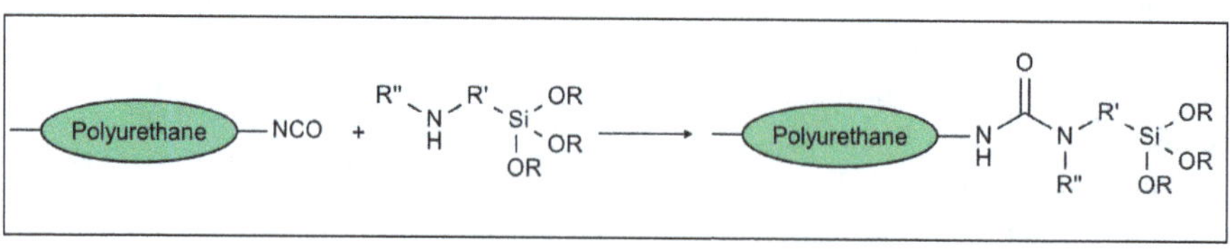

Figure 7.6: Silane-terminated polyurethane by reaction of NCO-prepolymer with aminoalkyl alkoxy silane

R=Methyl, Ethyl, R'=CH_2 (alpha) or $(CH_2)_3$ (gamma), R"=Alkyl

Table 7.1: Silane-terminated polyethers and silane-terminated polyurethanes: comparison of some properties

Property	Silane-terminated polyethers	Silane-terminated polyurethanes
Unformulated raw material:		
Functionality	~ 4	~ 6
Viscosity	lower	higher
Reactivity	lower	higher
In the crosslinked formulation:		
Elongation at break	higher	lower
Tensile strength	lower	higher
Elasticity	lower (higher plasticity)	higher (lower plasticity)

do not contain any urethane structures, however, due to their close relationship to silane-terminated polyurethanes, comparison of the two systems is required for a complete description of the resins.

Silane-terminated polyurethanes

Silane-terminated polyurethanes[7, 8] have also been available on the market for several years. The basic structure is shown in Figure 7.4. Of the various production methods, essentially two processes are established:

- The reaction of NCO prepolymers based on high molecular weight polyether polyols (e.g. "Acclaim" type) and diisocyanates (e.g. isophorone diisocyanate) with secondary aminoalkyl alkoxy silanes (see Figure 7.5).[9, 10]
- The reaction of long-chain polyether polyols with an alkoxyl-bearing isocyanato silane (see Figure 7.6). [11, 12]
- There are two basic types of silane end groups – alpha and gamma that can be used for either process . Gamma-silane groups have three methylene groups between the silicon atom and the amine atom, while alpha-silane groups have one methylene group. Due to the high reactivity of the alpha-silane, only silanes with two methoxy groups are used commercially.[13]

The combination of polyurethane building blocks with the crosslinking and adhesion mechanisms of silicones combines the strengths of both chemistries: the cohesion strength and elasticity of the polyurethane backbone with the blister-free moisture curing and adhesion properties of trifunctional alkoxy silane end groups.

Table 7.2: Typical composition of an NCO-reactive one-component polyurethane sealant

Function	Product examples	Parts by weight
NCO prepolymer	"Desmoseal" M	30
Plasticizer	DIDP, DINP	30
Filler	Chalk (ground types), pigments	30
Additives	Catalysts: e.g. tin and/or amine catalysts	10
	Stabilizers	
	Adhesion promoters: e.g. 3-glycidyloxypropyltrimethoxy silane	
		100

DIDP = diisodecyl phthalate, DINP = diisononyl phthalate

Thus, the prevalent types of silane-modified polyethers in the market consist of a highly flexible polypropylene polyether chains with two reactive alkoxy silane groups at each end. In contrast, the silane-terminated polyurethane has a more rigid polyether-polyurethane chain as the polymer backbone, and at each end, there are three reactive alkoxy silane groups. In principle, the crosslinking mechanism of both product groups is identical, but the higher functionality of the silane-terminated polyurethanes yields greater reactivity and a higher crosslinking density in the resulting polymer.

Some of the differences that result from the various polymer backbones and functionalities are summarized in Table 7.1. The more rigid polymer chains of the silane-terminated polyurethanes have a higher viscosity at the same chain length and tend to yield crosslinked end products with a higher shear modulus and hardness. It is thus possible to use them in formulating both sealants and flexible adhesives (harder than sealants). The flexible silane-terminated polyethers are lower in viscosity (yet have the same chain length), are less reactive as a result of their lower functionality, and yield less highly crosslinked sealants that tend to be softer.

7.3 Formulation

7.3.1 NCO-reactive one-component polyurethane sealants

The typical composition of a one-component polyurethane sealant used in construction applications is shown in the following Table 7.2. In such a formulation, the simpler ground chalk powders can be used as fillers. Silane-terminated polyurethanes, on the other hand,

Table 7.3: Typical composition of a one-component sealant based on silane-terminated polyurethane

Function	**Product examples**	**Parts by weight**
Polymer binder	Silane-terminated polyurethane binder, i.e. "Desmoseal" S	25
Plasticizer	DINP, DIDP	20
Filler	Chalk (mixtures of ground types, precipitated types), pigments	50
Additives	Catalysts: e.g. strong amines, tin compounds, mixtures of both	5
	Antioxidants, stabilizers	
	Adhesion promoters: e.g. aminosilanes (i.e. aminopropyltrimethoxysilane)	
		100

DIDP = diisodecyl phthalate, DINP = diisononyl phthalate,

require reinforcing fillers with a large surface area to develop adequate strength. Due to the presence of NCO groups in one-component polyurethane sealants the use of some additives comprising NCO-reactive moieties is not recommended. This includes certain antioxidants, rheology modifiers and adhesion promoters that are used for example in formulations containing silane-modified polyurethanes.

Formation of carbon dioxide from the moisture cure reaction can create bubbles or a foamed product as noted previously. Several techniques can be employed to mitigate the formation of carbon dioxide. First, the NCO content of the formulation can be reduced as low as possible. Second, an oxazolidine can be added to the formulation. The highly reactive oxazolidine reacts first with water to produce an amine which reacts with the polyisocyanate before the polyisocyanate can react with water.

7.3.2 Silane-terminated polyurethanes

The formulation of sealants based on silane-terminated polyurethanes requires harmonization of a large number of additives that all play a significant role in optimizing the processing and end properties (see Table 7.3).

The storage stability of the formulations depends greatly on achieving the lowest possible moisture content. The amount of water scavenger, usually vinyl trimethoxy silane, is important for the elimination of any possible traces of residual moisture. The amino silanes, added as adhesion promoters, also have a stabilizing effect, because their alkoxy silane groups are present in the mixture in much higher proportions than those bonded to the silane-terminated polyurethane. This excess consequently prevents the alkoxy silane groups bonded to the polyurethane from reacting with trace quantities of reactive components. However, high amounts of silane water scavenger or amino silane adhesion promoter can be detrimental especially for soft elastic systems. These silanes react in the subsequent crosslinking process, and after crosslinking, can cause an increase in the modulus of the sealant.

The viscosity and rheological properties of the formulations are determined by the addition of thixotropic agents, fillers, plasticizers and, if necessary, solvents. The mechanical properties of the crosslinked formulations are determined by the silane-terminated polyurethane, plasticizers, fillers and silanes. Their adhesive properties greatly depend on the amino silane adhesion promoter. It should be noted here that the basicity of these amino silanes can increase the crosslinking rate. Plasticizers, fillers, the amount of catalyst, and light stabilizers[14, 15] also affect the adhesive properties.
Silane-terminated polyurethane prepolymers (e.g. "Desmoseal" S) can be used to produce low-modulus sealants with a high elongation at break or high-modulus adhesives, depending on the formulation.

7.4 Processing

One-component, ready-to-use sealants are supplied in cartridges or flexible pouches and applied to the joint using a manual or compressed-air gun. Sealants are also supplied for the DIY home improvement market in plastic tubes or special cartridges that can be emptied without the use of special guns. In industrial applications, e.g. on a production line, the sealants are supplied in large containers and pumped through hose lines to metering and application units. Appropriate personal protective equipment should be worn and recommendations on safe handling provided in safety datasheets should be followed.

Small volume applications of two-component sealants are applied through hand held, manual, electric or compressed-air guns. The resins are stored in two-component plastic cartridges and do not mix until entering a static mix tube. The formulation is then dispensed at the other end of the static mix tube into the joint.

Large volume applications of two-component sealants are applied through cart mounted electric or compressed air powered meter mix equipment. The resins are stored separately on tanks mounted on the cart. Gear pumps push the material to a hand held static mix tube. For DIY applications two-chamber cartridges combined with static mixers are available.

7.5 Outlook

Sealants and elastic adhesives based on polyisocyanate prepolymers are widely used in the industry and continue to grow in volume. Growth of the more recently introduced silane-terminated polyurethanes is expected to be higher, especially after the recent introduction of special products like "Desmoseal" XP 2774 which allow the formulation of low modulus sealants with high elastic recovery. Due to growing importance of workplace hygiene, sealants and elastic adhesives with reduced content of volatile compounds, i.e., monomeric silanes are expected to see increased market relevance.

7.6 References

[1] R. M. Evans, Polyurethane Sealants, CRC Press Inc, Lancaster, 1993

[2] M. Pröbster, Baudichtstoffe, Springer Fachmedien, Wiesbaden

[3] I. R. Panek, I.-P. Cook, Construction Sealants and Adhesives. John Wiley and Sons, Inc., New York 1991

[4] W. Endlich, Kleb- und Dichtstoffe in der modernen Technik, Verlag W. Giradet, Essen 1998

[5] M. Pröbster, Moderne Industriedichtstoffe, Vulkan Verlag, Essen, 2006

[6] E. Baust, W. Fuchs, Praxishandbuch Dichtstoffe, Herausgeber: Industrieverband Dichtstoffe e.V. (IVD), Public Relations Verlag und Werbung GmbH, 2004

[7] US 3,632,557 (1972) Union Carbide Corporation

[8] B. A. Ashby, (General Electric Company), Moisture curable siloxy terminated polyether, Patent Application, US 3,408,321, 1968

[9] EP 0 596 360 A1 (1993) Bayer AG

[10] EP 1 124 872 (2001) Covestro Deutschland AG (form. Bayer MaterialScience)

[11] EP 1 924 621 (2006) Covestro Deutschland AG (form. Bayer MaterialScience)

[12] EP 0 372 561 A2 (1989) Asahi Glas Company Ltd. Japan

[13] V. Stanjek, R. Weidner, Alpha Silanes. In: Oye H, Brekken H, Nygaard L, eds. Silicon for the Chemical and Solar Industry X. Trondheim 2010

[14] G. Mathur, J. E. Kresta, K. C. Frisch, Stabilization of Polyether-Urethanes and Polyether (Urethane-Urea) Block Copolymers. Advances in Urethane Science and Technology Vol. 6, Technomic Publishing Co., Inc., 1978

[15] H. J. Fabris, Thermal and Oxidative Stability of Urethanes. Advances in Urethane Science and Technology Vol. 6, Technomic Publishing Co., Inc., 1978

8 New areas of application for polyurethanes

The most important areas of application for polyurethanes in the coating, adhesive and sealant segments have been demonstrated in the foregoing chapters. Chapter 8 intends to illustrate the wide variety of possible uses of polyurethanes as starting points for new applications, including those outside established areas of use. This spin-off effect will be described below on the basis of a few selected examples.

8.1 Polyurethanes for medical application

Polyurethanes are generally considered to be skin-friendly and biocompatible. This is a basic requirement for use on injured skin. Therefore, polyurethanes are versatile raw materials for many medical applications, including wound care, hydrogels, coatings of medical textiles, bandages, sealant materials in medical devices and even implants.

8.1.1 Polyurethanes for wound care

One major application field for polyurethanes in medical applications is wound care. Today, polyurethane raw materials are playing an important part in professional wound treatment. The objective of wound care besides protection is to relieve pain and shorten treatment times. For advanced wound care, moisture management of wounds is the key to a fast and sustainable healing process without further infection of the wound. Such wound dressings usually consist of an inner absorbent core and a backing film layer that ensures breathability, protection from dirt and infection (see Figure 8.1). Non-adherent dressings are directly applied on the wound and can be fixed with a secondary dressing (bandage or tape) whereas other dressings contain a skin-contact adhesive for direct fixation on the body.

Nowadays, hydrophilic foam dressings are the major type of dressings used in the therapy of chronic wounds (see Figure 8.2), where the wound healing process may take several weeks or even months. In those dressings, thermoplastic polyurethanes are used in backing films. They can be manufactured in extrusion processes as well as by solvent-borne casting. Aqueous dispersions, which offer a more sustainable alternative, can also be used. Polyurethane foams are the most wide-spread materials used for absorbent dressings, some of which use a super absorber layer to increase the absorption capacity.

For that purpose, polyurethane based foams have been in use already for many years.[1] Currently, MDI and TDI-based foams are state of the art, but new aliphatic prepolymers offer a disruptive alternative for the market. Recent achievements target the development of aliphatic hydrophilic polyurethane foams for wound care with optimized moisture management, high fluid retention, non-yellowing properties and compatibility to gamma-irradiation sterilization processes, which is unprecedented in existing state-of-the-art wound care foams. These new foams can be processed by casting the raw material mixture between two liners, followed by the direct raise of the foam layer to the intended thickness. They can also be easily functionalized with antimicrobials or super absorbent polymers. Manufacturing has evolved to allow for direct foaming between different substrates (films, non-wovens), eliminating additional materials such as adhesion layers and processing steps in wound dressing manufacturing.[2]

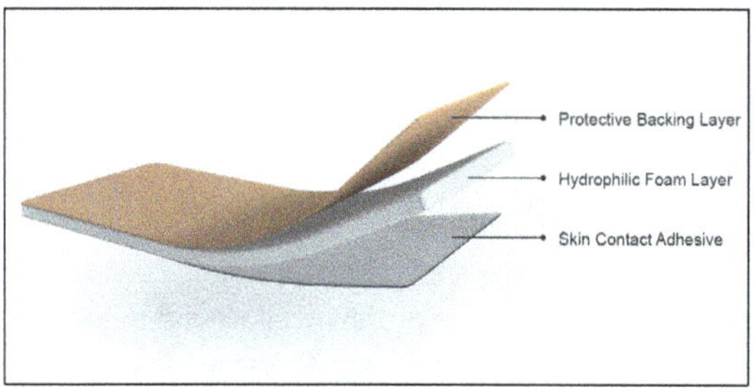

Figure 8.1: Schematic view of a typical wound dressing

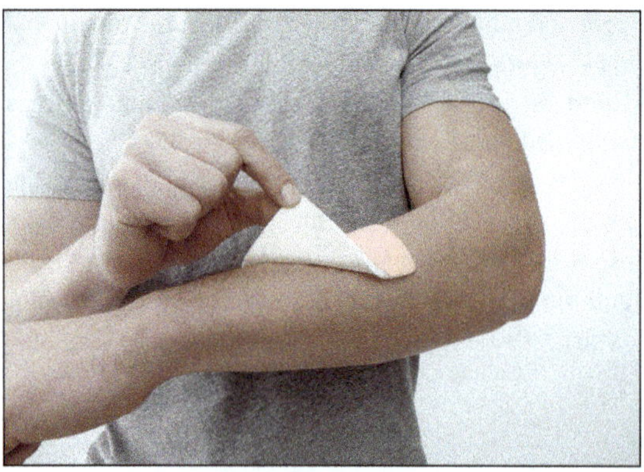

Figure 8.2: Application of a wound dressing

In addition to foams, polyurethane adhesive gels are used in wound care. They exhibit a remarkable degree of adhesion on uninjured skin but are completely non-sticky on contact with the moist wound. Such dressing sheets can therefore be removed easily and painlessly, without causing any new damage to the healing wound.[3, 4] It has also been shown that their use can minimize scar formation.[5, 6]

Manufacturing of polyurethane adhesive gels takes place in a solvent-free, sustainable process. Due to their hydrophilic nature, unique breathability and the adjustable adhesive strength not only applications in modern wound care, but also as skin-friendly low trauma adhesives for device fixation, ostomy and NPWT (negative pressure wound therapy) become available.

Another application of polyurethane materials features wearable devices for monitoring or therapy control. See Figure 8.3, here, electronic devices are connected to sensor patches attached to the skin. Tailored dressing materials such as thermo-formable foams, skin-friendly adhesives with long wear time such as polyurethane gels mentioned above and polyurethane barrier films enable the production of easy to use patches in an efficient roll to roll process.[7] Thermo-formable foams can be obtained by physically frothing suitable polyurethane dispersions and casting them onto a release liner. Stable foams are obtained after drying at temperatures such as 120 °C, see Figure 8.4.

8.1.2 Polyurethane as latex substitute

The prophylactic measures taken worldwide in the early 1990s to guard against HIV, hepatitis B and C, and later also bird flu, led to a sharp rise in the use of latex-based protective gloves. However, glove users frequently complained of allergic reactions. Latex contains allergens that lead to skin hypersensitivity. This fact triggered an intensive search for alternative materials.

Approximately 2 % of the global population suffers from natural latex sensitization with employees in the medical area being especially affected.[8] The market is constantly looking for alternatives for natural latex satisfying the desired properties for high-end applications in medical, such as probe covers and surgical gloves. Trials to replace natural

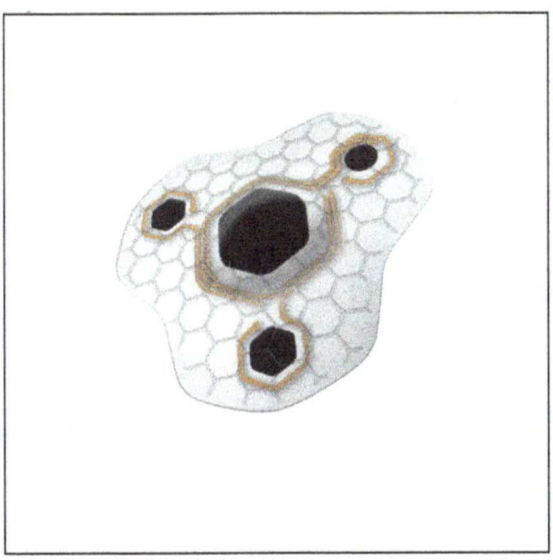

Figure 8.3: Schematic image of wearable patch

Figure 8.4: Sensor embedment on thermo-formable foam

latex by synthetic rubber were only partly successful, as an increasing number of persons are not only allergic to the proteins from latex milk, but also to the required rubber curing additives.[8]

Such allergic reactions are not known to be caused by films based on polyurethane dispersions as they avoid the use of accelerators and sulfur, which are particularly related to Type I and IV hypersensitivities.[9] Apart from their non-allergic properties, polyurethane gloves are comparable to latex gloves in elastic properties, feature improved wearing comfort and are odor-free. In recent years, targeted optimization of polyurethane dispersions for use as latex substitute (see Figure 8.5) has produced a significant improvement in the resistance of such gloves to alcohol-based disinfectants.[10] Similar polyurethane dispersions are also used for manufacturing other dipped articles for sensitive skin-contact applications such as probe covers and condoms.

8.1.3 Outlook

Driven by the aging population in the western world, progressing wealth in the emerging countries and new medical device innovations, the demand for innovative new material solutions in medical applications will increase. At the same time cost pressure in the governmental health care systems will call for improvements in processing and cost efficiency. Tailor made polyurethane material solutions offer a very attractive combination of product differentiation as well as processing efficiency potential mainly in the areas of wound care, surgery and disposable devices.

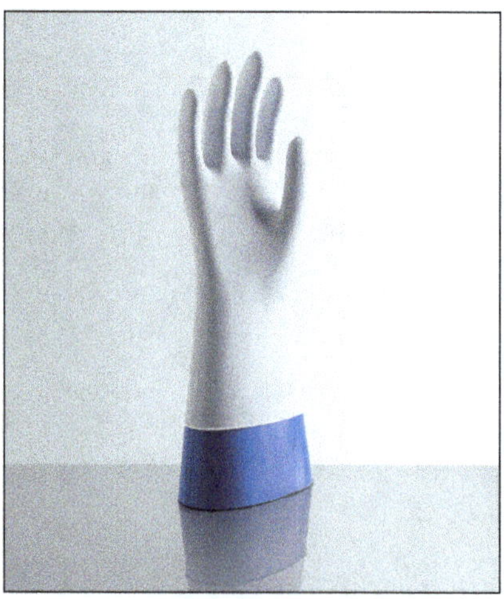

Figure 8.5: Example of a polyurethane based dipped glove

Digitalization is the second important growth driver in medical applications. Here, innovative wearable devices for diagnostic and therapy will create a large new market segment driven by new electronic and sensor technologies. The need for new sensor patches will create a materials market where polyurethanes will play a significant role due to the versatility and performance in skin contact applications.

Furthermore, new regulations on workers' and patients' safety will create a significant push for sustainable material solutions such as water-based and low monomer or low solvent containing polyurethanes in the medical device market.

8.2 Polyurethanes in cosmetic applications

At the end of the 1990s, aqueous polyurethane dispersions were launched into cosmetic applications. In comparison to other existing technologies, polyurethane products perform with excellent film-forming properties and have proven skin friendliness characteristics. They allow skin to breathe freely due to the film's very high water vapor transmission. Water-borne polyurethane dispersions form highly flexible films that mimic the movement of skin without any tightening sensation.

Cosmetics formulators typically use a film former in sun care and color cosmetic formulations to impart long lasting properties in UV protection and color respectively, such as water resistance, sebum resistance or transfer resistance. Water-borne polyurethane dispersions also make durable water-based nail polish formulations that can be easily peeled off.

8.2.1 Hair styling

In hair styling products polymeric film formers are mainly responsible for the styling performance and the durability of the hairdo in humid environments.[11]

Typically used film formers in cosmetic products are based on vinyl-pyrrolidone copolymers, vinyl acetate or acrylate chemistry.

While providing strong hold, these types of polymers still show some disadvantages. The main drawbacks of these three film forming technologies are the tacky and greasy skin feeling imparted onto the skin and the difficulties with formulating suitable products, e.g. high viscosity, neutralization often required. Used on hair these polymers typically form a rigid and brittle film, imparting a strong hold to the hair but, not durable in time.

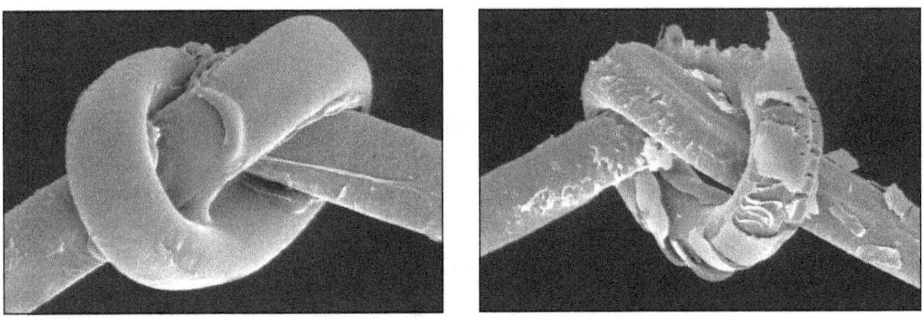

Figure 8.6: Scanning electron microscope (SEM) images of single knotted hair fiber coated with a polyurethane dispersion on left and an acrylate copolymer on right. Under stress, the acrylate copolymer becomes brittle, while polyurethane dispersion film shows no cracking.

Modern hair styling trends reveal consumer desire to combine high performance, e.g. durability, strong hold, with natural feel and look. The hair fixative polymers based on polyurethane dispersions contribute numerous benefits to hair styling formulations compared to other film-forming agents currently in use. These benefits include superior resistance to high humidity and durable styling/fixative effect (see Figure 8.6) even after mechanical action as well as no adverse effect on tactile properties of the hair. The hair remains silky and shiny after application of the styling products.[12]

Unlike currently used film formers, aqueous polyurethane dispersions are significantly easier to use. They do not have to be neutralized and don't require high energy input rendering them suitable for cold production processes. The absence of organic solvents and free emulsifiers in the aqueous polyurethane dispersion combined with the proven skin friendliness of the polyurethanes, make this chemistry a great choice for cosmetic products.

8.2.2 Skin care

Polyurethane dispersions also open a new route of formulating high-SPF (sun protection factor), water-resistant sun protection products.[13] The polymer structure may contained by design both hydrophilic and hydrophobic segments which allow for a unique combination of great water resistance and high rate of water vapor transmission.[17] The obtained film is thus more breathable and less occlusive than conventional film formers. The challenge of obtaining high water resistance can be further met by reducing the emulsifier concentration of formulations thanks to the co-emulsifying properties of polyurethane dispersions, avoiding both the re-emulsification and the removability of the product film in contact with water. Table 8.1 shows a basic water-borne polyurethane formulation of a sun care spray of a sun protective factor (SPF) 50.

The hydrophobic nature of polyurethane dispersion films imparts an outstanding water resistance to products containing pigments. Hence, not only excellent water resistance but also non-transfer properties[18] can be achieved by adding polyurethane dispersion in oil-in-water color emulsion. These properties make them particularly suitable for formulating water-resistant mascara and foundation products.[14] Due to their ability to form a uniform film, polyurethane dispersions, impart even color coverage on skin. As a result of the highly elastic nature of polyurethanes, formulations are comfortable to wear, reflecting consumer's need for colored cosmetic products having a natural feeling on the skin (i.e. non-tacky skin feeling, or non-tightening effect on skin).

In a new cosmetic application the polyurethane film acts like an invisible barrier, protecting skin against environment pollution such as particular matter.[15] The outstanding elasticity of the polymer film makes polyurethane dispersions suitable for peel-off mask as replacement of polyvinyl alcohol film formers. After applying water to the treated skin,

Table 8.1: Sun care spray based on a polyurethane dispersion

	Part by weights
Aqua	32.54
Microcrystalline cellulose (and) algin	0.10
Xanthan gum	0.05
Glycerol	5.00
Disodium EDTA	0.05
C12–15 alkyl benzoate	12.00
Dibutyl adipate	10.00
Propylheptyl caprylate	5.00
Bis-ethylhexyloxyphenol methoxyphenyl triazine	6.00
Ethylhexyl triazone	1.00
Diethylamino hydroxybenzoylh benzoate	6.00
Diethylhexyl butamido triazone	6.00
Sodium stearoyl glutamate	0.20
Acrylates/C10–30 alkyl acrylate cross-polymer	0.05
Alcohol denat.	10.00
Polyurethane dispersion ("Baycusan" C 1000),	6. 00
Sodium hydroxide	0.01
	100 %

the product film can be easily peeled off in one piece, more gently compared to polyvinyl alcohol-based peel-off products.

8.2.3 Outlook

Polyurethane film formers are still in the development growth phase compared to the traditional film formers based on polyvinyl pyrrolidone or polyacrylate chemistry. However, their market share is expected to grow in view of the current consumers' needs for cosmetic products having more natural and comfortable wear with long-lasting properties. From the sustainability standpoint, the environment friendly nature of polyurethanes will definitively reinforce their market position due to the biodegradation of polyurethane dispersion film formers being superior to traditionally used polyacrylate and polyvinylpyrrolidone based film formers in cosmetics. Furthermore, first generation polyurethane dispersions with more than 50 % renewable carbon are already available for cosmetic applications.[16]

8.3 Polyurethanes for light guiding applications

Polyurethane coatings display superior optical properties and, therefore, they are ideally suited for advanced functionalities, e.g. their use for light guiding applications. In order to meet this requirement, further functionalization must be introduced and, in the case of light guiding applications, is achieved by using radiation curable polyurethane-based photopolymers.

8.3.1 Photopolymers

Photopolymers are polymers that change their properties when exposed to light. These changes are often manifested structurally, for example undergoing a second cross-linking step of monomers orthogonal to the initial monomers when exposed to light. A wide variety of technologically useful applications rely on photopolymers, e.g. several coatings and varnishes depend on photopolymer formulation for proper hardening upon exposure to light.

Light guiding in thin films may be realized by e.g. surface modifications, pigments or particles or spatially distributed refractive index modulations (i.e. volume holographic recordings). The major challenges in these application fields are a high optical transparency, chemical and physical robustness, an easy processability, reliable manufacturing and efficient light guiding. A plethora of material concepts has been evaluated to fulfill the mentioned requirements and polyurethane based systems have been proven as a suitable and versatile approach.[19]

8.3.2 Polyurethane photopolymers

Properties like high transparency and superior mechanical robustness of aliphatic polyurethanes qualify them for their utilization in advanced optical applications. Among other examples, holographic data storage and holographic light guiding elements based on aliphatic polyurethanes as matrix material for photopolymers have emerged into several industrial applications over the last decade.[20]

Photopolymers yielding robust, reliable and scalable recording medium for the storage of different kinds of information can be manufactured in large scale and with an unmatched overall flexibility and supply chain robustness. The versatility of polyurethane based photopolymers in the area of these highly functionalized systems originates from the excellent means to design and control the mechanical and optical properties of the photopolymer material.

When combining e.g. a difunctional polyether-polyol or a polyether-polyester polyol with a low viscous crosslinker like, e.g. "Desmodur" N 3900 the glass transition temperature of the resulting polyurethane can be varied from significantly below room temperature (< -60 °C) to room temperature. The formation of locally different refractive indices within the film volume requires a chemistry that is orthogonal to the formation of the polyurethane. The polyurethane thus hosts the active ingredients to modify the refractive index spatially within the material by holographic recording. This writing chemistry at first consists of a photo initiating system (e.g. dye and photo initiator) tuned for the recording wavelength of the respective writing laser light wavelength and secondly of reactive monomers to chemically amplify the primary recording information.[21] The most prominent examples for these amplifiers are epoxy or acrylic monomers.

Furthermore, all active compounds, e.g. dye and photo initiators and monomers, must be compatible with the polyurethane matrix to guarantee high transparency. After the recording step (see below), an interpenetrating network of the polyurethane matrix and the spatially distributed polyacrylic or polyepoxide network will be formed. By this the information that is copied from the interference pattern during holographic recording into a refractive index pattern in the photopolymer will be maintained and is able to tolerate harsh environmental conditions.

8.3.3 Manufacturing process

A very competitive way for the manufacturing, subsequent recording and copying of photo polymer films on industrial scale is ideally realized in a continuous roll-to-roll approach (see Figure 8.7).[22]

Thus, the 2K polyurethane coating process starts with polyol and crosslinker being pumped in from two independent OH and NCO reservoirs that also contain the writing chemistry, additives, solvents and catalysts. Hereafter, precise dosing is applied and efficient mixing of the OH and NCO parts in a static or dynamic mixing unit forms the coating formulation. This formulation is precisely dosed and subsequently applied on a transparent substrate film. A variety of optical clear thermoplastic film substrates (e.g. polycarbonate film, PET film, polyamide film or triacetate film) can be coated (see Figure 8.8). In the next step, the solvent containing wet layer is heated in the dryer section. Here, the solvent is mostly evaporated, and the polyurethane matrix is formed by cross-linking. Residual solvent – especially higher boiling ones – are further evaporated at last and the tack-free photopolymer film is finally wound up and protected with a cover film.

8.3.4 Holographic recording

As depicted in Figure 8.9 the holographic recording process first of all requires coherent laser light to operate. The two coherent laser beams interfere and generate areas of high intensity, where the polymerization and the spatial rearrangement by diffusion of the monomers occurs. Hereby e.g. a 3D image of a cup comprising all optical information available for replay under the suitable conditions is generated. The 3D image itself (or any other optical function) is equivalent to a pattern of spatial modulation of the refractive index in the volume of the photopolymer, here visualized as alternating dark and light blue areas. Not only patterns equivalent to 3D images can be recorded in the photopolymer but also many optical functions inherent to that of a lens, a prism or a mirror.

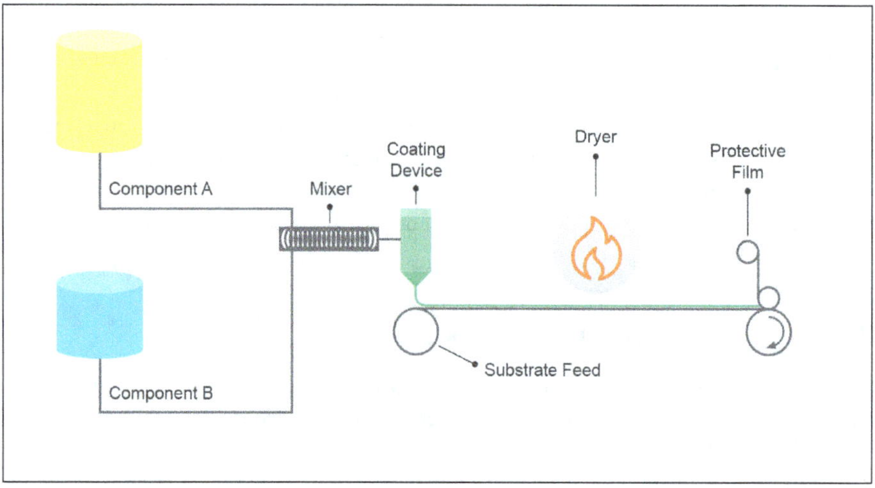

Figure 8.7: Manufacturing sketch for photopolymer films in a continuous process starting with the liquid precursors (OH and NCO lines) including the writing chemistry to the coating on thin polymer substrate films and finally product isolation

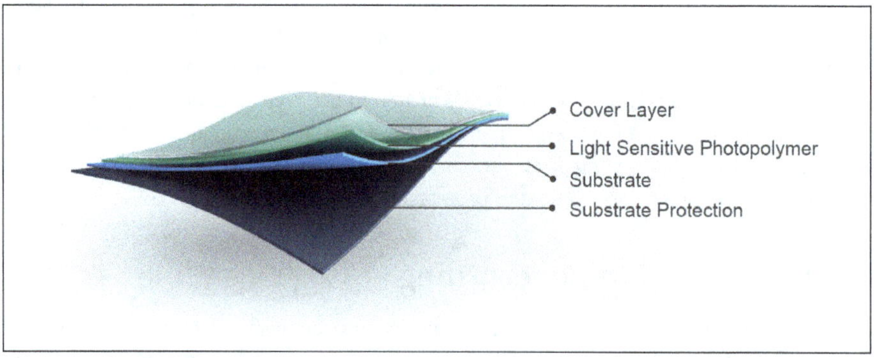

Figure 8.8: Typical layer stack of a photopolymer film: The light sensitive photopolymer is supported on a substrate film and protected with a cover layer

The unique combination of low glass transition temperature polyurethanes with the writing chemistry allows for an efficient holographic recording process. The rather rapid migration and diffusion of the incorporated writing chemistry is best described and understood by a non-local photo-polymerization driven diffusion (NPDD) model. As a further benefit, the polarity of the polyurethane can be modified over a certain range thus enabling one to load significant amounts of writing chemistry while maintaining the high optical transparency.

The three-dimensionally crosslinked polyurethane matrix further hosts the holographic interference pattern generated by laser light sources. The formation of the holographic information may thus be only generated by light. In a post processing step some additional UV/VIS light reacts residual monomers and bleaches the residual components of the photo initiators. This duel chemistry approach can be utilized in volume holographic recording to generate some highly selective optical functions ranging from recording for optical data storage to holographic optical elements useful for numerous applications like selected augmented reality applications or notch laser wavelength filtering for eye protection.

After holographic recording, some measures have to be taken to protect the polyurethane based holographic device from harmful environmental influences to enjoy a long-lasting function. Here, liquid coating as well as transfer films can be applied.

A special advantage of this polyurethane based photopolymer film is the option to also combine some functions, e.g. a lens and a prism in a very lightweight fashion via volume holographic recording. This will open several new options for the design of devices requiring optical functions that are not known in refractive optics.

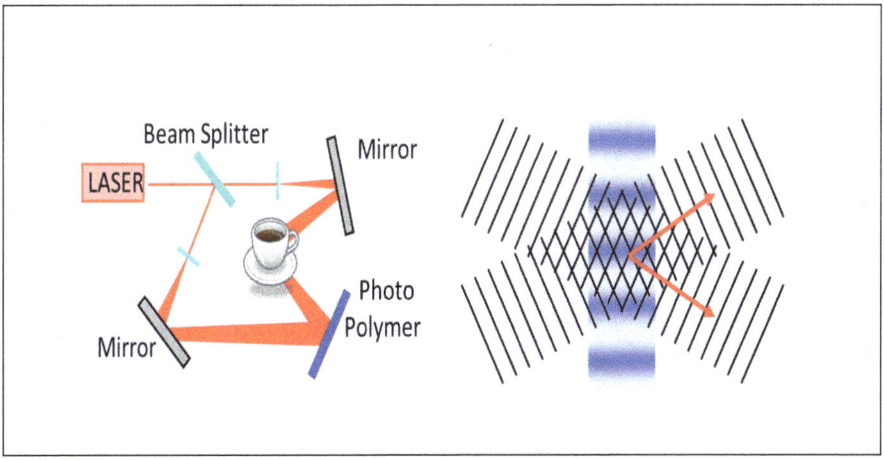

Figure 8.9: Transmission type holographic recording principle in a photopolymer film

8.4 References

[1] US 3 861 993 (1975) J. L. Guthrie
[2] EP 2 300 510 (2011) Covestro Deutschland AG (form. Bayer Material Science); EP 2 470 580 (2012) Covestro Deutschland AG (form. Bayer Material Science); EP 2 632 501 (2014) Covestro Deutschland AG (form. Bayer MaterialScience)
[3] J. Klode, L. Schöttler, I. Stoffels, A. Körber, D. Schadendorf, J. Dissemond, Journal of the European Academy of Dermatology and Venereology 2011_25_933
[4] O. Ganster, Bayer-Forschungsmagazin Research, 11: „Das Gel für schwere Wunden“, 1999
[5] R. Klopp, W. Niemer, M. Fraenkel, A. von der Weth, J Wound Care 2000_9_7_319
[6] W. Wigger-Alberti, M. Kuhlmann, D. Wilhelm, U. Mrowietz, K. Eichhorn, J. Ortega, A. Bredehorst, K.-P. Wilhelm, J Wound Care 2009_18_5_208
[7] EP 2 468 496 (2012) Covestro Deutschland AG (form. Bayer MaterialScience)
[8] Source: www.daab.de/allergien/latexallergie/
[9] Accelerators in Rubber Surgical Gloves by Milt Hinsch (http://latexallergyresources.org/news/accelerators-rubber-surgical-gloves)
[10] H. Suzuki, S. H. Khoo, M.-S. Weiser, MARGMA Conference abstract and publication, 2018
[11] S. Viala, SÖFW, (9), 36, 28, 2010
[12] S. Viala, P. Xiong, M. Krause, Household & Personal Care Today, vol.11, 40, 2016
[13] Y. Berezkin, A. Wulfinghoff, S. Viala, Personal Care Europe, (2), 64, 2018
[14] S. Viala, S. Dörr, Y. Berezkin, Cosmetic Science Technology, 160, 2009
[15] WO2017147739 (2017) Covestro Deutschland AG (form. Bayer MaterialScience)
[16] Covestro Press release, 05.04.2018, “Sustainable polyurethane film formers for cosmetic formulations”
[17] Baycusan, Covestro edition 2018-03, Order-No: COV00084332
[18] Handbook of Cosmetics Science & Technology, Chap. 37: Decorative Products, R. Korichi, J.-F. Tranchant, 3rd Edition, 2010, p. 401–402
[19] B. A. Kowalski, R. R. McLeod, Journal of Polymer Science, Part B: Polymer Physics, 54, p. 1021–1035, 2016
[20] F.-K. Bruder, T. Fäcke and T. Rölle, Polymers, 9, 472, 2017; DOI:10.3390/polym9100472
[21] C. Ley, C. Carré, A. Ibrahim and X. Allonas (March 22nd, 2017). Application of High Performance Photoinitiating Systems for Holographic Grating Recording, Holographic Materials and Optical Systems Izabela Naydenova, IntechOpen; DOI: 10.5772/66073
[22] F.-K. Bruder, T. Fäcke, F. Grote, R. Hagen, D. Hönel, E. Koch, C. Rewitz, G. Walze, B. Wewer, Mass production of volume holographic optical elements (vHOEs) using “Bayfol” HX photopolymer film in a roll-to-roll copy process. Proc. SPIE, 10127, 101270A, 2017

9 From combinatorial chemistry to lab automation and data sciences

9.1 Introduction and history

Combinatorial and high throughput technologies have been established and implemented during the last two decades of the past century and mainly applied in the area of drug discovery. Design of experiments techniques, highly automated and parallelized seamless handling/testing of small-scale samples and the use of high-performance data-mining software were essential to the success realized in pharmaceutical research and development.[1]

Even though first ideas were reported much earlier[2], the transfer of concepts for combinatorial and high throughput technologies into material sciences started to become successful in the middle of the 1990s.[3] Due to the complex requirements of material sciences, combinatorial and high throughput approaches are consisting of even more different and complex stages than in drug discovery. Adequate design of experiments at the beginning, appropriate data handling and finally efficient data mining and assessment are the critical backbone of potentially successful experimental workflows. The broad range of formulations, application and curing procedures of miniaturized samples turned out to be of most importance and hence, very critical for coatings, adhesives and sealants related preparation workflows (see Figure 9.1).

In the first decade of combinatorial and high-throughput approaches, the combinatorial (small scale) syntheses of components also required miniaturization creating significant challenges in designing equipment that could blend formulations, apply them to a substrate, and test them. Moreover, these miniaturized technologies required extended validations especially in coatings workflows to assure that the results are comparable with conventional testing procedures.

In the early phases of combinatorial and high throughput developments, two different approaches were used: On the one side, material-science based companies, such as Bayer MaterialScience, now Covestro, BASF[4], Dow[5], GE, DuPont[6] started building their own facilities internally to address company specific developments. On the other hand, technology providers, including start-up companies, such as Symyx Technologies Inc. (its hardware business evolved into Unchained Labs Ltd.), hte GmbH (Heidelberg, Germany) and Bosch (Waiblingen, Germany) started developing flexible workflows that incorporated powerful software tools, automatized (and often miniaturized) building blocks for synthesis, formulation, and physical-analytical testing (cf. first patents of Symyx Technologies Inc.[7]). In many cases technology-oriented companies and big innovative material science companies started co-operating to further accelerate and optimize specific developments.[8]

Figure 9.1: Dosing unit for high viscous liquids

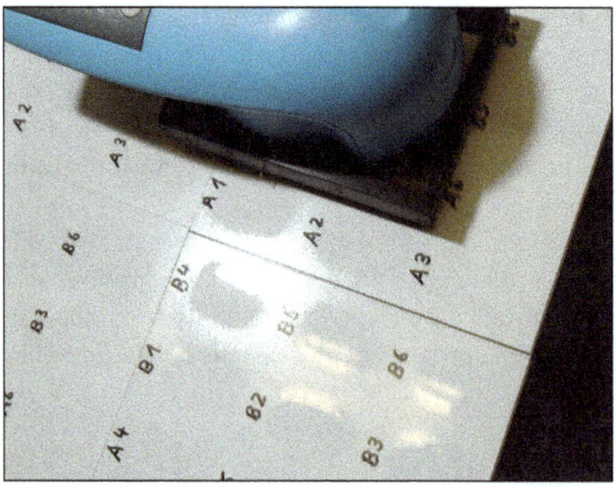

Figure 9.2: Miniaturized spray applied coatings ready for optical analysis

Applications of combinatorial and high-throughput techniques during the past decades revealed – especially in coatings workflows – several limitations in the initial approach.

In many cases the parallelized combinatorial small-scale synthesis of components turned out to be less important for coatings and adhesives and could be omitted (see Figure 9.2).

This led the providers of high-throughput equipment (such as Chemspeed Technologies, Bosch, Labman, and others) to build platforms that forgo miniaturization. Their equipment now provides, besides a comprehensive software solution, automated preparation and handling of formulation, application and preparation of samples, and flexible testing procedures all at normal full scale[9] (see Figure 9.3). Materials manufacturers[5, 10] as well as small and medium-sized enterprises (SME's)[8] have adopted their use in their R&D and quality insurance processes[11].

9.2 Use cases in polyurethane coatings and adhesives

Material development applying combinatorial approaches has now been in use for more than 15 years. *Potyrailo* et al.[12] summarized in their review the basic concept related to materials research and especially to coatings.

In the early phase, polyurethane researchers started with topics where the analysis of a high number of experiments appeared as the more straightforward approach to success, e.g. catalyst research in one-component reactive coatings, development of novel blocking agents, and the development of guide formulations.[13]

Majoros et al.[14] report on the preparation of prepolymers based on polyether diols and triols and toluene diisocyanate "Desmodur T 80". They investigated more than 100 compositions measuring their reaction kinetics in solvent and in bulk. *Nasrullah*[15] reports on a multistep synthesis of polyurethane dispersions involving the preparation of an isocyanate-terminated prepolymer followed by neutralization, dispersion in water and chain extension. *Oh et al.*[16, 17] investigated how formulation changes in low-VOC PU dispersions affect end-use property such as hardness, flexibility, and water and chemical resistance.

In addition to these combinatorial studies, in the early phase of this approach, researchers focused on developing specific automation techniques to evaluate coatings. *Bach* et al.[18] reported on a dye elution technique to monitor crosslinking of thermally activated PUR-hardeners, while *Majumdar*[19] implemented pot-life measurements. *Sormana* [20, 21] investigated automated mechanical analysis methods and *Chisholm* et al.[22, 23] reported on abrasion measurements. *Fadel* et al.[24] studied abrasion resistance in corrosion protection and *He* et al.[25] developed high-throughput electrochemical impedance to analyze polyurethane coatings. High throughput impact resistance tests were explored by *Sormana*[26] that showed a dynamic impact response of multiple positions on a combinatorial library film.

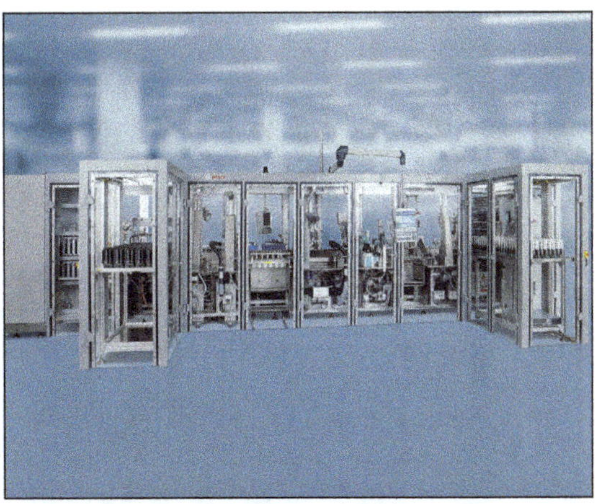

Figure 9.3: Equipment for paint preparation[62, 63]

Even automation methods for measuring anti-fouling

coatings have been developed and used in combinatorial studies. *Casse*[27] developed a spin-jet apparatus that delivered a jet of water of controlled, variable pressure into the wells of 24-well plates. Similarly, *Stafliens* et al.[28] report on a 24-well parallel test approach using the marine bacterium *Cytophaga lytica*, the analysis relies on automated image analysis[29, 30]. *Dean Webster's* group at North Dakota State University has investigated PMDS-Urethane hybrid systems. The North Dakota team screened di- and tetra-functional polydimethylsiloxane oligomers and copolymers with polycaprolactone blocks in siloxane-polyurethane coatings[31, 32]. The best performing coatings resisted fouling when exposed to algal (Ulva), bacterial (*Cytophaga lytica, Halomonas pacifica*), and barnacle (*Balanus amphitrite*) outperforming commercial silicone systems. 3-Aminopropyl-terminated polydimethylsiloxanes and polycaprolactone-polydimethylsiloxane triblock copolymers were used in several studies[33, 34], where surface energy and thus water contact angles were controlled by adjusting the molecular weight of the siloxane block. They surprisingly found that the siloxane polymer level in the coating formulations does not have a significant effect on the surface energy of the coatings[35], as they first thought. In more recent studies the North Dakota team investigated the effect of composition on surface roughness[36, 37]. Their results indicate that coatings made with polycaprolactone polyol appeared to be somewhat rougher compared to coatings made with acrylic polyols. This is important because the adhesion strength of sporelings of Ulva increased with an increase in surface roughness. *Majumdar* et al.[38] investigated antifouling and fouling release effects. They looked into tethered quaternary ammonium salts functioning as chemically bound biocide combined with a silicone-PU hybrid coating that acts as a release matrix.

Investigations on adhesion in vivo systems to PU coatings through a combinatorial approach are growing in number. Polyurethanes and polyacrylates-acrylamides based coatings have been found to resist binding by major bacterial pathogens, making them suitable for medical devices, reducing the occurrence of implant-associated infections.[39]

Tests that assess weather resistance have also been automated. *Lowry* et al.[40] discuss the required UV protection by nano-ZnO in a polyurethane/acrylic clear topcoat and analyzed the influence of nano-ZnO concentration and dry film thickness on the film's optical properties, properties such as UV permeability and visible transmittance of the coating.

Gambino et al.[41] facilitated the development of direct-to-metal UV coatings that due to their strong shrinkage are prone to poor adhesion. They identified the required selection and combination of resins that, after being UV-cured, adhere with excellent strength to various grades of iron and steel substrates.

Also for adhesives new analysis technologies have been implemented e.g. for peel tests[42] and adhesion analysis.[43] More advanced are techniques to measure phase behavior of polymer blends using optical small angle light scattering technique applicable for high-throughput testing.[44] So while literature on adhesive development is rather sparse in comparison to coatings applications, the challenges around adhesion and its influences high in number make adhesive development rather straightforward to apply high-throughput approaches combined with data-driven analysis techniques to it as shown by multiple publication bio-medical field.[45] Also work on glass-fiber reinforced polymer composites are done that investigate the profile's behavior of adhesive joints.[46]

All these studies have in common that they deal with larger sets of data. In the following paragraph we discuss in more detail the implications of managing data.

9.3 Data-driven coatings development

In recent years, with the advent of big data, the internet of things, and new developments in machine learning, an ever-increasing effort towards data-driven development of new materials has emerged. Supporting this trend, the governments of three of the world's major economies have each funded major initiatives towards this goal: the USA in 2011 launched the Materials Genome Initiative[47], Japan, in 2015, launched the Materials Research by Information Integration initiative[48], and the European Union in 2015 launched the Novel Materials Discovery (NOMAD) Laboratory[49]. They share a common conviction that data-driven research will help uncover structures and correlations which will lead to the discovery of new and novel materials and will help to reveal new properties of known materials.

This data-driven approach to material development, especially polymers, is still in its infancy. Its application to coatings, adhesives and sealants and specifically to polyurethane formulations, are embryonic. A case in point: *Winkler* et al.[50], in 2012, reviewed quantitative structure-property relationships modeling for diverse materials, among these polymers, no mention, however of polyurethanes, much less of coatings, adhesives or sealants. So to speak of successes of a data-driven approach to polyurethane coatings, adhesives and sealants, at this point, is premature, although inevitable. What is not premature is the need to address many of the issues and challenges that data-driven development, and the digitalization of coatings, will bring.[51] To set the stage for a review of the challenges ahead, first, a definition: Data-driven approach to material development is known as quantitative structure-property relation (QSPR).[52, 53] This is when subject matter expert trains algorithms, using appropriate data to makes an accurate (enough) prediction of the performance of a yet to be made formulation, or yet untested properties of existing formulations. To train algorithms, the expert requires both detailed, accurate and relevant digital descriptors of the coating structure at the nano-, micro-, and millimeter scale, and detailed, accurate, reproducible performance measurements, properties, of the same polymers.

The issues that need to be addressed for a successful data-driven coatings development program fall into three categories:

1. data quality (and quantity),
2. data scope, and
3. model building.

It is important to note that not only quantitative descriptors of the ingredients, but also their order of addition, mixing parameters, thermal history, and any other subtle yet important determinants of the polymers structure should be collected.[48]

Likewise, the quantitative units that measure performance must be measured with high accuracy and reproducibility, the method of measurement documented as well as all relevant information. Even with careful documentation of all important contextual information, the challenge remains for many labs in assembling all the data efficiently. Fortunately, the solution to this challenge is the main topic of this chapter, automated high-throughput testing, particularly the workflow described in the last paragraph of Chapter 9.1. A full-scale automated assembly line that blends the formulations, applies them to substrates, prepares the samples, and finally tests them, assures that all the data is standardized, reproducible, and documented. Furthermore, with automated testing, new formulations can be made and tested 24/7, so that both high quality and quantity[54] of data can be assured. Only with such standardization and documentation can a data library serve to build on the promises of data-driven development of coatings, adhesives and sealants.

The second issue is data scope. The data used to train a predictive model must define a technology space unambiguously. The answer to the question a study sets out to uncover should be interpolated from the data not extrapolated.[55] Therefore, a simple brute-force combinatorial approach will not suffice when building a database of descriptors and their properties. Statistically designed experiments are required.[56] Although these techniques were originally designed for agricultural studies, techniques have been developed and extended to most fields[57, 58] and have found extended use in coatings[59–61]. A statistically correct designed experiment will assure that conclusions are significant, robust and interpretable, and the risk of your model contaminated by unidentified factors minimized.

The third important issue in building a data-driven development program is the area of selecting and training machine learning algorithms. To follow best practices is of critical importance. The most common danger is overfitting, that is to say, that the prediction model can predict the results of the coatings used to train it but fails in predicting new data. Because this topic falls outside of the focus of this book, see e.g. *Kuhn* and *Johnson's* book "Applied Predictive Modeling" [55].

9.4 Outlook

The need to access more data in a data-driven economy will foster a continuous implementation of high throughput-testing and combinatorial approaches in experimentation. The industry will continue to move towards more sustainable, environment and climate-friendly solutions and the efficiency requirements in product development will continue to rise as well. Polyurethane chemistry is the most versatile and performant materials technology and is therefore predestined to be beneficially investigated with such a data-driven approach. Therefore, it is expected to see more studies from materials manufacturer as well as in the formulating industry and lab automation will continue to widen its capabilities supported by machine builders that see customization as a key part of their offering.

9.5 References

[1] J. R. Broach and J. Thorner, Nature, 384 (6604), p. 14–16 (1996)

[2] W. F. Maier, K. Stöwe and S. Sieg, Angewandte Chemie International Ed., 46, p. 6016–6067, 2007

[3] X.-D. Xiang, X. Sun, G. Briceno, Y. Lou, K.-A. Wang, H. Chang, W. G. Wallace-Freedman, S.-W. Chen, P. G. Schultz, Science, 268, p. 1738–1740, 1995

[4] N. Gruber, Ch. Jung, W. Schrof, Farbe&Lack, 113 (11), pp. 20, 22, 24–26, 2007

[5] T.-Ch. Kuo, N. A. Malvadkar, R. Drumright, R. Cesaretti, and M. T. Bishop, ACS Combinatorial Science, 18 (9), p. 507–526, 2016

[6] J. D. Hewes, Economic Impact of Combinatorial Material Science on Industry and Society, In High-Throughput Analysis, A tool for combinatorial Materials Science, Eds. R. A. Potyrailo and E. J. Amis, Kluwer Academic, New York, 2003

[7] US 2003138025 (2003) Symyx Technologies

[8] D. Gagro, High volume, high quality – Thanks to HTS, European Coatings Journal, 11, p. 10–13, 2014

[9] B. Reckter, VDI Nachrichten, 23, p. 10–11, 2018

[10] Coatings Xperience – Accelerating Coatings Development with High Throughput Technology, American Coatings Society, Paint. Org, Coatings Tech, 19–21, August 2017

[11] G. Meichsner, R. Hiesgen, T. Etzl, M. Vanderhoeven, T. Brinz, Color Res Appl. 1–10, 2018

[12] R. Potyrailo, K. Rajan , K. Stoewe, I. Takeuchi, B. Chisholm and H. Lam, ACS Combinatorial Science, 13 (6), p. 579–633, 2011

[13] H. Bach, Ch. Gürtler, S. Nowak, Farbe&Lack, 108, p. 30, 2002

[14] L. I. Majoros, B. Dekeyser, R. Hoogenboom, M. W. M. Fijten, J. Geeraert, N. Haucourt, U. S. Schubert, Journal of Polymer Science, Part A: Polymer Chemistry 48(3), p. 570–580, 2010

[15] M. J. Nasrullah, J. A. Bahr, C. Gallagher-Lein, D. C. Webster, R. R. Roesler, P. Schmitt, ICE 2007, International Coatings Expo: Clean-Lean-Green: Innovative Solutions for the Global Coatings Community, Toronto, ON, Canada, Oct. 3–5, 75/1–75/28, 2007

[16] J. K. Oh, B. Erdem, J. Anderson, K. Nanjundiah, J. Sweeney, Proceedings of the International Waterborne, High-Solids, and Powder Coatings Symposium 37th, p. 83–95, 2002

[17] J. K. Oh, J. Anderson, B. Erdem, R. Drumright, G. Meyers, Progress in Organic Coatings, 72(3), p. 253–259, 2011

[18] Test method for optimizing film forming polymers, PCT Int. Appl. WO 02/42745 A2 20020530, 2002

[19] P. Majumdar, D. A. Christianson, D. C. Webster, PMSE Preprints 90, p. 799–800, 2004

[20] J.-L. Sormana, J. C. Meredith, Macromolecular Rapid Communications 24(1), p. 118–122, 2003

[21] J.-L. Sormana, J. C. Meredith, Macromolecules 37(6), p. 2186–2195, 2004

[22] B. Chisholm, R. Potyrailo, R. Shaffer, J. Cawse, M. Brennan, C. Molaison, Progress in Organic Coatings 47(2), p. 112–119, 2003

[23] R. A. Potyrailo, B. J. Chisholm, D. R. Olson, M. J. Brennan, C. Molaison, Proceedings

of SPIE-The International Society for Optical Engineering 4578 (Fiber Optic Sensor Technology and Applications 2001), p. 145–152, 2002

[24] E. Fadel, S. K. Ghosh, J. Oomen, FATIPEC Congress 29th, p. 146–151, 2008

[25] J. He, J. Bahr, B. J. Chisholm, J. Li, Z. Chen, S.N. Balbyshev, V. Bonitz and G.P. Bierwagen, Journal of Combinatorial Chemistry, 10 (5), p. 704–713, 2008

[26] J.-L. Sormana, J. C. Meredith, Materials Research Innovations, 7(5), p. 295–301, 2003

[27] F. Casse, S.J. Stafslien, J. A. Bahr, J. Daniels, J. A. Finlay, J. A. Callow, M. E. Callow, Biofouling 23(1/2), p. 121–130, 2007

[28] S. Stafslien, J. Daniels, B. Mayo, D. Christianson, B. Chisholm, A. Ekin, D. Webster, G. Swain, Biofouling 23(1/2), p. 45–54, 2007

[29] E. Ribeiro, S. J. Stafslien, F. Cassé, J. A. Callow, M. E. Callow, R. J. Pieper, J. W. Daniels, J. A. Bahr and D. C. Webster, J. Comb. Chem. 10 (4), p. 586–594, 2008

[30] S. J. Stafslien, D. Christianson, J. Daniels, L. Vander Wal, A. Chernykh, B. J. Chisholm, Biofouling 31(2), p. 135–149, 2015

[31] A. Ekin, D. C Webster, Abstracts of Papers, 230th ACS National Meeting, Washington, DC, United States, Aug. 28–Sept. 1, PMSE-276, 2005

[32] A. Ekin, D. C. Webster, J. W. Daniels, S. J. Stafslien, F. Casse, J. A. Callow, M. E. Callow, J. Coat. Technol. Res., 4 (4) p. 435–451, 2007

[33] A. Ekin, D. A. Christianson, D. C. Webster, PMSE Preprints 93, p. 896–897, 2005

[34] R. J. Pieper, A. Ekin, D. C. Webster, F. Cassé, J. A. Callow, M. E. Callow, J. Coat. Technol. Res., 4 (4) p. 453–461, 2007

[35] A. Ekin and D. C. Webster, J. Comb. Chem. 9, p. 178–188, 2007

[36] R. B. Bodkhe, S. E. M. Thompson, C. Yehle, N. Cilz, J. Daniels, S. J. Stafslien, M. E. Callow, J. A. Callow, D. C. Webster, J. Coat. Technol. Res., 9 (3) p. 235–249, 2012

[37] R. B. Bodkhe, S. Thompson, M. E. Callow, D. C. Webster, PMSE Preprints 99, p. 376–377, 2008

[38] P. Majumdar, E. Lee, N. Patel, K. Ward, S. J. Stafslien, J. Daniels, B. J. Chisholm, P. Boudjouk, M. E. Callow, J. A. Callow & S.E.M. Thompson, Biofouling, 24:3, p. 185–200, 2008

[39] S. Venkateswaran, M. Wu, P. J. Gwynne, A. Hardman, A. Lilienkampf, S. Pernagallo, G. Blakely, D. G. Swann, M. P. Gallagher, M. Bradley, Journal of Materials Chemistry B: Materials for Biology and Medicine 2(39), p. 6723–6729, 2014

[40] M. S. Lowry, D. R. Hubble, A. L. Wressell, M. S. Vratsanos, F. R. Pepe, C. R. Hegedus, Journal of Coatings Technology and Research 5(2), p. 233–239, 2008

[41] C. Gambino, R. Subramanian, M. Dvorchak, FutureCoat Proceedings, New Orleans, LA, United States, Nov. 1–3, 073/1–073/7, 2006

[42] A. Chiche, W. Zhang, C. M. Stafford, A. Karim, Measurement Science and Technology 16 (1), p. 183–190, 2005

[43] A. M. Forster, C. M. Stafford, A. Karim, E. J. Amis, PMSE Preprints 88, p. 490–491, 2003

[44] B. Steinhoff, Adhäsion Kleben & Dichten 60 (4), p. 46–47, 2016

[45] Y.-S. Hwang, J. Kim, H. J. Yoon, J. I. Kang, K.-H. Park, H. Bae, Macromolecular Research 26(1), 1–8, 2018

[46] A. Orefice, G. Mancusi, V. P. Berardi, L. Feo, G. Zuccaro, Composites Part B 144, p. 237–253, 2018

[47] Material Genome Initiative, www.mgi.gove (Accessed March 2018)

[48] Materials Research by Information Integration, www.nims.go.jp/MII-I/en/ (Accessed March 2018)

[49] The NOMAD Laboratory: a European Centre for Excellence, http://nomad-coe.eu/ (Accessed March 2018)

[50] T. Le, C. Epa, F. R. Burden and D. A. Winkler, Quantitative Structure-Property Relationship Modeling of Diverse Materials Properties, Chem. Review, 112, p. 2889–2919, 2012

[51] D. J. Audus and J. J. De Pablo, AACS Maro Lett. 6, p. 1078–1082, 2017

[52] T. Fujita, D. A. Winkler, J. Chem. Inf. Model. 56, p. 269–274, 2016

[53] N. Adams, P. Murray-Rust, Macromolecular Rapid Communications 29, p. 615–632, 2008

[54] A. Halevy, P. Norvig, F. Pereira, The Unreasonable Effectiveness of Data. Intelligent Systems, IEEE. 24.08–12.10.1109/MIS.2009.36, 2009

[55] M. Kuhn, K. Johnson, Applied Predictive Modeling. 1_{st} ed., New York, 2013

[56] G. E. P. Box, J. S. Hunter, W. G. Hunter, Statistics for Experimenters: Design, Innovation, and Discovery, 2nd Edition, Wiley-Interscience, 2005

[57] J. Cornell, Experiments with Mixtures: Designs, Models, and the Analysis of Mixture Data. Wiley, 2002

[58] J. N. Cawse, Acc. Chem. Res. 34, p. 213–221, 2001

[59] R. Snee, R. Hoerl, Strategies for Formulations Development: A Step-by-Step Guide Using JMP. SAS Institute, 2016

[60] A. Rössler, Design of Experiments for Coatings, Vincentz Network, Hanover, 2014

[61] P. Whitcomb, M. Anderson, Design of Experiments for Coatings. in Coatings Technology Handbook, 3rd Edition (ed. A. Tracton) 15-1–15-6, CRC Press, 2005

[62] Robert Bosch Packaging Technology GmbH

[63] Chemspeed Technologies AG

10 Occupational hygiene in the manufacture and processing of PU systems

10.1 Occupational health and safety

10.1.1 Monomeric and polymeric isocyanates

Typical starting materials used to produce higher molecular weight polyurethane coatings and adhesives raw materials are toluene diisocyanate (TDI), 1,6-diisocyanatohexane (hexa-methylene diisocyanate, HDI), isophorone diisocyanate (IPDI), diphenylmethane diisocyanate (MDI), 4,4'-dicyclohexylmethane diisocyanate ($H_{12}MDI$), pentamethylene diisocyanates (PDI) and xylylen diisocyanates (XDI).

Overexposure to monomeric diisocyanates can cause skin, eye, nose, throat, and lung irritation. They also can lead to skin or respiratory sensitization. A sensitized individual may react to future exposures of diisocyanates at very low levels, even below the levels that might be considered safe for others. Chronic exposure to monomeric diisocyanates at levels above the exposure limits has been reported to result in diminished lung function. Exposure of the skin could result in tanning and irritant effects. Skin sensitization and the occurrence of allergic symptoms (e.g. reddening of the skin, swelling and inflammation) are also possible.

Further, animal tests and other research indicate that skin contact with diisocyanates can play a role in causing isocyanate sensitization and respiratory reaction.

With the exception of MDI, which is accorded special status on account for its low vapor-pressure, other monomeric diisocyanates have a significant volatility. Table 10.1 shows vapor-pressures and other physical data of the monomeric diisocyanates named above.

For occupational health reasons, monomeric diisocyanates are generally not used as raw materials for coatings, adhesives and sealants. To overcome the hazardous potential related to monomeric diisocyanates and to achieve low volatility, they must first be converted

Table 10.1: Physical data of selected monomeric diisocyanates

Starting material	Type	Molecular weight	Boiling point [°C/mbar]	Melting point [°C]	Vapor pressure at 25 °C [mbar]	Saturation concentration at 25 °C [ppm]	[mg/m³]
TDI (isomer mixture)	Aromatic	174.1	252–254/ 1013	< 14	2.5×10^{-2}	24.3	176
HDI	Aliphatic	168.2	255/1013 112/7	-67	1.4×10^{-2}	13.6	95
IPDI	Cyclo-aliphatic	222.3	158/13	Approx. -60	1.2×10^{-3}	1.2	10.8
MDI	Aromatic	250.3	> 350/1013	39–43	$< 1 \times 10^{-5}$	< 0.01	< 0.1
H_{12}MDI	Cyclo-aliphatic	262.4	Decomp./ 1013	Approx. 25	2.1×10^{-5}	0.02	0.23

TDI = Toluene diisocyanate, HDI = 1,6-Hexamethylene diisocyanate, IPDI = Isophorone diisocyanate, MDI = 4,4'-Diphenylmethane diisocyanate, H_{12}MDI = 4,4'-Dicyclohexylmethane diisocyanate;

into higher molecular weight polyurethane hardeners, using suitable modification reactions. Due to the irritating properties of monomeric diisocyanates, efforts are made during manufacture to ensure the lowest possible residual monomer content in polyurethane hardeners for coatings, adhesives and sealants. This applies particularly to products based on TDI, HDI and IPDI because of the relatively high volatility of these monomers.

Classification and labeling

European Union

On January 20th, 2009 the Regulation (EC) No. 1272/2008 on the classification, labelling and packaging of substances and mixtures (CLP)[1] entered into force. The CLP Regulation aligns existing EU legislation to the United Nations Globally Harmonized System (GHS) of Classification and Labelling of Chemicals. Different to GHS, CLP considers also EU harmonized classification of substances according to CLP Regulation, Annex VI, Part 3, Table 3.1.

Regulation (EC) No. 1272/2008 (CLP) is primarily aiming at the self-classification of substances and mixtures by the industry, which places these chemicals on the market. In general, classification is based on test data.

If a substance has a harmonized classification in accordance with CLP Regulation Annex VI, application of this classification is mandatory. As outlined in CLP Annex VI Part 1, "for certain hazard classes, including acute toxicity ..., the classification according to the criteria in Directive 67/548/EEC [author's note: formerly also known as Dangerous Substances Directive] does not correspond directly to the classification in a hazard class and category under this Regulation [author's note: i.e. CLP regulation]. In these cases, the classification in this Annex shall be considered as a minimum classification." Such a minimum classification is marked with an "*". If there is information available that leads to classification in a more severe category compared to the minimum classification, then classification in the more severe category must be applied.

Further, if the substance also meets the criteria for hazard classification, which are not taken into account by the harmonized classification, the harmonized classification should be supplemented by the classifications in the form of a self-assessment.

Table 10.2 summarizes the classification and labelling of common diisocyanates according to CLP. Precautionary statements, which are also part of the labelling, are not displayed in the table.

Typically, polyisocyanates are classified: Sensitization of the skin, category 1 (H317) (i.e. May cause an allergic skin reaction). Specific polyisocyanates may bear additional classification and labelling due to irritating properties and/or harmful effects to the environment. Labelling of commercial grades of polyurethane hardeners is further influenced by differences in the level of residual monomers or by supply in solvents. As an example, Table 10.3 (see page 402) displays the typical classification and labelling of an HDI-based polyurethane hardener.

For mixtures containing isocyanates, CLP Regulation requires a supplementary labelling element:

EUH204 Contains isocyanates. May produce an allergic reaction. The PU industry is committed to a high safety culture to prevent workers from sensitization, who deal with diisocyanates or polyurethane hardeners, which contain small amounts of residual monomers. Represented through the trade associations ISOPA and ALIPA safety initiatives like "walk the talk" (2006)[2] and "alipa safeguard we care that you care" (2010)[3] were successfully launched and helped already to reduce the number of cases significantly. Both campaigns raised the awareness of workers dealing with hazardous substances and refining measures to avoid the contact with diisocyanates and polyurethane hardeners via skin and breathing system. Nevertheless, a few cases of sensitization still occur every year.

In order to further increase the level of occupational health and safety for people working with diisocyanates and diisocyanates containing mixtures, the European competent authorities are in the process to launch a REACH use restriction for such products with a cumulated content of diisocyanates equal or above 0.1 % throughout the whole value chain, which limits the use of such products in industrial and professional applications to those cases where a combination of technical and organizational measures as well as a minimum standardized training package have been implemented. This measure will ensure that only trained people work with diisocyanates or diisocyanates containing mixtures that are well aware of the hazards and the suitable measures to avoid exposure. To further improve the industrial hygiene standard, some selected raw material suppliers already offer ultra-low monomer containing polyurethane hard-

Table 10.2: Classification and labelling of diisocyanates according to CLP (classification, labelling and packaging of substances and mixtures)

Diisocyanate		**HDI**	**IPDI**	**H_{12}MDI**	**TDI**	**MDI**
Classification	Acute toxicity, oral, category 4 (H302)	X				
	Acute toxicity, inhalative, category 1 (H330)	X	X		X	
	Acute toxicity, inhalative, category 2 (H330)			X		
	Acute toxicity, inhalative, category 4 (H332)					X
	Skin irritation, category 2 (H315)	X	X	X	X	X
	Eye irritation, category 2 (H319)	X	X	X	X	X
	Sensitization of the respiratory airways, category 1 (H334)	X	X	X	X	X
	Sensitization of the skin, category 1 (H317)	X	X	X	X	X
	Carcinogenicity, category 2 (H351)				X	X
	Specific target organ toxicity (single exposure), category 3 (H335)	X	X	X	X	X
	Specific target organ toxicity (repeated exposure), inhalative, category 2 (H373)					X
	Chronically hazardous to the aquatic environment, category 2 (H411)		X			
	Chronically hazardous to the aquatic environment, category 3 (H412)				X	
Labellingpictogram						
Signal word		Danger	Danger	Danger	Danger	Danger

Table 10.2: Continue

Diisocyanate		HDI	IPDI	$H_{12}MDI$	TDI	MDI
Hazard Statement	H302: Harmful if swallowed	X				
	H330: Fatal if inhaled	X	X	X	X	
	H332 Harmful if inhaled					X
	H315: Causes skin irritation.	X	X	X	X	X
	H319: Causes serious eye irritation	X	X	X	X	X
	H334: May cause allergic or asthma symptoms or breathing difficulties if inhaled	X	X	X	X	X
	H317: May cause an allergic skin reaction	X	X	X	X	X
	H335: May cause respiratory irritation	X	X	X	X	X
	H373 May cause damage to organs through prolonged or repeated exposure if inhaled					X
	H351 Suspected of causing cancer.				X	X
	H411: Toxic to aquatic life with long lasting effects		X			
	H412 Harmful to aquatic life with long lasting effects.				X	

eners with a cumulated content of diisocyanates below 0.1 %. This will help to further reduce the number of sensitization cases and by the same time, keep the superior technical performance of polyurethane technology available in the European market.

United States of America

In the United States, labelling requirements are outlined by the Occupational Safety and Health Administration (OSHA) in HazCom 2012 (29 CFR 1910.1200). The standard does not dictate classifications, rather it provides the criteria for how to classify chemicals according to the UN Globally Harmonized System of Classification and Labelling of Chemicals (GHS) as adopted in the US.

In the US, all isocyanates are classified as irritating to the respiratory system. MDI and $H_{12}MDI$, and TDI are classified as irritating to the skin and eyes, while HDI and IPDI are classified as corrosive to the skin and eyes.

Table 10.3: Classification and labelling according to CLP of "Desmodur" ultra N 3300 as an example for a common aliphatic polyurethane hardener[21]

Hexamethylene-1,6-diisocyanate homopolymer (oligomerisation product of isocyanurate type) residual monomer < 0.1 % HDI		
Classification		
Acute toxicity, inhalative, category 4 (H332)		
Sensitization of the skin, category 1 (H317)		
Specific target organ toxicity (single exposure), category 3 (H335)		
Labelling		
Pictograms	Signal word	Hazard statement
	Warning	H317 May cause an allergic skin reaction H332 Harmful if inhaled H335 May cause respiratory irritation

In the US, the American Coatings Association (ACA) is one resource for downstream users of coatings products. They have developed a manual (ACA's "HMIS" Implementation Manual) for converting a GHS classification to a workplace "HMIS" label designation, among other topics.

Similar to the EU, in the US the PU Industry is also committed to a high safety culture to ensure workers are informed on the safe use and handling of diisocyanates or polyurethane hardeners, which contain small amounts of residual monomers. The American Chemistry Council's Aliphatic and Aromatic Diisocyanate Panels provide literature and information for downstream users of these technologies. They also promote outreach and awareness to workers and the general public on working with polyurethane chemistries. These trade associations have also partnered with OSHA to create a voluntary alliance, which provides members, occupational physicians, stakeholders, and others within the polyurethanes value chain with information, guidance, and access to training resources that will help them further protect the health and safety of workers.[4]

The labelling regulations for chemicals differ widely around the world. With the introduction of the Globally Harmonized System of Classification and Labelling of Chemicals (GHS), the hope was that labelling would be more harmonized around the world. However, the reality is that while it is closer aligned than it was previously, each region has varying requirements which could cause a product to be labelled differently in different regions. The benefit is that the base building blocks are the same, so recognition and understanding of the general system of pictograms and hazard statements is now more uniform on the global scale.

China

China has fully implemented GHS for all hazardous substances and mixtures from May 1st, 2011. Decree 591 requires manufacturers of hazardous chemicals to classify, label and package chemicals as well as prepare safety data sheets (SDS) in accordance with relevant national standards. Decree 591 also requires distributors to provide SDSs and labels if selling hazardous chemicals. In addition, GHS Safety data sheets (SDS) and labels are required by new substance notification under Ministry of Ecology & Environment (MEE) Order 7 and HazChem registration under Ministry of Emergency Management (MEM) Order 53.

Chemical classification criteria and the content and format of SDSs and labels are set by several compulsory national standards (prefixed with "GB") and recommended national standards (prefixed "GB/T") issued between 2008 and 2013. In 2013, the Standardization Administration of China (SAC) issued 28 compulsory chemical classification standards (GB 30000.2-2013 to GB 30000.29-2013 Safety rules for classification and labelling of chemicals) and each standard corresponds to one hazard class under GHS. Those standards are fully aligned with UN GHS Rev. 4 and have adopted all building blocks under UN GHS Rev. 4.

For chemicals listed in the Catalogue of Hazardous Chemicals, Ministry of Emergency Management (MEM) has published harmonized chemical classifications of those chemicals. Industry must use the classifications given in the guidance or more severe ones to classify their chemicals and prepare SDSs and labels.

In this context, it should also be emphasized that all kind of reactive chemistries and even solvents and additives used for coatings and adhesives bear a risk of negative health effects if workers dealing with such products do not implement proper measures to protect themselves from skin contact or inhalation. Depending on technical and organizational measures in place, the personal protection equipment used by the workers dealing with chemically different products should typically look very similar.

Exposure limits

The European Union and its member states define maximum workplace exposure limits for hazardous gases, vapors and, increasingly, also for substances that form dust. Such a limit gives the maximum permissible concentration of a substance in ppm (i.e. milliliters of gas or vapor per cubic meter of air) or, in the case of suspended particles, in mg/m^3 which, in the light of available knowledge, does not constitute a health risk or unreasonable burden during long-term work with the product, even for many years, based on an average for a working period of an eight-hours working day and up to 42 hours per week. The exact definition of the exposure limit can vary from country to country. In addition to the eight-hour limit, there are also short-term exposure limits, which restrict the maximum exposure levels, e.g. in discontinuous processes.

Table 10.4: Typical occupational exposure limits for selected diisocyanates

Diisocyanate	**Occupational exposure limit** [ppm]	[mg/m³]
Toluene diisocyanate (TDI)	0.005	0.035
Hexamethylene diisocyanate (HDI)	0.005	0.035
Isophorone diisocyanate (IPDI)	0.005	0.046
4,4'-Dicyclohexylmethane diisocyanate ($H_{12}MDI$)	0.005	0.054
4,4'-Diphenylmethane diisocyanate (MDI)	0.005	0.050

While a majority of European states have specified individual workplace exposure limits for monomeric diisocyanates, some states in the Asia Pacific region decided to follow a more generic approach and specify limit values based on mg NCO groups per m^3 (see Table 10.4).

For exposure assessment, some countries apply the generic approach for specifying workplace exposure limits on basis of NCO groups to polyurethane hardeners. However, this method only takes into account the significant differences in the toxicological properties of high molecular weight oligomeric or polymeric polyurethane hardeners, compared to those of monomeric diisocyanates. A possible consequence is an unjustified overestimation of the risk of exposure to polyurethane hardeners.

In the course of the REACH regulation, various polyurethane hardeners are registered as substances. Toxicological data available for these substances allow establishing so called "derived no effect levels" (DNEL) which can act as limit values in exposure assessments.

In practice, employees are usually not exposed to single substances but to mixtures, e.g. different paint components and impurities. Per definition, occupational exposure limits are valid only for exposure to single substances. Their suitability for the toxicological assessment of mixtures is limited, as mixture components can strengthen or weaken each other. The model of an evaluation index (IEL) is often used as a basis assessment of mixtures (Figure 10.1). The IEL is defined as the sum of the quotients calculated from the measured air concentrations (C_i) of the individual substances and the corresponding exposure limits (EL_i) and should be reliably below 1.

$$I_{EL} = \frac{C_1}{EL_1} + \frac{C_2}{EL_2} + \frac{C_3}{EL_3} + \ldots\ldots\ldots + \frac{C_i}{EL_i}$$

Figure 10.1: Evaluation index

Isocyanate analysis

Derivatization methods are primarily used to determine the isocyanate concentration in the air at the workplace. In these methods, a suitable reagent, generally an amine is used to convert the isocyanate into a stable derivative that can be identified and quantified by means of chromatography. The trapping reagent is present as a solution either in a wash bottle (impinger) or on impregnated glass fiber filters. A typical method following the requirements of DIN EN 482 is the accredited method no. 7120 from BGIA[5] which is used for the detection of e.g. TDI, MDI, HDI, IPDI and PDI – an exemplary selection of diisocyanates industrial used. The method allows measuring the mentioned aliphatic and aromatic diisocyanates in the air as gas or dust as a stationary or person related measurement:

- Using a suitable pump, air is drawn through two glass fiber filters impregnated with 1-(2-methoxyphenyl)piperazine.
- After the test period, the glass fiber filters are extracted with a solution of 1-(2-methoxyphenyl)piperazine in acetonitrile.
- Quantitative analysis is done using high performance liquid chromatography (HPLC). The diisocyanates are detected with diode array or fluorescence.
- Measurement uncertainty: 20 %
- Detection limit, absolute: 0.0005 mg

Workplace measurements which meet the requirements of occupational safety and health regulations are typically performed by accredited laboratories. The mutual recognition of accreditations among member states warrants the acceptance of conformity assessments conducted by these laboratories around the world.

Depending on the targeted application, it should be considered as well that polyurethane hardeners also may contain traces of processing chemicals, like catalysts, stabilizers or even residues of possible side reactions, which might be relevant to decide about the suitability of a product for a specific use. Responsible suppliers will provide their customers with the relevant information to help them to decide on their specific use cases.

10.1.2 Co-reactants for polyurethane hardeners

The standard co-reactants for polyurethane hardeners are polyols such as polyester, polyacrylate or polyether polyols that may be free of intentionally added solvents or dissolved in solvents commonly used in coatings and adhesives or which are supplied in the form of an aqueous dispersion. Such raw materials are often not classified as hazardous according to the CLP classification criteria. If supplied in solvents, then classification of the coatings or adhesives raw materials is typically related to the solvent content of the product.

For the United States, the classification and labelling of the co-reactants is governed by OSHA HazCom2012, the same standard as for the isocyanates.

In any case safety data sheets (SDS) for co-reactants should be reviewed prior to use.

10.1.3 Processing of polyurethane coating, adhesive and sealant systems

General protective measures

Independent of whether national legislation requires an assessment of the workplace and documentation of the results, each workplace should be investigated and assessed in terms of present or potential hazards to the skin and respiratory tract. Valuable guidance for individual activities in this regard is offered by the exposure scenarios described in the extended safety data sheets for the individual products.

The suitability of the various options for ensuring the best possible respiratory protection depends on the object. In accordance with EU Directive 98/24/EC[6], the employer has the obligation to ensure that the risk to the health and safety of the employee from a hazardous chemical substance is eliminated or reduced to the minimum. The directive specifies a ranking of the protective measures. The use of personal protection equipment should only be mandatory if other measures such as the appropriate design of the workplace, extraction and ventilation are not sufficient to protect employees from contact with hazardous substances. It goes without saying that spray coating operations, where excessive ventilation would be counter to the aim of the process, should be performed wearing personal protective equipment (see Figure 10.2).

The conditions governing the suitability of the individual safety measures and types of equipment are described in the information provided by the manufacturers of respiratory equipment.

In addition, the following requirements must be considered when handling reactive polyurethane coating materials:

- In addition to following the general safety measures appropriate for coatings and adhesive applications, it must be ensured that people with allergies – especially asthmatics – and those who suffer from bronchial catarrh or other chronic respiratory diseases are not involved in work that might additionally impair their respiratory function. People with allergies must not be exposed to reactive polyurethane coatings. If there are doubts as to the suitability of an individual to work with these coatings, a medical examination must be required. If there are any problems, a physician must be consulted.
- In the event of coughing, pressure in the chest or asthma-like symptoms during or after working with the coatings, avoid any further exposure and consult a physician.
- Safety glasses are mandatory. If, nevertheless, spray mist gets into the eyes, rinse immediately with plenty of water and then consult a physician.
- Skin contact with reactive polyurethane coatings has to be avoided. Suitable protective clothes and protection gloves have to be worn when working with such coatings as recommended in the product safety data sheets.

- To avoid exposure to solvents, freshly coated objects should be left to dry in well-ventilated rooms.
- When baking thermal curing polyurethane coatings, the gases from the baking oven must be extracted to prevent the build-up of harmful vapors in the work area. If necessary, the exhaust gases from the oven must be scrubbed in accordance with the air pollution control regulations (e.g. GMBl. No. 25 to 29, p. 511[7]).

Suitable respiratory protection includes half-masks with replaceable filter cartridges and air supplied hoods.

Suitable overalls closed at the neck and wrists and gloves are strongly recommended to prevent skin contact with coating products. Suitable materials for gloves as well as

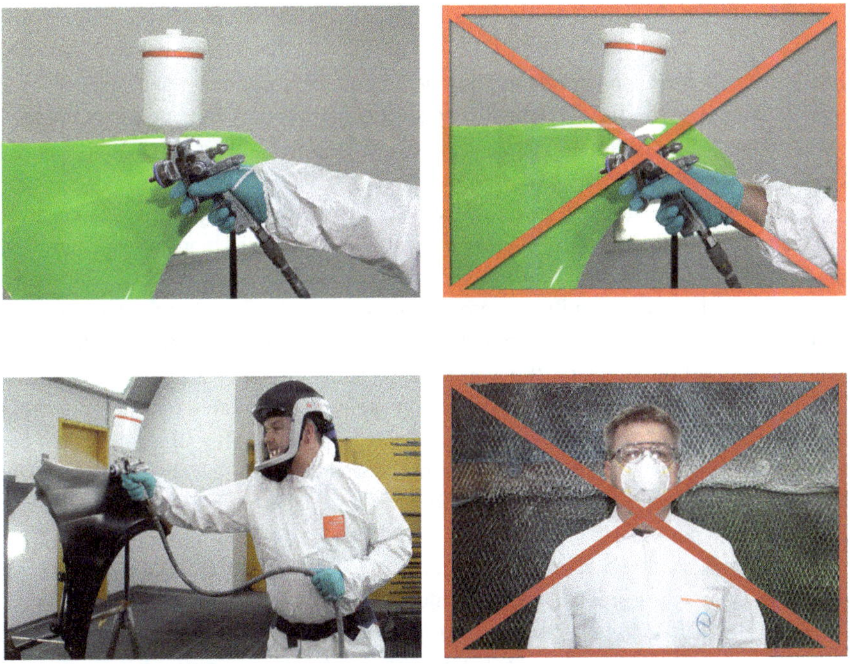

Figure 10.2: Protective measures during coating and adhesive spray application

Table 10.5: Occupational exposure limits for solvents and dusts in European countries (2018)

Substance	Occupational exposure limit [mg/m³]
Xylene	108 – 440
Ethylbenzene	20 – 442
C3-Alkyl benzene (cumene)	50 – 245
Ethyl acetate	500 – 1461
Butyl acetate	300 – 724
2-Butoxy ethanol	49 – 120
1-Methoxy-2-propyl acetate	270 – 550
Methyl isobutyl ketone	80 – 410
General dust exposure limits	
Inhalable dust	10
Respirable dust	1.25 – 5

first-aid measures are described in the material SDS for the individual products.

Spray application

When spraying polyurethane coating and adhesive systems, a number of hazardous circumstances must be considered. First, volatile components such as solvents and small amounts of residual monomers are present in much higher concentrations in the air during a spray operation as compared to that of simple evaporation. Additionally, the non-volatile components such as polyols, polyurethane hardeners, pigments, extenders and auxiliaries, are distributed finely in the air in the form of an aerosol (spray mist). Depending on the spraying conditions, a proportion of the coating droplets are smaller than the threshold particle size of 7 µm assumed for lung penetration.[8, 9]

In response to the fact that respirable aerosols, i.e. those that can penetrate the alveoli of the lungs, and even inert fine dusts represent a health hazard, general exposure limits have been defined for inhalable and respirable dusts respectively. Table 10.5 summarizes the occupational exposure limits for some common solvents as well as the particulate exposure limits. Because of varying limits in the individual European states, the table gives ranges of the limiting values for the workplace for individual substances.

In the assessment of hazard potential, inhalable and respirable particles are ranked higher than solvents. The hazard to spray applicators from solvents and spray mists is common to all coating systems, independent from the chemical nature. Polyurethane coatings present a specific potential hazard in the form of residual monomeric diisocyanates and their sensitizing potential.

Even in case of water-borne coatings, inhalation of spray mist presents a risk. This means that personal safety measures must be in place, even when spray-applying water-borne systems.

Due to the many different components in spray mists, it is in general not yet possible to quantify concentrations of spray mists that may be harmful to man. It is therefore essential to require the best possible protection against the inhalation of spray mists. This requirement applies not only to spray mists of two-component polyurethane coatings, but also for spray mists derived from all coatings, even if the hazard potential of the different coating systems varies.

Water-borne one-component and two-component reactive systems

It has already been stated that water-borne reactive polyurethane systems in terms of occupational hygiene should be treated in the same way as solvent-borne systems. Even though the risk presented by solvents has been eliminated or at least minimized (most water-borne coating systems contain a low percentage of co-solvents), some monomeric diisocyanates and polymeric polyurethane hardeners still present a health risk, in the same way as they do in organic media. Even when using a water-borne system, workers dealing with such materials still may need personal protection equipment (PPE). Reference to the product SDS for appropriate information has always to be taken.

Baking urethane resins

A special type of polyurethane coating system contains thermally activated polyurethane hardeners. These are basically polyisocyanate resins where the isocyanate groups are chemically converted into less hazardous compounds like urea, urethanes or amides. Other than polyisocyanates, such resins are inert and non-reactive at room temperature.

The chemical conversion of the isocyanate groups is typically done by using monofunctional compounds which contain a reactive hydrogen atom in the molecule and react with isocyanates, e.g. ε-caprolactam, butanone oxime, 3,5-dimethylpyrazol or diethyl malonate. The resins can be combined with suitable co-reactants to formulate baking coatings with good storage stability. Even true one-component baking coating resins with good long-term storage stability are feasible. Some resins also contain both, hydroxyl groups and inactivated isocyanate groups in the same molecule, which eliminates the risk of mixing errors at the formulator.

During thermal curing of such coating systems, e.g. can coating, coil coatings, industrial baking coatings, automotive primer surfacers and stone-chip primer surfacers, a small amount of the blocking agent is released. It should be considered that traces of the monomeric diisocyanates, on which the product is based, may be found in gases emitted from the baking oven. In terms of occupational hygiene and air pollution control, gases from the oven have to be extracted to prevent a build-up of hazardous substances in the air near the oven. Conformity with emissions control legislation must be ensured by an adequate air supply to the oven and the implementation of any necessary waste gas purification measures, e.g. incineration or absorption.

10.1.4 Spill response, handling containers and waste disposal

In preparation for accidental spills, it is advisable to have a written procedure for dealing with such an emergency and a trained emergency response team in place. A number of factors will affect the extent of hazard associated with a spill: the volume of the liquid spilled, volatility and flammability of the material, temperature of the material and location of the spill.

Anticipating how, when and where spills are likely to occur is critical to spill prevention and clean-up efforts. Clean-up of polyisocyanate residues may be extremely difficult if a spill occurs on a porous surface. Therefore, non-porous flooring is recommended for high-potential spill areas. A suitable coating should be applied to porous surfaces such as wood or concrete to reduce porosity.

The overriding principle in spill response is to protect people first, then prevent or minimize the impact of the release on the environment. Only trained personnel should conduct a response to any release.

The following procedure should be understood as a rough guideline:

- Implement the site emergency response plan and evacuate non-emergency personnel. The magnitude of the evacuation depends upon the quantity released, site conditions and the ambient temperature.
- Consult with the safety data sheet (SDS) of the material for guidance on how to handle the spill and to dispose of the waste appropriately.
- Wear necessary personal protection equipment PPE as specified in the SDS or the site emergency response plan.
- Ventilate and remove ignition sources.
- Control the source of the leak.
- Contain the released material by damming, diking, retaining, or diverting into an appropriate containment area.
- Absorb or pump off as much of the spilled material as possible. When using absorbent, completely cover the spill area with suitable absorbent material (e.g. vermiculite, kitty litter, "Oil-Dry", etc.). Allow for the absorbent material to absorb the spilled liquid.
- Shovel the absorbent material into an approved metal container. Do not fill the container more than 2/3 full to allow for expansion and do not tighten the lid on the container. Repeat application of absorbent material until all liquid has been removed from the surface.
- For spills involving a solid product, remove mechanically (sweep up, vacuum, shovel etc.) and collect and place into an approved metal container. Decontaminate the spill surface area using a neutralization solution (see SDS of the polyisocyanate or Table 10.6); scrubbing the surface with a broom or brush helps the decontamination solution to penetrate into porous surfaces.
- Wait at least 15 minutes after first application of the neutralization solution.
- Cover the area with absorbent material and shovel this into an approved metal container. Residual surface contamination can be checked using a wipe test pad to verify decontamination is complete (e.g. "CLI Surface Swype"). If the wipe test pad demonstrates that isocyanate remains on the surface (red color on pad), repeat

applications of neutralization solution, with scrubbing, followed by absorbent until the surface is decontaminated (no color change on wipe pad).

Table 10.6: Preparation for the neutralization of residues of polyurethane hardeners

Mixture A	**Parts by weight**
Water	88
Soda (Na_2CO_3)	10
Detergent	2
Mixture B	
Water	90
Conc. ammonia	8
Detergent	2
Mixture C	
Industrial alcohol (ethanol, isopropanol or butanol)	50
Water	45
Conc. ammonia	5

Mixture B reacts faster than Mixture A, but should only be used with adequate ventilation present due to the ammonia content (TLV). Mixture C is flammable and may only be stored and handled in explosion-proof rooms.

- Apply the lid loosely to the metal waste container. Do not tighten the lid because CO_2 gas and heat can be generated from the neutralization process. With the lid still loosely in place, move the container to an isolated, well-ventilated area to allow release of CO_2. After 72 hours, seal the container and properly dispose of the waste material and any contaminated equipment (i.e. broom or brush) in accordance with existing federal, state and local regulations.
- Incineration of organic materials in appropriately designed and licensed incinerators is the preferred method of disposal (of liquid product). However, when incineration or any other type of disposal is considered, it is recommended that the operation be inspected to determine whether the disposer can properly and safely handle the container and dispose of any waste. The disposal of the waste must be in accordance with existing federal, state and local regulations. If possible, the incineration temperature should be above 1,000 °C. Adequate dwell times and air excess must be ensured. Water vapor must not be used in the incinerator feed.

10.2 Consumer protection aspects

10.2.1 Polyurethane coatings, adhesives and sealants – indoor air quality

Repeated experience over many years confirms that properly formulated and cured polyurethane coatings and adhesives yield no harmful or hazardous emissions after evaporation of the commonly used solvents.[10, 11] For example, in the comment on the specification of German reference values for indoor air quality in reference to diisocyanates it is stated: "The working group did not consider establishment of a guide value for diisocyanates useful, for despite initial higher concentrations in the air during application of paints and adhesives containing diisocyanates (concentrations within the OEL range (Occupational Exposure Limits)), they quickly decrease and long-term exposure is not likely after hardening".[12]

Water-borne polyurethane coating systems – whether aqueous preparations of reacted polyurethanes or water-borne two-component reactive coatings – still often contain higher boiling solvents as flow promoters to control film formation. Examples of such additives include ethers of diethylene glycol, 1,3-butane diol or 1,4-butane diol. Their boiling points are much higher than those of solvents normally used in coatings, which is why they may be emitted from the coating over a prolonged period.

In this context it should be mentioned that solvents are also subject to changes in occupational health evaluation and chemical legislation. In the past, e.g. N-methyl-pyrrolidone (NMP) has been used as an efficient and versatile co-solvent in the formulation of water-borne coatings and printing inks. Some years ago, NMP has been classified as reprotoxic (cat. 1B) and therefore mixtures containing 0.3 % or more of this solvent must be labelled as being "H360D May harm the unborn child". So, such formulations are today recognized as CMR (Convention on the Contract for the International Carriage of Goods by Road) substances which will significantly reduce the acceptance of such products in the market. Innovative raw material suppliers started at an early stage to develop water-borne resins which do not require NMP as co-solvent for proper film formation. Modern water-borne coatings and adhesives avoid such ingredients to protect workers applying these systems and to eliminate such substances from indoor emissions after application.

Formulators need to be aware of the application and later use of their products, taking into consideration the classification of their products and requirements relative to volatile organic compounds (e.g. indoor emissions). Especially in applications like construction and automotive interior, the sensitivity of consumers towards possible risks resulting from chemical ingredients has grown during the last years. Competent and innovative raw material suppliers support formulators to select raw materials suitable for the intended use.

Rarely a hazard but occasionally a nuisance, odor problems may occur when fatty-acid-modified polyester resins (alkyd resins) are used as the polyol components in two-component reactive systems. These problems can be attributed to the presence of free fatty acids in the resins or to their oxidative degradation ("becoming rancid"). The only form of prevention is careful selection of the raw materials and quality control. More details can be found in the specialist literature e.g.[13, 14].

Polyurethane coatings – both one- and two-component systems – have proven themselves over many years as decorative and protective finishes for applications with very high requirements to low indoor emission and low elution.

In this context it is worth mentioning that other widely used components and additives of coatings and adhesives are subject to restrictions through legislation and restriction lists applied by end users in specific branches. Two examples of such restriction lists include the Global Automotive Declarable Substance List (GADSL) from the automotive industry[15] and the Apparel and Footwear International Restricted Substance List (RSL) Management group (AFIRM) which is supported by a variety of apparel and footwear producers[16]. As a result, organotin compounds used as catalysts will fade out of coatings and adhesives in the future.

Innovative coatings and adhesives raw material suppliers will undoubtedly play a role in helping to cope with changes in regulations and provide the market with products that fulfill future requirements.

10.2.2 Do-it-yourself and polyurethanes

As discussed above, polyurethane hardeners exhibit irritating and sensitizing properties. Therefore, specific protective measures are required to avoid contact to skin, eyes and prevent inhalation, especially in case of spray systems. As it is not feasible to implement such measures in the homeworkers use, companies represented in the European Aliphatic Isocyanates Producers Association (ALIPA)[17] explicitly exclude the do-it-yourself (DIY) application of aliphatic di- and polyisocyanates.

In the US, industrial applicators are protected by the classification and labelling requirements of OSHA's HazCom 2012 and consumers by labelling requirements designated by the Consumer Product Safety Commission (CPSC).

The use of MDI-based products is regulated on EU level by Regulation (EC) No. 1907/2006 (REACH), Annex XVII (Restrictions on the Manufacture, placing on the market and use of certain dangerous substances, mixtures and articles)[18].

However, any new use should be reviewed against the product safety data sheet for applicability. In China there is currently no specific regulation on DIY in force.

10.2.3 Relevant legal provisions covering raw materials for coatings and adhesives in contact with foodstuffs

In order to protect the consumer, the use of materials in contact with foodstuffs is increasingly regulated by specific legal provisions. Details are provided for the provisions that apply to coatings and adhesives in the European Union, with the special example of Germany, and the regulations applied in the USA.

European Union

The desire for binding European Union legislation covering food-contact applications was implemented in the 1980s. In 1989, the Council of Ministers passed the framework directive 89/109/EEC[19] concerning materials and articles intended to come into contact with foodstuffs. The principle underpinning of this legislation is that articles in the finished state must not transfer to foodstuffs any constituents harmful to human health. This framework directive was replaced by the more precisely worded Regulation (EC) 1935/2004. This also places no specific requirements on individual product groups but formulates general principles under Article 3: the food contact materials shall not endanger human health, bring about an unacceptable change in the composition of the food or deteriorate its organoleptic characteristics. Article 5 of this Regulation also permits implementing measures for 17 specific product categories to be established, including plastics, coatings, adhesives and printing inks.

Regulation (EU) No. 10/2011

Whereas for coatings and adhesives no specific measures have been adopted, Regulation (EU) No. 10/2011 (often called the Plastics Implementation Measure, PIM) covers materials and articles made of plastic which come into contact with foodstuffs. Furthermore, the EU Commission is planning an implementing measure for printing inks.

Once a specific measure such as the PIM for plastics has been adopted, this automatically replaces, in so far as they existed. Indeed, this goes further: since the Commission is working on a specific measure for printing inks, this effectively prohibits Member States from adopting their own measures, such as the planned (21st) change of the German Consumer Goods Ordinance.

The PIM includes detailed positive lists of permitted monomers and additives (for the latter, national exceptions remain valid). These positive lists are regularly updated by modifying regulations. Moreover, the directive specifies migration limits for the transfer of constituents to the foodstuff. Besides a general migration limit (the sum of all the migrating substances) of 60 mg/kg food, the positive list also contains specific migration limits for many substances. Conformity with these migration limits must be proven by analytical testing. If adhesives, coatings and printing inks are used on plastics articles, then the provisions of the PIM apply them, too.

European Council Resolutions AP (96)5 and AP (2004)1

The European Council (a supranational body without jurisdiction) passed resolution AP(96)5 in 1996 in respect of coatings that are intended to come into contact with foodstuffs and updated it with the resolution AP(2004)1 in 2004. The European Council resolution also contains inventories of monomers and additives. These are similar to the positive lists of the Regulation (EU) No. 2011 but also differ from them in part. The European Council resolutions are only recommendations and are not legally binding. Nevertheless, they constitute a supranational body of rules aligned with the specific needs of coatings.

Germany

In Germany, the Federal Institute for Risk Evaluation (BfR) established a Commission to draft individual recommendations[20] for specific applications in the field of food handling and processing. These recommendations contained lists of starting materials and auxiliaries as well as requirements in respect of the migration of constituents from a defined surface into foodstuffs. At present, there is no general recommendation for coatings. The adhesive sector is specially covered by BfR recommendation XXVIII "Crosslinked polyurethanes as adhesive layers for packaging materials for foodstuffs".

United States of America

In the USA, the regulation of plastics in contact with foodstuffs is the responsibility of the Food and Drug Administration (FDA). The relevant legal provisions are contained in the Code of Federal Regulations (CFR, Title 21, Parts 170 to 199). The FDA regulations are structured completely differently from EU legislation and cover a great number of individual application areas and product groups. Therefore, it is not surprising that significant discrepancies sometimes exist between US and EU regulations.

Depending on their composition, polyurethane resins can have FDA clearance for use in food-contact applications through the adhesives regulation (21 CFR 175.105), the polyurethane resin regulation (177.1680) for contact with bulk quantities of dry solid foods, and the rubber articles regulation (177.2600) for repeated-use applications with all food types. In contrast, polyurethane resins are not listed in the Resinous and Polymeric Coatings regulation (175.300) and are therefore not generally allowed for use in food-contact coatings applications.

As an alternative to a listing in a relevant FDA regulation, individual products in the USA can also receive clearance for use in food contact applications through the submission of a Food Contact Substance Notification (FCN) which must demonstrate that a food contact substance is safe for its intended use. FCN submissions are reviewed and approved by the FDA.

China

In China, National Health Commission (NHC) is in charge of safety of food contact materials, including additives, plastics, coatings, adhesives, rubbers, etc. NHC has established a comprehensive standard framework to regulate food contact materials. The materials, which are no listed by these standards, are not allowed to use in food contact applications. These standards include GB 9685 for additives, GB 4806.6/GB 4806.7 for plastics and GB 4806.10 for coatings. Furthermore, these standards are still evolving and new standards, like adhesive standard, are expected to set-up.

Generally, the global food contact legislations are far from harmonization. For instance, these China standards list specific polymers rather than monomers or monomer groups. Therefore, there are significant discrepancies among US, EU and China. Users should carefully evaluate the compositions of polyurethane coatings and adhesives and make sure the final finished polymers are listed in these standards.

10.2.4 Polyurethane coatings and drinking water

European Union

In the European Union there is at present no binding regulation in force for materials in contact with drinking water, with the result that each product must be evaluated and approved in each country. However, the EU is working on a European Acceptance System (EAS) for construction products that come into contact with drinking water, which in the medium term is intended to harmonize the regulations for drinking water contact across the EU without mutual recognition.

To achieve a degree of harmonization on a voluntary basis, Germany, France, The Netherlands and UK (so-called 4 Member States; "4MS") have been working on a common assessment process based on positive listing of raw materials and testing procedures. Other countries, notably Italy, Portugal and Denmark have shown interest, but not formally joined. The 4MS published its common approach in March 2016 and its positive list of approved raw materials in May 2017.

In Germany, the Federal Environment Agency is responsible for regulating drinking water contact materials. Cornerstone is the Ordinance on the Quality of Water intended for Human Consumption, last revised in 2016. Previously, the various types of drinking water materials, including plastics and coating, were regulated through numerous individual guidelines. The Federal Environment Agency is now converting all the organic material-related guidelines into a new ordinance, the "Attestation of Conformity relating to suitability of products in regard to drinking water hygiene" which adopts the 4MS common approach. The section on coatings includes various material types, including polyurethanes and epoxy resins.

In a new development, the EU Commission made a proposal to recast the Drinking Water Directive 98/83/EC in February 2018. This Directive sets quality standards for drinking water and empowers Member States to adopt measures to ensure that these standards are met. The proposal also includes mandating harmonized CEN Norms for testing products in drinking water contact under the Construction Products Regulation (EU) No. 305/2011. Until these norms are adopted, the national regulations and 4MS approach will continue to be the most important requirements for drinking water.

United States of America

In the United States, the Food and Drug Administration (FDA) and the Environmental Protection Agency (EPA) are both responsible for the safety of drinking water. EPA regulates public drinking water (tap water), while FDA regulates bottled drinking water. The regulation of bottled water falls under the FDA Food Safety Modernization Act. FDA monitors and inspects bottled water products and processing plants under its food safety program. When FDA inspects plants, the Agency verifies that the plant's product water and operational water supply are obtained from an approved source; inspects washing and sanitizing procedures; inspects bottling operations; and determines whether the companies analyze their source water and product water for contaminants. Materials used in contact with bottled water are regulated under FDA indirect food additive regulations. Polyurethanes are not listed in the FDA Resinous and Polymeric Coatings regulation 21 CFR 175.300, therefore the use of a polyurethane coating in a water bottle application must be approved by the FDA through the submission of a food contact notification.

EPA regulates public drinking water under the Safe Drinking Water Act (SDWA), a federal law that protects drinking water supplies throughout the nation. Under this law, the Agency sets regulatory limits for the amounts of certain contaminants in water provided by public water systems. The National Primary Drinking Water Regulations (NPDWR) are legally enforceable primary standards and treatment techniques that apply to public water systems. Polyurethane coatings and thermoplastic polyurethane (TPU) articles may be used in public drinking water systems provided the water continues to meet NPDWR standards.

China

In China, National Health Commission (NHC) is responsible for drinking water safety. NHC sets regulatory limits for certain contaminants in drinking water in GB 5749. Polyurethane coatings or adhesives used as protective materials in drinking water system shall meet the criterial set in GB/T 17219.

10.2.5 Behavior of polyurethane coatings, adhesives and sealants in the event of fire

Polyurethane materials are organic materials that contain carbon, hydrogen, oxygen and also nitrogen. Other elements are brought into the coating system through the use of pigments, additives and auxiliaries. Such materials are inherently combustible. However, fire testing of multicoat polyurethane systems on mineral substrates has led to their classification as hardly flammable (fire class B1) to non-combustible (fire class A2) in accordance with DIN 4102. The corresponding EU classes are C, B and A2 respectively, according to EN 13501-1.

When carbon compounds decompose, they form carbon dioxide, water vapor and, depending on their composition and the fire conditions (e.g. temperature and availability of oxygen), varying amounts of carbon monoxide, hydrocarbons, lower and higher aldehydes and ketones, and soot-like effluents.

Products that contain nitrogen, including polyurethanes but also other natural and synthetic materials such as leather and textiles, e.g. wool and polyamide, also generate volatile nitrogen compounds such as ammonia, nitrogen oxide, nitriles and at temperatures of 800 to 1000 °C, hydrogen cyanide (prussic acid). The latter is a highly endothermic compound whose formation, like that of carbon monoxide, is promoted by high temperatures.

Polyurethanes have also been observed to release traces of diisocyanates or of unknown isocyanate-like compounds. This reaction starts at around 250 °C and reaches its peak at 400 to 600 °C, i.e. at temperatures that are characteristic for a developing fire.

This behavior is common to all products that contain nitrogen, including natural products also formulations based on acid-curing alkyd resins with melamine or urea formaldehyde resins as the crosslinkers as well as formulations based on amine-cured epoxy resins. In a real fire situation, the type and concentration of harmful combustion products depend on the prevailing conditions. These can only be simulated to a certain extent in laboratory tests. For example, the lack of oxygen in a smoldering fire favors the formation of hydrogen cyanide, provided the fire temperature is high enough. However, being a combustible carbon compound, it will continue to burn in fully developed fire conditions in the presence of oxygen.

The main risk factor presented by the combustion gases of all organic materials is carbon monoxide. In a fire situation, it is always present in harmful concentrations. It is therefore highly recommended that fire-fighters wear self-contained breathing apparatus.

10.3 References

[1] REGULATION (EC) No 1272/2008 OF THE EUROPEAN PARLIAMENT AND OF THE COUNCIL of 16 December 2008 on classification, labelling and packaging of substances and mixtures, amending and repealing Directives 67/548/EEC and 1999/45/EC, and amending Regulation (EC) No 1907/2006; https://echa.europa.eu/regulations/clp/legislation

[2] http://isopa.org/product-stewardship/walk-the-talk/

[3] www.alipa.org/index.php?page=alipa-safeguard---we-care-that-you-care

[4] ACC Aliphatic Ref: https://adi.americanchemistry.com/

ACC Aromatic Ref:
https://dii.americanchemistry.com/
OSHA Alliance Ref: https:
//polyurethane.americanchemistry.com/
ACC-and-OSHA-Alliance.html

[5] www.ifa-arbeitsmappedigital.de/ce/diisocyanate-monomer-2-4-tdi-2-6-tdi-2-4-mdi-4-4-mdi-hdi-ipdi-und-ndi/_sid/KLRP-402799-zVDI/detail.html

[6] Council Directive 98/24/EC of 7 April 1998 on the protection of the health and safety of workers from the risks related to chemical agents at work (fourteenth individual Directive within the meaning of Article 16(1) of Directive 89/391/EEC); http://eur-lex.europa.eu

[7] Erste Allgemeine Verwaltungsvorschrift zum Bundes-Immissionsschutzgesetz TA Luft – Technische Anleitung zur Reinhaltung der Luft, July, 24 2002; GMBl. No. 25–29 from July 30 2002 p. 511

[8] A. Bürkholz et al., Farbe und Lack 82 p. 693–698, 1976

[9] MAK- und BAT-Werte-Liste, Deutsche Forschungsgemeinschaft, Senatskommission zur Prüfung gefährlicher Arbeitsstoffe. Wiley-VCH Verlag GmbH & Co. KGaA, Weinheim; appears annually, www.wiley-vch.de/books/info/dfg/index.html

[10] G. Balle, M. Kuck, W. Wellner, Polyurethanlackierungen – eine Emissionsquelle für monomere Diisocyanate? Zur Analvtik monomerer Diisocyanate, insbesondere Toluylendiisocvanat, in der Raumluft, Staub – Reinhaltung der Luft 51 (1991) p. 231–236

[11] M. Fischer, E. Böhm, Erkennung und Erfassung von Schadstoffemissionen aus Möbellacken, Schriftenreihe Schadstoffe und Umwelt, No. 12, Erich Schmidt Verlag, Berlin 1994

[12] The Federal Environment Agency (Umweltbundesamt); Indoor Air Hygiene; www.umweltbundesamt.de/uba-info-daten/daten/irk.htm

[13] H. Kittel, Lehrbuch der Lacke und Beschichtungen, S. Hirzel Verlag Stuttgart Leipzig 1998

[14] Streitberger, Goldschmidt, BASF Handbook Basics of Coating Technology, 3rd Edition, Vincentz Network, 2018

[15] www.gadsl.org/

[16] www.afirm-group.com/afirm-rsl/

[17] The European Aliphatic Isocyanate Producers Association (ALIPA); Avenue Edmond Van Nieuwenhuyse Laan 4, B-1160 Brussels (Belgium); www.alipa.org

[18] Regulation (EC) No 1907/2006 of the European Parliament and of the council of 18 December 2006 concerning the Registration, Evaluation, Authorization and Restriction of Chemicals (REACH), establishing a European Chemicals Agency, amending Directive 1999/45/EC and repealing Council Regulation (EEC) No 793/93 and Commission Regulation (EC) No 1488/94 as well as Council Directive 76/769/EEC and Commission Directives 91/155/EEC, 93/67/EEC, 93/105/EC and 2000/21/EC; https://echa.europa.eu/regulations/reach/legislation

[19] Council Directive 89/109/EEC of 21 December 1988 on the approximation of the laws of the Member States relating to materials and articles intended to come into contact with foodstuffs; Official Journal L 040, 11/02/1989 P. 0038–0044

[20] Veröffentlicht in den jeweils aktuellen Ausgaben des Bundesgesundheitsblattes

[21] Covestro Deutschland AG

11 Sustainable development

11.1 The global context of sustainable development

The definition of sustainable development is probably best summarized in the Brundtland Report from 1987: "Sustainable development is development that meets the needs of the present without compromising the ability of future generations to meet their own needs."[1] This report achieved widespread public awareness following the Earth Summit on Environment and Development in Rio de Janeiro in 1992. Consequently the United Nations took up work towards a more sustainable future. The United Nations recognized climate change as a threat for human welfare in its 2015 Paris Agreement, which was signed by the majority of the parties of the United Nations.[2] In addition, the UN Sustainable Development Goals (SDG's) were agreed upon by the UN in 2016, which ambitiously strive for eradication of poverty and implementing peace globally by 2030.[3] This resulted in governments, the private sector and non-governmental organizations joining forces to find solutions for the challenges ahead.[4, 5] Likewise sustainable development will have an impact on the regulatory environment for businesses, thus influencing the coatings and adhesive market and its multitude of applications.

The rather complex field of Sustainable Development may be best characterized by two main aspects relevant to life cycle thinking:

- Reducing the negative impacts and fostering the positive impacts of mankind on environment and society (people, planet, profit) over the life-cycle of a product or service, and
- Decoupling economic growth from linear use of resources for generating and supplying a product/service to a circular model approach where products are designed to last longer and to be recycled.

While the field of sustainable development has existed for over 30 years, it still has to be considered a developing science. It is an evolving topic in which many more models (e.g., climate and waste recovery models) will need to be developed and scientific discussions continued.[6]

U. Meier-Westhues, K. Danielmeier, P. Kruppa, E. P. Squiller: Polyurethanes

11.2 Reducing the negative impact/ fostering positive impact

Products and services can have an impact on the:
- Society and communities (PEOPLE) – e.g. health and well-being, quality of life, social capital etc.
- Environment (PLANET) – e.g. water and soil, fossil depletion, waste generation etc.
- Economic system (PROFIT) – e.g. income and expenditures, employment, taxes etc.[7]

While there are well established tools to quantify the economic impact of human interactions, the measurement of environmental and societal impacts is still in development stages. The assessment of environmental impacts, however, is more advanced and commonly accepted than the evaluation of societal impacts.

An established methodology to measure environmental impact is the environmental Life-Cycle Assessment (LCA)[8] which is also widely applied within the chemical industry.[9, 10] Standard databases and methodologies are available and demonstrate that LCA data are gaining importance for regulations and industry sectors. A couple reference examples include the Product Environmental Footprint (PEF) initiative of the European Commission and the International "EPD" System.[11, 12] The most relevant LCA industry average data for polyurethane systems are available via industry associations such as ISOPA, ALIPA, EPDLA and Plastics Europe.[13–16] Polyurethane related Life Cycle Inventory data were published by the Plastic Division of the American Chemistry Council (ACC) in 2010.[17]

The measurement of social impacts can be achieved using a social Life-Cycle Assessment (S-LCA). A couple of frameworks are available to assess social impacts e.g. from the United Nations Environment Programme (UNEP) and Society of Environmental Toxicology and Chemistry (SETAC).[18] The Roundtable for Product Social Metrics, a cross industry working group, recently took effort to establish a pragmatic framework for S-LCA that is widely accepted and adopted among companies.[19]

In the chemical sector S-LCA is addressed by individual companies as well as within sector initiatives like the World Business Council for Sustainable Development (WBCSD).[20–22]

Being that social, environmental and economic interdependencies are not fully understood today, sustainability has to be seen as mentioned above – as an evolving science, with concepts and models still changing and developing based on ongoing scientific discussions and insights.[6]

11.3 Decoupling economic growth from linear use of resources

The circular economy concept, which was a focus topic during the World Economic Forum in Davos in 2013, is an approach to decouple economic growth from the linear use of resources.

The current practice of how we use resources is a conveyor-belt-type, linear 'take-make-consume-dispose' economic model.[23] Due to a growing world population and an increasing middle class, resource consumption is expected to triple until 2050 according to a 2011 UN report, hence the current economic model limits sustainable development.[24] The forecasted demographic trends call for a change in societal and economic patterns, especially regarding the use of resources for production, consumption and disposal of products and services. Regulatory frameworks will be the main drivers for future resource consumption models.[25]

Circular economy concepts are about the extension of a product's or service's lifetime and the recycling of materials at the end of the product life cycle. Here the use of high performance coatings and adhesives designed to last a long time in service and thus reducing the number of times a product and/or service needs to be produced, will show their strengths. Moreover the use of alternative building blocks, like CO_2 or bio-based raw materials, is a contribution to circular economy.

11.4 Sustainable development in the coating, adhesive and sealants industry

The chemical industry has acknowledged its responsibility for sustainable development for some time now by following the principles of Responsible Care in its activities worldwide.[26] Sustainable development is on the agenda of all major chemical industry associations. Additionally, the chemical industry is employing sustainable development strategies and the UN SDGs.[27–32]

Specifically, paints and organic coatings are generally used to ensure the effective protection of a substrate against corrosion, deterioration and mechanical damage. In this way, they contribute to prolonging the service life of goods for a broad variety of substrates, such as in transportation, electronic devices, parts used in industrial plants and building infrastructure. In other words, coatings can help to preserve resources. Aliphatic polyurethane based coatings are especially powerful for outdoor applications and due to their weather and corrosion resistance they help to prolong the lifetime of infrastructural parts, which are exposed to aggressive climate conditions, like off-shore windmills or oil-platforms. Resource efficiency is realized by an extended lifetime of the coating and lower maintenance cycles. The latter contributes positively to workers occupational health and safety.

Likewise, adhesives can be designed to preserve and efficiently use resources. In contrast to other joining technologies such as welding or soldering, adhesives were the first to provide a method for permanently joining diverse types of substrates and the most disparate substrate combinations. Adhesives nowadays are also contributing to light weight solutions which allow, for example, the use of renewable resources such as wood in building and construction, as well as, the saving of resources such as fuel or electrical energy in cars or planes.

As much as coatings, adhesives and sealants contribute to sustainable development, some unavoidable environmental impacts (e.g. emissions to air and water, waste, CO_2-emissions caused by energy consumption) occur during the life cycle of a coating, adhesive or sealant. This comprises the whole life cycle from the manufacture through application up to the recycling or disposal of the coated, glued or sealed object (see Figure 11.1). These environmental impacts and strategies for control, reduction and avoidance are largely independent of the chemistry of the binders or base polymers used in the coating or adhesive.

The ecological impact of coatings, adhesives and sealants, the mitigation of these impacts and the history of regulations in Europe is described in more detail in the 1st edition of this book.[33]

As the worldwide coating, adhesive and sealant industry grows, so does the number of regional regulations. This edition does not endeavor to cover all of the specific regulatory aspects among the different regions. Generally, regulations are a major driver for

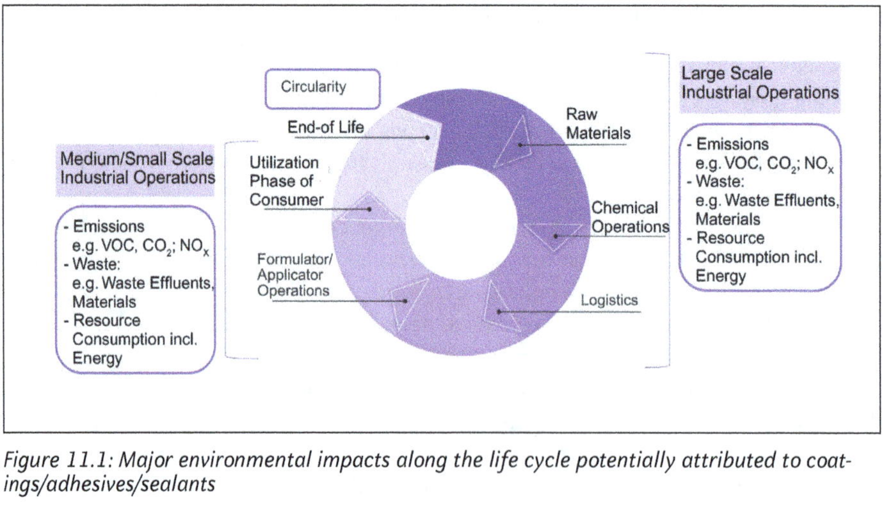

Figure 11.1: Major environmental impacts along the life cycle potentially attributed to coatings/adhesives/sealants

change that can lead to more sustainable solutions. The monitoring of existing and upcoming regulations is an essential activity for companies who desire to innovate sustainable solutions in a profitable way.

Coating, adhesive and sealant technologies are often based on reactive chemical compositions. Due to the reactivity of the molecules they could have an effect on human health if humans are exposed by direct contact. For example, while at the moment of the publication of this book methylene-bis(phenylisocyanate) is already restricted for consumer use under EU REACH, all diisocyanates are under discussion for a restriction for professional and industrial uses. Also, other chemicals that are used in the coating, adhesive and sealant industry, such as crosslinkers, catalysts, co-solvents, co-reactants, shelf-life extender, are observed by regulators. More and more Non-Governmental Organizations create lists of chemicals, which are proposed to be substituted.[34, 35] These lists are often based only on hazardous properties and don't consider the actual risk related to the use of such substances. In the view of industry hazardous properties alone should not lead to substitution if a substance can be handled safely. Therefore, the chemical industry has to take major efforts to ensure safe handling of the products they offer in the market. To get a better understanding about sustainability related risks and opportunities on the product portfolio and for a proactive, innovative product development, many chemical companies develop and apply methodologies for a sustainable product portfolio assessment, which includes not only regulatory trends, but also market perception. Such assessment creates a better understanding for innovation opportunities supporting sustainable developments.[36]

Polyurethane based coating, adhesive and sealant technologies help to reduce the impact to the environment and are contributing to sustainable development in some major ways:

- As a function provider for more durable and stress resistant materials, polyurethane coatings have been shown to extend service life-time of products e.g. in protective and marine coatings, in the automotive and transportation sector, in construction applications, in the good appearance, preservation of wood products, etc. and thus, save resources in the long term (see Chapter 5.2).
- As an enabler for more efficient application processes, polyurethane coatings can save energy and resources in industrial production processes by lowering curing temperatures and curing time requirements (e.g. polyaspartic technologies).
- By enabling low VOC applications with solvent-free polyurethane dispersions used in:
  - Water-borne primer surfacers and basecoats in automotive OEM applications (see Chapter 5.3)
  - Water-borne topcoats in wood, metal and architectural wall coatings (see Chapters 5.1, 5.2, 5.9)
  - Water-borne textile coatings for synthetic leather (see Chapter 5.7)
  - Water-borne adhesives (see Chapter 6.5)

  - Water-borne film formers for cosmetics (see Chapter 8.2)
  - High-solid polyaspartic coatings for protective coatings (see Chapter 5.2)
- By lowering the carbon footprint through the use of bio-based instead of fossil raw materials, e.g. pentamethylene diisocyanate-based polyurethane hardener, and polyesters using bio-based building blocks (see Chapter 5.3)
- By increasing energy efficiency of buildings with joint sealants (see Chapter 7)

11.5 Outlook

For a long time, sustainability related topics (often also stimulated by regulations) have been a driver for ground-breaking innovations in many industries and thus in the coating and adhesive industry as well. With increasing awareness about sustainability and with increasing pressure on the planet's boundaries, all stakeholders will likely be more and more active in directing the impact of the industry and society, considering the people, planet, profit approach. This will inspire further innovations such as circular economy models. This will result in new product designs and challenge current uses and applications which might in the long term have significant effects on how coatings and adhesives are being used.

Opportunities will be created by engaging into collaborations and partnerships along the value chain and cross-industry as well as with academia. The chemical industry will remain a strong partner for innovation in adhesives and coatings in the future. Due to its technical performance, durability and versatility polyurethane systems will play a major role.

11.6 References

[1] G. H. Brundtland, Report of the World Commission on environment and development: "our common future.", United Nations, 1987

[2] United Nations, The Paris Agreement, http://unfccc.int/paris_agreement/items/9485.php (accessed March 2018)

[3] United Nations, Sustainable Development Knowledge Platform, https://sustainabledevelopment.un.org/sdgs (accessed March 2018)

[4] WBCSD, Vision 2050, www.wbcsd.org/Overview/About-us/Vision2050 (accessed March 2018)

[5] United Nations Global Compact, Making Global Goals Local Business, www.unglobalcompact.org/sdgs (accessed March 2018)

[6] Robert W. Kates, What kind of a science is sustainability science? PNAS December 6, 2011. 108 (49) 19449–19450; https://doi.org/10.1073/pnas.1116097108 (accessed April 2018)

[7] T. F. Slaper, T. J. Hall, The Triple Bottom Line: What Is It and How Does It Work?, Indiana Business Review, www.ibrc.indiana.edu/ibr/2011/spring/article2.html (accessed April 2018)

[8] ISO 14040: Environmental Management – Life Cycle Assessment – Principles and Framework

[9] WBCSD, Life Cycle Metrics for Chemical Products, www.wbcsd.org/contentwbc/download/1886/23998 (accessed March 2018)
[10] ICCA, How to Know If and When it's Time to Commission a Life Cycle Assessment, www.icca-chem.org/wp-content/uploads/2016/05/How-to-Know-If-and-When-Its-Time-to-Commission-a-Life-Cycle-Assessment.pdf (accessed March 2018)
[11] European Commission, Single Market for Green Products Initiatives, http://ec.europa.eu/environment/eussd/smgp/index.htm (accessed March 2018)
[12] The International "EPD" System, www.environdec.com/ (accessed July 2018)
[13] ISOPA, Toluene Diisocyanate (TDI) & Methylenediphenyl Diisocyanate (MDI) ISOPA, 2012, https://isopa.org/media/1101/isopa-eco-profile-mdi-tdi-2012-04.pdf (accessed March 2018)
[14] ALIPA, Aliphatic Isocyanates for Polyurethane Products, 2014, www.alipa.org/uploads/Modules/Publications/ALIPA%20Eco-Profile%20%20Aliph%20Isocy%202014-12.pdf, 2014 (accessed March 2018)
[15] EPDLA, Life Cycle Inventory of Polymer Dispersions, 2013, https://hnlkg4f5wdw34kx1a1e9ygem-wpengine.netdna-ssl.com/wp-content/uploads/2017/07/4_EPDLA-Life-Cycle-Assessment-LCA-Summary-Report.pdf (accessed March 2018)
[16] Plastics Europe, Eco-profiles, www.plasticseurope.org/en/resources/eco-profiles (accessed April 2018)
[17] American Chemistry Council (2010), https://plastics.americanchemistry.com/LifeCycle-Inventory-of-9-Plastics-Resins-and-4-Polyurethane-Precursors-Rpt-Only/
[18] M. Kühnen, & R. Hahn, Indicators in social life cycle assessment: a review of frameworks, theories, and empirical experience. Journal of Industrial Ecology, 21(6), p. 1547–1565
[19] The Roundtable for Product Social Metrics, https://product-social-impact-assessment.com/roundtable-for-product-social-metrics/ (accessed April 2018)
[20] BASF, Assessment of three sustainability dimensions, 2017, www.basf.com/documents/corp/en/sustainability/management-and-instruments/quantifying-sustainability/seebalance/BASF_SEEbalance_Flyer.pdf (accessed March 2018)
[21] WBCSD, Social Life Cycle Metrics for Chemical Products, 2016, www.wbcsd.org/contentwbc/download/1918/24428 (accessed March 2018)
[22] DSM, Brighter Living Solutions Methodology, www.dsm.com/corporate/sustainability/brighter-living-solutions/brighter-living-solutions-methodology.html (accessed March 2018)
[23] Ellen McArthur Foundation, Towards the Circular Economy, 2013, www.ellenmacarthurfoundation.org/assets/downloads/publications/Ellen-MacArthur-Foundation-Towards-the-Circular-Economy-vol.1.pdf (accessed March 2018)
[24] International Resource Panel, Decoupling Natural Resource Use and Environmental Impacts from Economic Growth, 2011, www.resourcepanel.org/file/400/download?token=E0TEjf3z (accessed March 2018)
[25] Reed Smith Client Alerts, The implications of the transition to a 'circular economy', and overview of the content and current status of draft EU implementing legislation, www.reedsmith.com/The-implications-of-the-transition-to-a-circular-economy-and-overview-of-the-content-and-current-status-of-draft-EU-implementing-legislation-05-26-2016/ (accessed April 2018)
[26] ICCA, The Quest for performance excellence, www.icca-chem.org/responsible-care/ (accessed March 2018)
[27] ICCA, The Road to Sustainable Development, www.icca-chem.org/sustainable-development/ (accessed March 2018)
[28] CEFIC, Teaming up for a sustainable Europe, http://www.cefic.org/Documents/RESOURCES/PositionPapers/Cefic-Sustainability-Charter-TeamingUp-For-A-SustainableEurope.pdf (accessed April 2018)

[29] ICCA, Global Chemical Industry Contributions to the Sustainable Development Goals, www.icca-chem.org/wp-content/uploads/2017/02/Global-Chemical-Industry-Contributions-to-the-UN-Sustainable-Development-Goals.pdf (accessed March 2018)
[30] IPCC, Sustainability, https://ippic.org/advocacy-policy/#sustainability (accessed March 2018)
[31] CEPE, CEPE Sustainability Activities, http://www.cepe.org/sustainability/ (accessed March 2018)
[32] FEICA, Sustainable Development, http://www.feica.eu/our-priorities/sustainable-development.aspx (accessed April 2018)
[33] U. Meier-Westhues, Polyurethanes: coatings, adhesives and sealants., 2007, Vincentz Network GmbH & Co KG, ed.1 chapter 11
[34] Chemsec, http://chemsec.org/sin-list/ (Accessed July 2018)
[35] "Greenscreen" for safer chemicals, www.greenscreenchemicals.org/ (accessed July 2018)
[36] WBCSD, Framework for portfolio sustainability assessments, www.wbcsd.org/Projects/Chemicals/Product-Portfolio-Steering-Framework (accessed April 2018)

12 Outlook

As described in detail in the previous chapters, the field of polyurethanes has developed into an extraordinary success story over the past 80 years. Polyurethane raw materials have become well-established worldwide in many application areas, especially those requiring high-quality performance in coatings, adhesives, and sealants.

Polyurethane technologies have progressed over the years thanks to their strong technical profile and multitude of formulation options. The technology has enjoyed large growth in the coatings, adhesive, and sealant areas due to the displacement of conventional systems where there were performance, economic and sustainability advantages to be achieved by polyurethanes. Strong demand in all regions, particularly in the Asian market, has also accelerated its rapid growth over the last two decades.

On the technical side, above-average growth rates can be explained by the technology's superior quality over conventional systems: polyurethanes fulfill growing requirements for coating, adhesive, and sealant performance. In addition, they also noticeably extend product service life. This makes them more durable and more sustainable.

The market quickly recognized this potential. Both established and new raw material manufacturers, coatings, adhesive, and sealant producers, and processors and partners throughout the supply chain invested in new equipment, expanded their capacities, and continuously developed polyurethane technology through innovation.

Other applications were added over time, such as polyurethane products for medical wound care, cosmetics, and holography. Innovative polyurethane products should continue to offer new opportunities in these and other areas in the future.

Growth in the polyurethane area is likely to continue also due to stricter legal regulations of VOC emissions and solvents. These will continue to increase demand for high-solid coatings, aqueous, powder-based, and radiation curing systems, as well as, raw materials with low residual monomer content. This drive for change is advancing at different rates in different regions but will continue to develop around the world.

There are also other worldwide trends at play: users are choosing aliphatic polyurethane systems over aromatic systems more frequently due to current technical requirements that call for better outdoor durability. And, in the trade sector, one-component polyurethane systems with a high profile of technical quality are preferred. Innovative technologies that reduce the number of layers and work steps, increasing efficiency for final processors, are also in demand. Such technologies require multifunctional coating, adhesive, and sealant systems. As examples from automotive coating and protective coatings show, polyurethane chemistry offers a wide variety of formulating options that allows it to play a key role in this area.

State of the art computer-based methods to develop formulations and catalysts will help create even more customized polyurethane systems in the future; these will, in turn, help expand the range of customized recipes. Low-temperature and abbreviated curing times will continue to be a trend in coatings. These and other aspects offer outstanding opportunities for growth for manufacturers of innovative polyurethane-based products.

The trend towards sustainable solutions deserves special attention. Bio-based raw materials can serve as innovative building blocks for sustainable coating, adhesive and sealant systems, providing the basis for a range of resource-saving products. The authors expect that the increasing demand for such polyurethane raw materials will affect both polyols and polyisocyanates in particular, if the products provided have equal or better technical performance and value with the products already established in the market.

There are also pioneering developments in using CO_2 as the building block for raw material synthesis. Polyurethane chemistry makes a key contribution to greater sustainability with at least equal, and often greater technical performance.

Last but not least, organic decomposition will play an ever greater role in the future as part of the debate over more sustainable products and manufacturing. Coatings and adhesives based on polyurethane dispersions can be designed to maximize organic decomposition where this is a desirable feature.

In light of the new challenges posed by stricter regulations and legal specifications, it is expected that future demand will focus not only on products with the lowest possible residual monomer content, but also new alternative coating characteristics achievable with polyurethane building blocks.

Innovative technologies and processing procedures such as digital printing or direct coating/direct skinning will help create entirely new product manufacturing processes, and will result in new requirements for polyurethane coatings, adhesives, and sealants. Innovators all along the supply chain are working together in close partnerships to drive innovation.

In considering future development, we can conclude that the polyurethane chemistry will continue to make a key contribution to production and infrastructure preservation and will create valuable goods. The chemicals industry will continue to serve as an innovative partner and drive development forward.

Authors

Dr Ulrich Meier-Westhues studied Chemistry at the RWTH Aachen University, Germany. He was with Herberts GmbH four years then moved to Bayer AG in 1989. In the business unit Coatings, Adhesives, Specialties (CAS) he had several positions in Marketing & Business Development. Before his retirement in 2015 he was responsible for Innovation and a member of the global CAS Management Committee in Bayer MaterialScience (today Covestro).

Dr Karsten Danielmeier is head of R&D for Covestro´s business unit Coatings, Adhesives, Specialties since June 2015 and a member of the global CAS Managemant Committee. After finishing his PhD in Synthetic Organic Chemistry at Bonn University, he joined Bayer AG in 1996 and has since worked in various functions with increasing responsibility in R&D, Product Management, Supply Chain Management and Technical Marketing in both Germany and the USA.

Peter Kruppa is head of Application & Technology Development for Covestro´s business unit Coatings, Adhesives, Specialties since July 2016 and a member of the global CAS Management Committee. He holds a diploma in process/chemical engineering from TU Dortmund and started at Bayer AG in 1988. He has worked in different roles, business units and corporate functions with increasing responsibility in engineering, technology and production, strategic planning and business planning both in Germany and Spain.

Dr Edward P. Squiller is a director in Application & Technology Development for Covestro´s NAFTA business unit Coatings, Adhesives, Specialties since 2015. After finishing his PhD in Physical Organic Chemistry at The Pennsylvania State University he joined the Mobay Chemical Company in 1984 (which was later named Bayer Corporation in the USA) and has since worked in various technical functions with increasing responsibility in the USA.

Index

Symbols

A

B

C

D

E

F

G

H

I

K

L

M

N

O

P

Q

R

S

T

U

V

W

www.ingramcontent.com/pod-product-compliance
Lightning Source LLC
LaVergne TN
LVHW021726021026
849161LV00004BA/732
* 9 7 8 3 8 6 6 3 0 7 8 2 7 *